Next Generation
3G and I

Next Generation Mobile Systems
3G and Beyond

Edited by

Minoru Etoh

DoCoMo Communications Laboratories USA

John Wiley & Sons, Ltd

Other Wiley Editorial Offices

John Wiley & Sons Inc., 111 River Street, Hoboken, NJ 07030, USA

Jossey-Bass, 989 Market Street, San Francisco, CA 94103-1741, USA

Wiley-VCH Verlag GmbH, Boschstr. 12, D-69469 Weinheim, Germany

John Wiley & Sons Australia Ltd, 33 Park Road, Milton, Queensland 4064, Australia

John Wiley & Sons (Asia) Pte Ltd, 2 Clementi Loop #02-01, Jin Xing Distripark, Singapore 129809

John Wiley & Sons Canada Ltd, 22 Worcester Road, Etobicoke, Ontario, Canada M9W 1L1

Wiley also publishes its books in a variety of electronic formats. Some content that appears in print may not be available in electronic books.

Library of Congress Cataloging-in-Publication Data

Next generation mobile systems 3G and beyond / edited by Minoru Etoh.
    p. cm.
ISBN-13 978-0-470-09151-7 (cloth)
ISBN-10 0-470-09151-7 (cloth)
1. Wireless communication systems – Technological innovations. 2. Cellular telephone systems – Technological innovations. 3. Mobile communication systems – Technological innovations. I. Etoh, Minoru.

  TK5103.2.N4453 2005
  621.3845'6 – dc22

                                                    2005003372

British Library Cataloguing in Publication Data

A catalogue record for this book is available from the British Library

ISBN-13 978-0-470-09151-7 (HB)
ISBN-10 0-470-09151-7 (HB)

Typeset in 10/12pt Times by Laserwords Private Limited, Chennai, India
Printed and bound in Great Britain by Antony Rowe Ltd, Chippenham, Wiltshire
This book is printed on acid-free paper responsibly manufactured from sustainable forestry in which at least two trees are planted for each one used for paper production.

To all the people who are collaborating with us

Contents

III   Middleware and Applications   189

Foreword

Thoughts on the XG system

The growth of the Internet, which I have been involved in, and the growth of mobile telephone services, which NTT DoCoMo has been a leader in, have been peculiarly inter-linked and at the same time separate. The Internet is a lab experiment that broke free, and finds itself in a world of its own contriving, which it often does not understand. The mobile telephone world was developed for commercial purposes, and is in many respects the son of its father, the wired telephone system. Each seeks to bring innovative services to its users.

The two intertwine in interesting ways. Research done in the 1990s developed a way for Internet end systems to break free of their wired moorings, resulting in what we call IP Mobility, mobility in the Internet layer. Two logical users of this soon developed: the nomadic laptop driving down the street might connect to a wireless LAN, and the mobile telephone that acquired services reminiscent of that same laptop.

Both move, and both have a need to be able to participate in peer-to-peer sessions, but they serve different needs. The laptop is, in the final analysis, something we use because its screen size and general ubiquity make it a reasonable replacement for the desk-mounted system left behind, but we would not think of it as a personal communications device. If nothing else, if I attach mine to my belt, it will soon either be damaged itself or damage my belt. The mobile telephone fits my belt well, and is very appropriate as a personal communications device. But if I were to try to write these thoughts on my telephone, I would soon go crazy. The instrument is not suited to the application. Both do electronic mail, but one is for megabyte attachments and the other for pithy messages or instant messaging. Both do calendaring, but one easily lets me see a month at a glance, while the other is more suited to managing my day. They are tools, both of them, suited to their own uses, and using in many respects common technology.

The development of mobility has worked its way back into the wired Internet in inter-esting ways. As we develop and deploy the concepts of Anycast Routing, in which a set of computers collaborate to offer a service, and I use Internet routing to attach to whichever happens to be closest at the moment, IP Mobility solves a problem for a stationary system. If I open a TCP connection to the nearest computer in an Anycast service, and then either I move or routing changes, I might find myself talking to another server without warning. Since the state I shared with the first is unknown to the second, I lose everything I have done to that point and must start over. But, if I treat the Anycast address as the Home Address of a mobile node and then use Optimized Routing to tie the mobile session to the particular server as a Care-of Address, my session remains stable even as routing changes. Treating the stationary node as a mobile node solves a difficult application problem.

To understand the network, one must, I think, grasp it at many layers simultaneously. One must understand the implications of the transmission layers, physical and link, and the intranetworking layer that builds a network of like systems under a common administration. One must go on to understand how these intranets interconnect at the Internet layer, and how various transports use them and respond to their anomalies. In the end, one must understand and be prepared to deal with the requirements of the application and the user who uses it. This is true because, in the end, the networked device, and the user, looks no further than the application concerned. The network must make that application work well for the user, or find another reason to exist.

I found myself, recently, thinking of ways to use the wired, wireless, and mobile Internets together. In December 2004, a tsunami swept hundreds of thousands of people – we may never really know how many – to their deaths, and the Internet community asked "how might we have helped?" The answer turns out to use interesting aspects of both types of communication systems. A Tsunami Warning Center, had one existed (as one does in the mid-Pacific) in the Indian Ocean, might have encoded a message using the Common Alerting Protocol, which is an information model for agency-to-agency distributions regarding disasters. They would have sent this message in any number of ways to subscribing centers run by appropriate authorities in various countries. Many of those would be by Internet – the web, RSS feeds, authenticated electronic mail, and so on. These centers would then consider what areas are likely to be affected, the level of urgency, certainty, and severity, and what the appropriate message might be, and then sent the message on to citizens likely to be affected. An obvious way to locate the people in a locality is to ask if their mobile telephone is registered in the cell in the locality. An obvious way to get their attention is to call each such registered telephone with a voice message, or to send a text message using cell broadcast, as is available in GSM and being deployed in Europe. The Internet and the mobile telephone system each are then used to distribute a message from a regional center through a crisis management agency, and finally to a person sleeping on the beach or in the projected path of a storm.

The authors collaborating on this work have tried to step aside from the marketing terminology of "next generation", to think about what kinds of systems they would really like to deploy, and why they would deploy them. To their credit, they have not thought of their system as taking over the world, as many in our industry do when they dream up new technologies. It is enough to find one's place in the world and serve a targeted set of needs well. In this book, they have explored the kinds of applications that will adapt well to a personal communication device. They have considered the kinds of transports these applications will use, how they interact with the rest of the Internet, and how they interact with the peculiar transmission systems – radios of various kinds – used in the mobile telephone world, and the aspects and algorithms of mobility that enhance that experience.

They point us in an interesting direction. And for that, I thank them.

Fred Baker
Cisco Fellow, California

Preface

The aim of this book is not to describe new wireless access technologies that will replace the third generation (3G) technologies, but to describe a complete ecosystem of technologies, including access technologies that will be essential for the development of a new mobile environment beyond the 3G era. The mobile applications currently envisioned for this future platform will be augmented by others that will arise because of the innovative opportunities offered by the new environment, and will change the nature of the mobile communications business.

Daily life is becoming increasingly dependent on mobile communication. This trend began with the first mass uptake of cellular telephones in the mid-1980s and continues with evermore diverse services being offered by a growing number of operators. There is every indication that this trend will continue and even accelerate as networks become more powerful and devices become more ubiquitous.

Throughout the evolution of the mobile network industry, mobile networks have been nominally characterized by the generation of their wireless access technology. At the time of publishing, year 2005, it is commonly accepted that there have been three generations of wireless technology, called 1G, 2G, and 3G.

The third generation was officially launched by NTT DoCoMo, Inc. in October 2001, and is, in effect, the "current generation," although we are early in this generation's predicted life cycle. Third-generation wireless technology is well defined, and devices that use 3G technology are entering, and even becoming commonplace in, the market.

Wireless connectivity, simple speech services, and computing devices are becoming commodities. The user community now demands evermore powerful functionality and continuously improving applications. This new and more sophisticated user demand is driving the research community to look toward the future of wireless networks. As a result, mobile communication researchers are starting to focus on the technology required for the next generation of mobile systems. This new focus creates a need for the research community to begin a dialogue about the future of mobile networks and communications. Throughout this book, we use the term "Next-generation (XG) mobile systems" to describe a complete mobile communication system beyond 3G that includes the whole technology "value chain" of future wireless networks. Our definition of XG is complex and inclusive, from future heterogeneous service platforms to the core network and from the heterogeneous access network to the user terminals. On the other hand, the term "Fourth Generation" (4G) mostly implies the fourth-generation radio access networks (RAN). The technologies required to realize XG systems are clearly not limited to new wireless access methods, as is sometimes proposed. We must pay attention to a wide range of emerging and existing research topics such as IP backbone networks, open and heterogeneous service platforms, terminal software, and multimedia applications. It is our belief that these technologies will be implemented and will evolve continuously, rather than suddenly, supplanting 3G technology in a revolutionary way. In this sense, although the next generation can be seen as a logical evolution of 3G, the XG image is very different and will require breakthrough technologies in many diverse areas. We firmly believe that the future mobile world will not be defined only by new wireless access technologies. We propose a clear distinction between two terms: the 4G Radio Access Network (RAN) and the XG mobile system. The 4G RAN part is clearly an important component of XG mobile systems, but it is insufficient to define it. We will discuss some existing and emerging technologies that we think are necessary for our definition of Next Generation in this book.

NTT DoCoMo, Inc. has created a research lab specifically to work on next-generation mobile system technology issues and to help lead the community in these discussions. The company believes that it is timely and in the best interests of the whole industry to share our vision of the future of the wireless networking industry, and the technologies required to define the industry beyond the current third generation.

This book examines the issues that are currently driving technology development in the wireless world. It surveys the technologies that are, in our opinion, most likely to become part of the foundation for mobile systems in the post 3G era.

Each chapter covers a different technology area. The current technology base is summarized, and the demands for new functionality and how these demands stress current 3G systems is discussed. Where appropriate, we employ existing standards as a tool to describe the current status of the industry, and emerging standards as a tool to anticipate the medium-term future. Emerging standards provide a comprehensive and commercially neutral indication of the most likely direction of mobile systems in the medium term (five to six years). Finally, current research is presented, including discussions about DoCoMo Labs USA group's research into future XG mobile system architectures.

A Note on Terminology: The world of future wireless networking systems is dogged by misunderstanding due to the confusing terminology that unfortunately must be used. The difficulties arise because the same terms have different meanings depending on which side of the Pacific or Atlantic you happen to be on. In this book, we use the following terms with the following meanings.

Next-generation mobile system. This refers to the whole (beyond 3G) mobile communication system in its entirety, including the whole technology "value chain" of the wireless ecosystem from the service platform, through the core and access networks to the user terminal and applications.

Next-generation mobile network. This refers to a subset of the "Next-generation Mobile System" defined above. The Next-generation Mobile Network includes the core network and radio access networks only. Please note that in keeping with established industry practice the shorthand "XG" may be substituted for the term "Next Generation".

4G radio access network (or 4GRAN). This refers specifically to the radio access network in the "Next-generation Mobile System". This is the radio/wireless network connecting the user terminal to the edge of the core network.

4G Wireless access technology. This is a reference to the technology employed in the 4GRAN and may occasionally be used in the same context as 4GRAN.

Acknowledgments

We gratefully acknowledge Takanori Utano, Chief Technical Officer in NTT DoCoMo, Inc., for giving us an opportunity to publish this book. We also acknowledge Candy Wong, a former employee, at present a research engineer for Mobile Software Lab, for her contribution. The following people were extremely helpful in contributing to the chapter on 4G Radio Access technology: Dr Keiji Tachikawa, a former president and CEO of NTT DoCoMo, Inc. and Dr Mamoru Sawahashi, a director of the IP Radio Network Development Department in NTT DoCoMo, Inc. We also thank John Wullert, a co-chair for the Parlay Emergency Telecom Services (ETS) working group, as well as Steve Crowley, a member of the 3GPP2 and IEEE 802 standards bodies, for their review of the chapter on an IP-Based 4G Mobile Network Architecture.

While it is the editor's name that appears on the cover of the book, no book would be possible without the combined efforts of the supporting team. We would like to extend a word of thanks to our management and Kayoko Fujita, a PR&CS Liaison at DoCoMo Communications Laboratories USA, Inc. for administrative support. Their encouragement and support throughout this process is appreciated immensely.

List of Contributors

Frank Bossen, Wai Chu, David Espinosa, Minoru Etoh, Xia Gao, Craig Gentry, Nayeem Islam, Ravi Jain, Moo Ryong Jeong, Toshiro Kawahara, James Kempf, Khosrow Lashkari, Ged Powell, Zulfikar Ramzan, Manuel Roman, Muhammad Mukarram Bin Tariq, Fujio Watanabe, Dong Zhou, 181 Metro Drive, Suite 300, San Jose, California 95110, USA

Xiaoning He, xiaoning@parawireless.com

Henry Song, cs_yus@yahoo.com

Gang Wu, wu@parawireless.com

Alper E. Yegin, Alper.Yegin@Samsung.com

Part I

A Vision for the Next Generation

The three generations of mobile networks deployed to date (1G, 2G, and 3G) have been defined by their technical characteristics. Even when the essential requirements, drafted to excite and initiate work on a new generation, were explicitly written in terms of services (e.g., IMT-2000), the resultant systems were defined (and, unfortunately, divided) by their access network technology.

While it is possible to characterize and differentiate the existing three generations of wireless networks by their service provision (and this is done in Chapter 1), this is not usually done. One of the reasons for this is the networks themselves generally offered services that appeared to be mobile versions of services traditionally available over the PSTN, with the emphasis on speech communication. Thus, the services themselves were not perceived as the differentiating aspect of each network. This is changing with the possibilities for enhanced services offered by current 3G networks and given the advanced services now deploying, describing a network by defining some set of technical aspects of its implementation is becoming almost meaningless.

We stand at the c08f012 of development for the next generation of mobile systems. We argue in this book that this mobile system will not be defined by the technology used in the access network, not least because the access network will be heterogeneous and includes many different technologies. Further, we argue that this heterogeneity will not be confined to the access network but will be an important aspect of the service platform layer. This argument is introduced in Chapter 1 and developed further in the second chapter. Chapter 1 lays out a brief review of the evolution of wireless network development before projecting the evolutionary trajectory to anticipate some key characteristics of the next generation, leading to the definition of imperative technologies for a successful next generation deployment.

Chapter 2 deals with the architectural aspects of the push toward a commercially viable next generation (XG) mobile network. A descriptive review of the development and implementation of the principal third-generation networks deployed today is given in the first part of this chapter. This provides an interesting and important perspective to the second part of the chapter, which develops some rationale and key requirements for the next generation. Important areas, such as the need for an all IP network, an open API approach and an accessible, ubiquitous service layer are discussed, culminating in a proposal for a general XG mobile system architecture. A comprehensive survey of relevant standards is presented, providing a view of current 3G technologies and allowing some educated speculation on future directions beyond 3G.

Next Generation Mobile Systems. Edited by Dr. M. Etoh
© 2005 John Wiley & Sons, Ltd

In this book, we do not focus only on new radio access networks. We anticipate that many other technologies, such as backbone networks, terminal software, and security systems will combine and synergistically evolve to create next-generation mobile systems. While development in the technology is evolutionary, the emergence of killer applications on the integrated systems will be revolutionary.

1

Evolution of Mobile Networks and Services

Minoru Etoh and Ged Powell

Before a book sets about describing the technologies required by the next generation of mobile networks, there is a clear need to address the question, "What is the next-generation mobile network?" A new generation of mobile technology is marked by a significant advance in functionality. It is commonly accepted that, so far, mobile networks have evolved through three distinct generations. These three generations are conventionally defined by their enabling (wireless access) technologies. So, it might seem that defining the technologies required for the next generation will, in fact, provide an answer to our question; however, this convention of defining each generation only by its enabling technologies breaks down when we consider the next or fourth generation. Another, more inclusive definition is required.

It is a commonly stated proposition that next-generation (XG) mobile systems will operate on Internet technology combined with various access technologies (such as wireless LAN) and run at speeds ranging from 100 Mbps in cell-phone networks to 1 Gbps in hot-spot networks (ITU-R Working Party 8F 2003). The technologies implied by this rather limited, technical definition are absolutely necessary, but not sufficient, to provide the required leap into a new generation. In order to produce the significant functional leap required for next generation, full and seamless convergence of mobile networks with the Internet is essential. Moreover, the efficacy (and therefore value) of the new network must exceed that of the current Internet. This could be achieved with enhanced capabilities, such as mobility support, real-time service provision, and reliable security.

If this increase in value is to be maximized, next-generation mobile systems will also need to be developed (and therefore defined) from the user perspective (Mohr 2002;

Next Generation Mobile Systems. Edited by Dr. M. Etoh
© 2005 John Wiley & Sons, Ltd

Tachikawa 2003a), where services and applications (more than technology) define success. The combination of these new technologies with this user-oriented perspective will allow the XG mobile systems to exceed the Internet's current utility. We consider these technologies and this user-oriented perspective to be the XG system imperatives. One purpose of this book is to refine and expand the definition of these imperatives. This chapter provides a brief introduction to each of the XG system imperatives. The following chapters deal with each of these technologies in more detail.

1.1 The Evolution of Mobile Networks

To date, there have been three distinct generations of mobile cellular networks. The first three generations of mobile networks are conventionally defined by air interfaces and transport technologies. However, it is worth noting that each generation clearly provided an increase in functionality to the mobile user, and could therefore be defined in those terms, rather than in transport technology terms. From this perspective, Figure 1.1 shows the generations, their transport technologies, and applications.[1]

In Figure 1.1, we summarize major functionalities of each generation as follows.

1G: Basic mobile telephony service

2G: Mobile telephony service for mass users with improved ciphering and efficient utilization of the radio spectrum

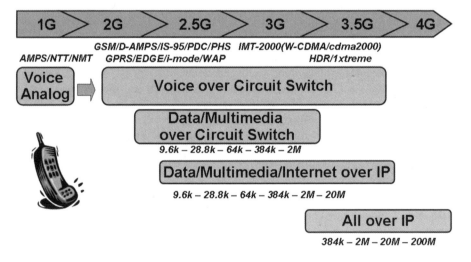

Figure 1.1 Generation of mobile networks

---

[1]The technical distinctions and definitions of 1G–3G mobile networks are more fully developed in several excellent publications including Kaaranen et al. (2001a) and Tachikawa (2002a).

2.5G: Mobile Internet services

3G: Enhanced 2.5G services plus global roaming, and emerging new applications (see the
next section)

The first generation (1G) is based on analog cellular technology, such as the American
Mobile Phone Service (AMPS) in the United States and the NTT system in Japan. The
second-generation (2G) technology is based on digital cellular technology. Commercially
deployed examples of the second generation are the Global System for Mobile Communications (GSM), the North American Version of the CDMA Standard (IS-95) and in Japan,
the Personal Digital Cellular (PDC). GSM also provides interregional roaming functionality.
Owing to this functionality, GSM continues to show outstanding progress by obtaining 1
billion customers worldwide in 2004.

Packet-switched networks were overlaid onto many of the 2G networks, in the middle of
the 2G period. Generally, 2G networks with packet-switched communication systems added
are referred to as 2.5G mobile networks. It is noteworthy that the functional leap between
2G and 2.5G networks delivered arguably the greatest user impact. The i-mode service,
first introduced by NTT DoCoMo in 1999, illustrates the significance of the transition
from 2G to 2.5G as well as the power of considering networks from a user, rather than
technology, perspective. i-mode was the world's first service that enabled browser-equipped
smart phones to access and navigate the web (Natsuno 2003).

This type of wireless data services offers a huge range of services over a variety of handsets. Its mobile computing functionality enables users to perform telephone banking, make
airline reservations, conduct stock transactions, send and receive e-mail, play games, obtain
weather reports and access the worldwide web. The current cell-phone's web-browsing
capability offers access to a wide and growing array of websites from internationally known
companies such as CNN to very local information. It has been a phenomenal commercial
success and continues to develop and expand.

Behind the success of the mobile computing lies the PDC-based Personal Digital Cellular-
Packet (PDC-P) system developed in 1997 (Hirata et al. 1995). PDC-P involves overlaying a
packet-based air interface on the existing circuit-switched PDC network, thus giving the user
the option of a packet-based data service. The i-mode service operates on this packet-based
data service and, as such, relies on the PDC-P network for its continued success. The PDC-P
network is world-leading technology that has clearly fulfilled its promise as an incubation
environment for killer applications. Following the success of PDC-P, the General Packet
Radio Service (GPRS) was introduced as a new nonvoice value-added service that allows
information to be sent and received across a mobile telephone network.

2.5G mobile networks facilitate instant connections where information can be sent or
received almost immediately and without any user activity required to establish a connection.
This is why 2.5G mobile devices are commonly referred to be as being *always connected* or
always on. The Internet service connection provided to cellular users by 2.5G and epitomized
by i-mode is a genuinely remarkable functional leap.

The third generation (3G) arrived in October 2001 when DoCoMo launched its W-
CDMA network. At the time of writing, 3G is beginning to be used in several countries
and is due to be launched very soon in many others. 3G mobile networks are characterized
by their ability to carry data at much higher rates than the 9.6 kbps (kilobits per second)

supported by 2G networks, and several tens of Kbps typically offered by 2.5G mobile technologies.

3G transport technology can be viewed as a mixture of "wireless N-ISDN" and an extended GPRS network. DoCoMo launched its W-CDMA network (this is DoCoMo's 3G network) in 2001. The network allows a 384-Kbps packet-switched connection for downlink and 64-Kbps circuit switch connection that is N-ISDN compatible. The functional leap implemented by 3G networks that differentiates them from 2G networks is also the Internet service connection. In this sense, it could be argued that 2.5G mobile networks and 3G mobile networks fall into the same generation. The 3G mobile networks, however, provide significantly greater bandwidth and can therefore accommodate new mobile services, such as enhanced multimedia applications that cannot be supported by 2.5G mobile networks. This important distinction is the first indication that applications and services, not only technologies, will determine future generation differentiations.

1.2   Trends in Mobile Services

The expansion of mobile communications so far has been led by the growth of voice usage. However, voice usage will certainly saturate in the near future, simply because the population growth is relatively low and the number of active hours in a day are limited. At the time of writing, about 80% of the world's mobile Internet users are in Japan. According to the forecast by Japanese government, the total number of Internet users in Japan is expected to reach 77 million by 2005, from a base of 60 million in 2003. This growth is primarily due to the projected increase in the number of mobile users accessing the Internet from a mobile device. Owing to the massive popularity of mobile Internet access realized by mobile services, the three major telecommunication operators in Japan (DoCoMo, KDDI, and Vodafone) all now operate mobile Internet services and realize significant revenue from data traffic.

In the following discussion, we will focus on Japan's market and technology trends. Japan is currently considered by market commentators to be at least two or three years ahead of most other regions in 3G development and is the clear world leader in mobile Internet usage. Japan therefore provides some interesting insights into potential markets and technologies. In the Japanese experience, there are some useful clues of how market strategists might choose to exploit this new mobile functionality.

Figure 1.2 illustrates the recent data-speed enhancement in Japan. Market research figures show that users are now replacing their 2G cell phones with 3G terminals. The number of 3G subscribers is projected to exceed more than 10 million in 2004. Japan has many competing mobile networks while GSM world (in Europe) uses the single standardized mobile network.

The current and anticipated mobile services in Japan can be viewed as follows:

E-mail: This is a killer application regardless of the mobile network generation. E-mail can be sent to other mobile phones or to anyone who has an Internet e-mail address. Mobile terminals can receive e-mail. There are two important observations to make about this application. First, it costs less than one cent per message. Second, interoperability with Internet mail is guaranteed; unlike the Short Messaging Service in GSM networks, mobile phone e-mail is fully compatible with normal Internet e-mail.

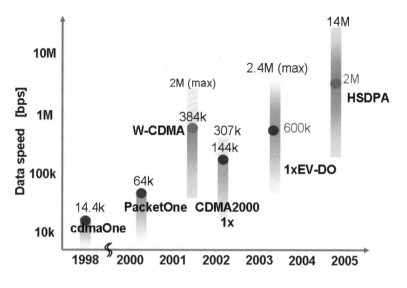

Figure 1.2 Recent data speed enhancement in Japan

Web Browsing: The legacy 2G network allows predominantly text-based HTML browsing with some limited and low-resolution graphics. In 2.5G and 3G networks, the JPEG standard is adopted and is commonly used in addition to the GIF standard. TFT displays with 262,144 colors, 2.4-in. 240 × 320 pixel resolution are typically used on 2.5G and 3G terminals, which is creating a convergence of mobile content and Internet content.

Location-dependent Services: DoCoMo launched a location-dependent web-browsing service in 2001. It is the first actual implementation of location-dependent services. The service delivers users a broad range of location-specific web content. Recognizing 500 different regions, the system pinpoints the locations of the subscribers according to their nearest base station and provides them with a content menu specific to that area. The location-estimation accuracy depends on cell size and the associated base station. The subscriber can then view cell-range information (such as restaurants and hotels), download relevant maps, and even access localized weather reports. Future location-dependent service systems will use both GPS and network information. A mobile device will report its precise location by accessing a GPS receiver, while a location server will instantly report the coarse location of a mobile device by triangulating on its signal. This higher resolution information will facilitate newer services.

Java Application: Users now can download and store a variety of dynamic applications via 2G and 3G networks. Most recent mobile phones are Java capable and able to run a wide variety of Java-based applications. These phones can also run the Secure Socket Layer (SSL) protocol, which provides secure transmission of sensitive information, such as credit card numbers. It is expected that the Java phones will be used for financial services and other e-commerce businesses, in addition to video games.

Videoclip Download: This 3G service enables users to obtain video content at enhanced speeds of up to 384 Kbps. Movie trailers, news highlights, and music files will be among the many types of rich content to be offered. Data will be provided in three formats: video with sound for promotional videos, news, and so on, still frames with sound for things such as famous movie scenes, and sound-only music files.

Multimedia Mail: Mobile picture mail services have proved a major hit in the Japanese market. A multimedia mail typically consists of a still picture or personal video content. In 2003, 80% of new cell phones are being sold with built-in cameras and those CCD resolutions have exceeded 1 mega pixels. Those mail services are extended to enable user to e-mail approximately several hundred KB video clips (of 10 or 20 s) taken either with the handset's built-in dual cameras or downloaded from sites. The phone shoots typically video content at a rate of up to 15 frames/s.

Video Phone: Visual phone service is a typical application on the top of 3G networks. This service utilizes a 64-Kbps circuit connection. Owing to this N-ISDN compatible connection, full interoperability of various video phones has been realized over ISDN, (Personal Handy phone System) (PHS), and 3G networks.

Notably, multimedia mail is expected to become an important or "killer" mobile application along with e-mail and web browsing. DoCoMo's 3G multimedia mail format is compliant to a 3GPP standard, Multimedia Messaging Service (MMS). This is discussed in Chapter 8. Java-enabled applications are anticipated to become very important in the next round of killer applications. Internet connectivity realized in 2.5G networks fostered e-mail and web-browsing applications. Wideband Internet connectivity realized in 3G networks is now fostering multimedia applications and Java-download applications.

In the next generation, it is very unlikely that mobile communication will be limited to enhanced person-to-person communication or broadband multimedia communication. As was the case in the evolution of Java applications, the cell phone is developing into an important mobile platform for daily life tools. This is currently occurring through interfaces such as IrDA, Bluetooth, contactless IC and Radio Frequency Identification (RFID) tag. Future wireless communication technologies will further fuel this movement. In addition to cell phones, built-in wireless modules that are dedicated for specific purposes and capable of directly communicating with 2.5G and 3G networks are becoming available. As a result, numerous new applications are emerging. Commuter pass, home-security services, product-delivery tracking, remote control of vendor machines, and telemetry are all examples of what is and what will become possible. These applications are depicted in Figure 1.3. These machine-to-machine services are now becoming commercially available, and it is reasonable to expect that this application area will grow into a significant component of XG services. Wireless machine-to-machine communication is clearly about enabling the flow of data between machines (nonhuman generators and consumers of data). However, it is important to understand that the ubiquity of machines is also an important assumption in this proposition. Work on sensor networks points very clearly to a world with ubiquitous machine data generators. It is not difficult to imagine these sensors communicating with each other in a local network before transmitting their (perhaps collective) data to some central point for action or storage. Given this kind of scenario, machines and machine networks communicating directly with humans is a very likely possibility. Notwithstanding the human

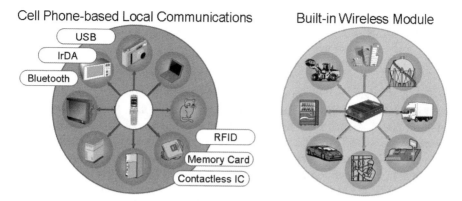

Figure 1.3 Beginning of machine-to-machine computing

dimension just introduced into this discussion, we will refer to this whole area as M-to-M computing. In the future, communication-capable devices and home electronic appliances will form various local device networks, and these local networks will interoperate with the global mobile networks. This description is typical of the functionality and ubiquity envisioned for next-generation mobile systems. The evolution of each network generation enriches mobile applications, and the applications bring the generation differences into very sharp focus. This is as it should be if we are to consider network generations from a user or services perspective.

1.3   Why Next-generation (XG) Mobile Systems?

When considering next-generation wireless access technologies, it seems clear that Japan, the United States, and Europe are diverging in their approach.

- In Japan, the focus is on pushing the technology envelope to achieve extremely broadband wireless, broader even than wired broadband in most of the world. We will discuss 4G wireless access technology in Chapter 3.

- In the United States, the focus is on developing a physical layer and media access that is not only a much better fit with IP than the current 3G protocols but can also cover a wide area and deliver to faster moving vehicles, which 802.11 cannot achieve. Standardization initiatives, such as 802.16 and 802.20, are proof of this focus.

- In Europe, the focus is on maximizing value from the existing 3G wireless access protocols, for example, by working on seamless intertechnology handover between GPRS and 802.11.

In the current US and European approaches, radically new wireless protocols and technologies may not be needed. The European and US approaches are referred to as "beyond 3" and neither appears at this time to need to employ radically new wireless access methods. In a commonly used narrow definition, 4G is taken to mean the next generation of wireless access networks that will replace 3G access networks in future. It is important to understand that throughout this book we use the term "XG" to avoid the confusion caused by 4G wireless perspective. In our definition, the XG system includes service overlay networks and terminal communication software with security enhancements.

Thus, we consider the description of XG mobile systems that we discussed in the introduction to this chapter. The kind of seamless, user-oriented experience required to make the transition into XG means that XG devices will be much more than Internet service connections (Forum 2001). As described in the introduction, a full integration of the Internet with wireless networks is a XG mobile system imperative. Please note that the integration is not straightforward, since the Internet was originally designed for fixed networks.

Currently, two distinct domains exist. These can be called the _wireless world_ and the _Internet world._ In 2.5G and 3G, these two domains are connected through network devices (gateways) that provide protocol, control, and other necessary translation functions. These gateways are one reason why current 2.5G and 3G mobile network Internet connections provide only a subset of the services available from the Internet. The other reason is the difference in bandwidth. The bandwidth gap is a legacy of the voice-centric evolution of wireless cellular networks and the scarcity of radio spectrum. The gateways are needed precisely because true convergence of the two worlds has not yet been realized. These two constraints result in a restricted or "watered down" user experience of wireless Internet access, as depicted in Figure 1.4.

The gateway connects the mobile network to the Internet. It counts the number of packets or e-mail transactions for billing purposes and also provides firewall protection to the mobile network. However, there is an important drawback in using this gateway model with current mobile networks, that is, its inability to provide IP-transparent, seamless connectivity. This

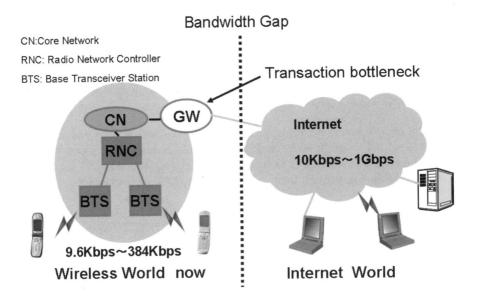

Figure 1.4 Divided world

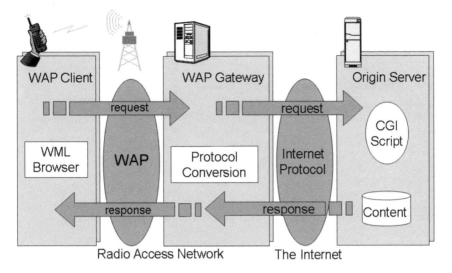

Figure 1.5 WAP architecture

issue is discussed in Wisely et al. (2003). The current gateway approaches provide two levels of connectivity: application level and transport level. Gateways provide this connectivity in a way that both connects and separates the Internet and wireless worlds.

Consider a wireless application protocol (WAP) architecture as an example of application-level connectivity. As depicted in Figure 1.5, WAP enables provision of a web browser for mobile terminals. The WAP gateway acts as a middleman or translator receiving user requests and reformatting them into HyperText Transfer Protocol (HTTP) requests. There are pros and cons in this method of communication between mobile terminals and the Internet. WAP clearly provides a workable content delivery capability for mobile terminals via a wireless link. On the other hand, access to the Internet is clearly not seamless and under certain conditions can be blocked by the gateway. The WAP approach also has serious implications for e-mail. All e-mail transactions to the Internet are relayed by the WAP gateway and are handled as SMS transactions (with the attendant charges). The same is true of e-mail messages from the Internet to the WAP terminal.

i-mode's Internet connectivity is fundamentally different from WAP. The gateway relays i-mode e-mail simply according to the Internet mail address. In supporting web access, unlike WAP, i-mode utilizes transport-level connectivity, where TCP/IP packets are relayed by the gateway. The transport-level connectivity has fostered a number of mobile applications outside the gateway. This connectivity provides more transparency to the mobile network, but there are two new issues: the heavy IP tunneling protocol stack and temporal connectivity. Figure 1.6 shows the protocol stack for the 3G standard. On the basis of the current 3GPP specifications, user IP datagrams are transferred via UDP tunneling between backbone switches. The tunnel is typically built on an ATM infrastructure, and IP over LAN emulation (LANE) over ATM carries the tunnel. Surprisingly, there are eight layers of protocol stacks. On one hand, an advantage of this methodology is that the Quality of Service (QoS) can be achieved by ATM adaptation layer. Voice traffic and switched circuit

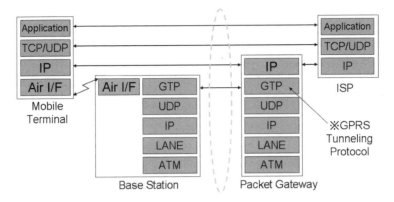

Figure 1.6 Media transport protocol on W-CDMA networks

multimedia traffic are directly mapped on ATM layer to guarantee QoS and also to reduce protocol overhead. When the voice traffic is dominant, this method is an extremely efficient and effective solution for a mixed traffic network. The tunneling protocol performs well so far in 3G networks for current and emerging "killer" applications, such as e-mail and web browsing, which use client–server communication. On the other hand, when the nonvoice traffic among numerous applications is dominant over ATM, that efficiency will be lost.

It is anticipated that future applications will employ peer-to-peer (P2P) communication to achieve functionalities such as file sharing as discussed in (Matei et al. 2002). Unlike the server–client communication, P2P is a communication model in which each application has the same capabilities and either application can initiate a communication session. In recent usage, P2P has come to describe direct file-sharing applications through a mediating server. We use the term P2P for a communication model that requires seamless connectivity. In addition to P2P applications, P2P networking technologies will facilitate M-to-M computing where significant amounts of M-to-M communication both internally (within the local network) and externally (to and from the local network) are generated. In this case, the gateway-centric tunneling protocol may face network addressing and congestion problems. The current gateway technique works in a similar way to a huge network address translation (NAT) system, where the terminal's IP addresses is dynamically assigned and temporally used for Internet connectivity. To facilitate P2P applications and M-to-M computing, it is clearly advantageous to provide IP-transparent seamless connectivity that is both internal and external. This will also avoid the possible transaction bottleneck caused by the gateway-centric model. If this can be realized, a simplified protocol stack can be employed in which all data is carried on native IP. A network with this simple protocol could also accommodate new application servers with direct connectivity to the mobile network. By employing a simpler IP routing without temporal address assignment, an extensible backbone can also be built and there are no heavy protocol problems. To maintain highly reliable communication link, and to support internal services such as itemized print billing and immediate credit, and to foster various new third-party applications, the extensibility is the key functionality. Extensibility also helps to accommodate heterogeneous radio access networks (RAN). However, several technical issues may emerge in network security, traffic control, mobility support, and so forth. They are discussed in Chapter 2 and subsequent chapters. Please note that some gateways are required to mitigate the effects of mobility and security threats. These may be functionally as well as physically distributed, rather than centralized. The issue is to find the optimal point of compromise between the seamless model and the confined one.

The previous discussion dealt with some of the difficulties associated with the gateway-centric nature of current solutions in providing connectivity between the Internet and wireless worlds, while recognizing IP-transparent seamless connectivity raises additional technical challenges.

The other major problem with the current situation is the difference in access network bandwidth between mobile networks and typical fixed-line access networks. To be precise, the problem is the limited bandwidth of the wireless access networks. The current DSL connection provides about a 12 Mbps permanent connection (Taniwaki 2003), while a typical 3G network provides several hundred Kbps connection bandwidth at most. There is a bandwidth gap, at least, of an order of magnitude. Discussions on this subject tend to be dominated by concerns about services such as streaming media, because that type of application typically requires high data bandwidth. However, it is also true that that overall network bandwidth and overall latency (the time lag between cause and effect) are inversely related. More bandwidth generally means less latency[2]. Reduced latency can be as valuable (from a service provider's perspective) as high data bandwidth. For services like Java program downloading, mobile payments, gaming and, in fact, any real-time application, latency may be the difference between a successful service and a failure. Fast delivery is essential in many typical mobile applications, for example users can view HTML content or download a Java application in a few seconds on the road. A higher bandwidth is also clearly beneficial to real-time applications (such as voice over IP) that require less than 250 ms latency in round-trip time.

It is clear that high system bandwidth is a desirable characteristic across the range of anticipated applications. ITR-R has predicted that by the year 2010 potential new radio interfaces will need to support data rates of up to approximately 100 Mbps for high-mobility situations, such as mobile access, and up to 1 Gbps for low-mobility situations, such as nomadic/local wireless access (ITU-R Working Party 8F 2003).

Radio access networks are conventionally characterized by two features: bandwidth and mobility. Another equally important feature is often neglected. This is connection ubiquity. Connection ubiquity is a generic term, used to describe service availability with radio access networks for which cell density and power consumption are important considerations. Figure 1.7 summarizes the important features of various radio access networks.

To ensure connection ubiquity together with high bandwidth and mobility, the network architecture must be heterogeneous rather than homogeneous (see Figure 1.7). The heterogeneity of the RAN will also provide a wider choice to customers, allowing providers to meet customers' individual requirements. As a result of this requirement for heterogeneity, the systems beyond 3G will be realized by a functional fusion of existing, enhanced, and newly developed elements of cellular systems, nomadic wireless access systems, as well as other wireless systems with high commonality and seamless interworking.

---

[2]In the wired Internet this may not hold. As wired bandwidth increases, the transmission time for packets will approach zero. However, the fixed latencies of propagation and variable latencies of router congestion will dominate.

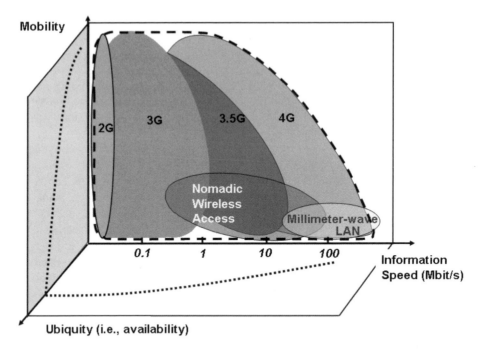

Figure 1.7   Radio access networks beyond 3G

As indicated in Figure 1.8, different radio access systems will be connected via the extensible IP backbone. The ITU-R also points out that interworking between these different access systems will be a key requirement, which can be handled in the core network or by suitable servers accessed via the core network (ITU-R Working Party 8F 2003). This interworking task includes management of handovers (horizontal and vertical) and seamless service provision with service negotiation including mobility, security, and QoS management.

The discussion has so far centered on two major technical problems: the bandwidth gap and the gateway-centric model. If the user-centric perspective is considered, Figure 1.9 summarizes the current situation. It is clear that essential XG components are missing from both the Internet and mobile networks. Some Internet-type applications, such as e-mail and web browsing, have been realized in 2.5G and 3G mobile networks; however, current mobile networks cannot provide true P2P applications based on seamless connectivity. Nor can they provide high-bandwidth applications, such as high-quality video streaming and real-time delivery of large volume of data. These applications are common on the Internet. On the other hand, the Internet does not sufficiently specify any integrated implementation of the AAA, mobility management, and ubiquitous connectivity support that is currently provided by mobile networks.

A XG mobile network will be the network that removes the gap between the wireless and Internet worlds in terms of connectivity and provides a superset of the current Internet's utility to mobile users. Thus we could define it as:

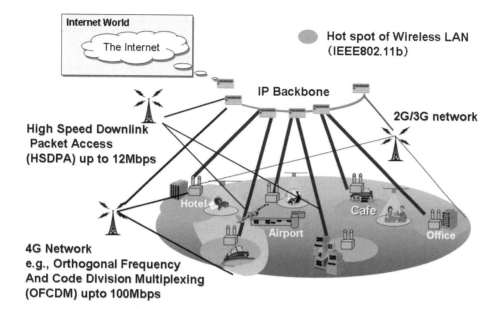

Figure 1.8 Integration of network access channels

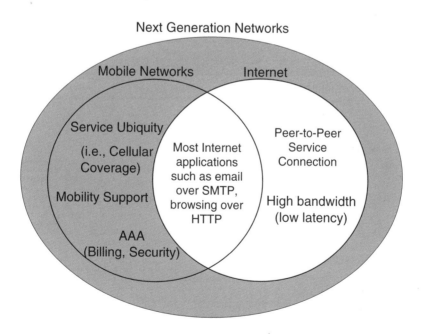

Figure 1.9 XG mobile network as a super set

(**XG:**) Seamless Mobile Internet Service Network that removes the gap between the wireless and the Internet worlds, and combines the positive aspects of the two worlds.

1.4 Next-generation Imperatives

As has been discussed in the previous sections, the next-generation network cannot be defined only by wireless technologies, air interface, IP backbone, or bandwidth. Many researchers, especially those involved in the wireless communication area, understandably tend to define XG mobile systems in these terms. We define a more comprehensive description of the XG network that recognizes that service ubiquity is an integral part of the XG description. A ubiquitous mobile Internet as described here will be an extremely fertile incubation environment for new and innovative applications. If the environment is open and accessible, innovative business models will drive countless new and diverse applications on top of these new technology environments. We use this description to derive a list of XG imperatives.

These imperatives are the existence of a RAN, an IP network, a ubiquitous service platform, and applications. In recognition of these imperatives, we have carefully selected technical topics for this book to cover the XG mobile systems and application technologies. Figure 1.10 briefly depicts the supposed XG mobile system architecture and each component is described in more detail in the following sections.

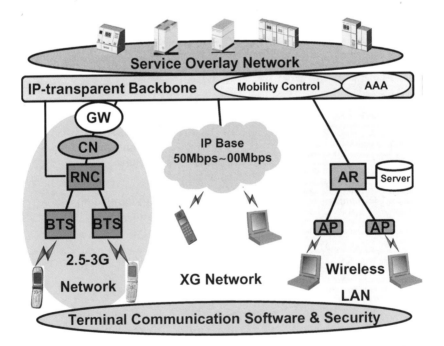

Figure 1.10 XG mobile system

1.4.1 Radio Access Networks (RAN)

We anticipate that various and complementary radio access networks will be used in combination to form the RAN in XG. Networks such as 2.5G, 3G, enhanced 3G, for example, high-speed downlink packet access (HSDPA) technology in Figure 1.2, 4G orthogonal frequency and code division multiplexing (OFCDM) technology and wireless LANs (WLAN), as we discussed in Figures 1.7 and 1.8. We anticipate that wireless LAN technologies will be one of the principal RAN technologies for the next-generation mobile communication system. These subjects are covered in Chapters 3 and 4.

1.4.2 IP Backbone

The main issues here are how to seamlessly integrate the heterogeneous radio access networks and to realize the two major functionalities on the top of RANs. The two functionalities are AAA (Authentication, Authorization and Accounting) and mobility support. The implementation of those functionalities must satisfy real-time constraints required by mobile applications and a large terminal base. Chapter 2 gives us a comprehensive discussion on the XG network architecture. Chapter 5 deals with the mobility and integration issues, and Chapter 11 addresses Authentication, Authorization, and Accounting.

1.4.3 Ubiquitous Service Platform

The term *Ubiquitous Service Platform* is used to cover terminal and application aspects of service delivery. The term also denotes a coherent set of characterizing concepts, as defined below:

(i) Heterogeneity in wireless access networks, backbone networks, mobile terminals and applications

(ii) Openness in terms of allowing and supporting third-party service providers to deploy and compose application services, such as web services

(iii) Ability to allow mobile users to engage in all kinds of Internet transactions and services with appropriate trust and security relationship management support and open interfaces support

We expect that the platform will consist of two major parts: the service overlay network at the network side and the terminal communication software at the terminal side. An XG network is expected also to support location-dependent information retrieval, Java-enabled services such as e-commerce, and media delivery as we saw in Section 1.2. Unlike the gateway-centric implementation, we expect that a service overlay network on the top of the IP backbone will accommodate these applications together with terminal communication software, especially when real-time operation and security are required. The vast majority of currently deployed systems assume a relatively simple client that is browser based and nonprogrammable. As we can see in Java applications, the advent of smarter phones with faster processors, more memory, and persistent storage is making it possible to provide a better user experience and new classes of applications. Note that terminal will be an essential part of the Ubiquitous Service Platform, where Open APIs and web protocols are also crucial technologies. The Ubiquitous Service Platform is a relatively complex concept and impacts many technology areas. For this reason, an understanding of Chapters 2, 6, 7, 9, and 12 is required in order to gain a full insight into this important XG component.

We also consider the AAA and the Mobility support at sub-IP layers of RAN, and in the service platform, since those functionalities are realized by well-harmonized coordination of networks and terminals. This discussion also relates to the system security that comprises network security and terminal security. Chapter 10 addresses this important area. Any security solution must be scalable without practical limit and flexible but robust. Not only will terminal base be huge, but terminal networking environments may be heterogeneous and rarely in a stable state.

Multimedia traffic is increasing far more rapidly than speech, and will increasingly dominate traffic flows. Since XG will effectively remove the limitations on bandwidth, the network will provide the user with the ability to more efficiently discover and receive multimedia services including e-mail, file transfers, messaging and multimedia distribution services. These services can either be symmetrical or asymmetrical, real time or not real time. They may consume data at rates requiring high bandwidths and low latency. With this forecast, we identify application technologies (for example, media coding technology), in addition to network and terminal technologies, which are essential to foster new applications. These new multimedia technologies are discussed in Chapter 8.

The following chapters cover and expand upon the essential technologies discussed in this chapter. It is our belief that these technologies, developed and deployed as we describe in this book, will result in a commercially viable and life-enhancing next-generation communication network.

2

The All-IP Next-generation Network Architecture

Ravi Jain, Muhammad Mukarram Bin Tariq, James Kempf, Toshiro Kawahara

2.1 Introduction

What is the next generation (XG) of mobile networks? One way of classifying generations of mobile communications technology is by the protocols or data rate over the air interface, ranging from 9.6 kbps for 1G to 384 kbps for 3G. Thus, XG could be defined in terms of air interface data rate also (say, Internet Protocol over the air or 100 Mbps downlink). However, the difficulties that 3G deployment is currently facing outside Japan, while probably temporary, clearly indicate that data rates alone are not enough to motivate many users to adopt this new technology.

In contrast, we consider the shift to XG fundamentally in terms of the innovative services and applications that users will have available and be willing to pay for. This orientation leads to several design choices. The first is that the next-generation architecture is based on supporting the Internet Protocol (IP) as a fundamental construct in all parts of the system, that is, an all-IP network, and, in particular, one based on IP version 6 (IPv6). While this choice is now becoming widely accepted in the technical community, it is important to isolate and critically examine the reasons for it. The second design choice is that the architecture is defined by a layered family of Application Programming Interfaces (APIs), some public and some private, but all designed to facilitate access to the network resources in a secure, useful, and billable manner. The third is that the need for rapid and flexible application deployment is causing migration of intelligence from the core toward the periphery of the

Next Generation Mobile Systems. Edited by Dr. M. Etoh

system, in both IP-based networks as well as the Public Switched Telephone Networks (PSTNs), and the XG architecture must be consistent with this trend.

This discussion starts with a review of the main 3G architectures, those developed or proposed by the industry for the 3rd Generation Partnership Project (3GPP), 3rd Generation Partnership Project-2 (3GPP2), and Mobile Wireless Internet Forum (MWIF), and briefly discusses their limitations in terms of both network and service architecture. Section 3 describes an approach to developing an XG architecture, starting by elaborating the rationale for key design choices. The last section also presents a high-level view of our proposed XG architecture, including its separation of functionality into four basic layers.

2.2   3G Architectures

When third-generation (3G) systems were initially considered, the goal was to enable a single global communication standard that could fulfill the needs of anywhere and any-time communication. International Telecommunications Union's (ITU) International Mobile Telecommunications (IMT-2000) vision (ITU-T 2000a) called for a common spectrum worldwide (1.8–2.2 GHz band), support for multiple radio environments (including cellular, satellite, cordless, and local area networks), a wide range of telecommunications services (voice, data, multimedia, and the Internet), flexible radio bearers for increased spectrum efficiency, data rates up to 2 Mbps in the initial phase, and maximum use of Intelligent Network (IN) capabilities for service development and provisioning. ITU envisioned global seamless roaming and service delivery across IMT-2000 family networks, with enhanced security and performance as well as integration of satellite and terrestrial systems to provide global coverage. Although some of the technical goals have been achieved, the dream of universal and seamless communication remains elusive. As a reflection of the regional, political, and commercial realities of the mobile communications business, the horizon of third-generation mobile communications is dominated by two largely incompatible systems.

One realization of IMT-2000 vision is called the *Universal Mobile Telecommunications System* (UMTS), developed under 3GPP.[1] This system has evolved from the second-generation Global System for Mobile Communications (GSM) and has gained significant support in Europe, Japan, and some parts of Asia. The system is sometimes simply referred to as the 3GPP system; however, we will refer to it as the UMTS network in this chapter.

The second version of the IMT-2000 vision continues to be standardized under 3GPP2[2] and is referred to as the CDMA2000 or 3GPP2 system. This system has evolved from the second-generation IS-95 system and has been deployed in the United States, South Korea, Belarus, Romania, and some parts of Russia, Japan, and China, that is, mostly the regions that had IS-95 presence. This chapter refers to this system as the CDMA2000 system.

These two systems are similar in functional terms, particularly from a user's point of view. However, they use significantly different radio access technologies and differ significantly in some of their architectural details, making them largely incompatible. This section

---

[1] 3GPP Organizational Partners include: Association of Radio Industries and Businesses (ARIB) of Japan, China Communications Standards Association (CCSA), European Telecommunications Standards Institute (ETSI), T1 of USA, Telecommunication Technology Association (TTA) of Korea, and Telecommunication Technology Committee (TTC) of Japan.

[2] 3GPP2's organizational partners include ARIB, CCSA, TIA, TTA, and TTC.

provides an overview of the architectural aspects of the UMTS and CDMA2000 systems. It also briefly discusses the architecture developed by the MWIF, as a proof of concept for all-IP mobile communications networks, and which contains many architectural approaches that will be important for next-generation systems. While MWIF itself has disbanded, work is being continued under the aegis of the Open Mobile Alliance (OMA). This chapter presents the 3GPP architecture in some detail, but for CDMA2000 and MWIF, we focus on the similarities and differences with the UMTS network.

2.2.1   UMTS

When 3G standardization efforts began in the latter half of the 1990s, a conscious effort was made to align 3G with the existing 2G GSM solutions and technologies. GSM at that time was, and for the most part still is, the dominant mobile communications standard through much of Europe and Asia. The decision to base 3G specifications on GSM was motivated by widespread deployment of networks based on GSM standards, the need to preserve some backward compatibility, and the desire to utilize the large investments made in the GSM networks. As a result, despite its many added capabilities, the UMTS core network bears significant resemblance to the GSM network.

So far, 3GPP has produced three releases. The first was released in March 2000 and is called *3GPP Release 99 or 3GPP-R99*. This release carries a very strong GSM flavor. For example, the core network design for circuit-switched traffic is almost identical to the GSM network. Japan became the venue for the first deployment of 3GPP-R99 when NTT DoCoMo rolled out its full commercial 3G service, referred to as Freedom of Mobile Multimedia Access (FOMA) in late 2001. 3GPP has since published two more, Release 4 (3GPP-R4) in March 2001, and Release 5 (3GPP-R5) in mid-2003. Release 6 (3GPP-R6) is expected in the spring of 2004. While the overall architecture in each of these releases is derived from GSM, there are certain important differences. These are summarized in Table 2.1 and described briefly below. This section provides a general overview of the UMTS network architecture as it stands in Release 5.

Network Architecture

The network architecture for 3GPP-R5 is described in documents from its Technical Specification Group (3GPP 1999b). 3GPP uses the term Public Land Mobile Network (PLMN) for a land mobile telecommunications network. The PLMN infrastructure is divided logically into an access network (AN) and a core network (CN). On top of the network infrastructure is a service platform, which is used for creating services. Figure 2.1 shows the very high level organization of the UMTS network.

The network supports two types of access networks, namely, the Base-station System (BSS) and the Radio Network Subsystem (RNS). BSS is the GSM access network, whereas RNS is based on UMTS, in particular the Wideband Code Division Multiple Access (W-CDMA) radio link. The Radio Access Network RAN specifications in Release 99 only include UMTS Radio Access Network (UTRAN), but allude to other alternative radio access networks. However, later releases have standardized a GSM/EDGE-based RAN, called *GERAN*.

Table 2.1   Evolution of 3GPP specifications

3GPP Release	Freeze Date	Highlights
3GPP-R99	2000	Creation of UTRAN both in FDD and TDD CAMEL phase 3 Location services (LCS) New codec introduced (narrowband AMR)
3GPP-R4	2001	GERAN concept established Separation of MSC into a MSC server and media gateway for bearer independent CS domain Streaming media introduced Multimedia messaging
3GPP-R5	March–June 2002	Introduction of IMS; IPv6 introduced in the PS domain IP transport in UTRAN Introduction of high-speed downlink packet access (HSDPA) Introduction of new codec (wideband AMR) CAMEL phase 4 OSA enhancements
3GPP-R6 (expected features)	Expected March 2004	Multiple input, multiple output antennas IMS stage 2 WLAN-UMTS interworking MBMS

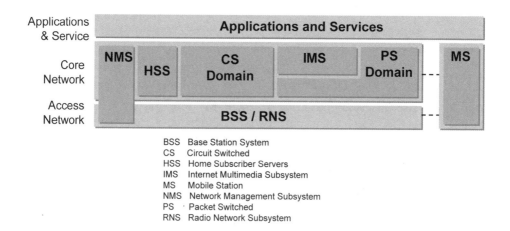

BSS   Base Station System
CS   Circuit Switched
HSS   Home Subscriber Servers
IMS   Internet Multimedia Subsystem
MS   Mobile Station
NMS   Network Management Subsystem
PS   · Packet Switched
RNS   Radio Network Subsystem

Figure 2.1 High-level architecture of UMTS network

While both types of AN provide basic radio access capabilities, UMTS provides higher bandwidth over the air interface and provides better handoff mechanisms, such as soft handover for circuit-switched bearer channels.

The CN primarily consists of a circuit-switched (CS) domain and a packet-switched (PS) domain. These two domains differ in how they handle user data. The CS domain offers dedicated circuit-switched paths for user traffic and is typically used for real-time and conversational services, such as voice and video conferencing. The PS domain, on the other hand, is intended for end-to-end packet data applications, such as file transfers, Internet browsing, and e-mail.

3GPP-R5 also includes the IP Multimedia Subsystem (IMS). Its function is to provide IP multimedia services, including real-time services, in the PS domain, including those that were previously only possible in the CS domain. A CN based on 3GPP-R5 can contain a CS domain, PS domain, IMS on PS domain, or a combination of these.

In addition, the core network has a logical function called the Home Subscriber Server (HSS) that consists of different databases required for the 3G systems, including the Home Location Register (HLR), Domain Name Service (DNS), and subscription and security information. It also provides necessary support to different applications and services running on the network. Network management is provided by the Network Management Subsystem (NMS), which essentially forms a separate vertical plane.

Figure 2.2 presents a more detailed view of the network architecture. A brief description of different subsystems follows, starting from the mobile station (MS), shown at the bottom of the figure. For full details, please refer to the specification (3GPP 1999b) and references therein.

The user's terminal is called a *mobile station* (MS) and logically consists of mobile equipment (ME) and an identity module. The ME consists of equipment for radio communication, while the identity module contains information about the user identity. The separation of MS and identity module achieves separation of the user and the device that, in principle, allows a user to switch to a different device by merely plugging in an identity module. The network supports two types of identity modules, the Subscriber Identity Module (SIM), similar to GSM systems, and the UMTS SIM (USIM), based on whether the station belongs to the older GSM-based system or to the newer UMTS-based system.

The RNS consists of a radio network controller (RNC) that controls radio resources in the access network. The RNC performs processing related to macrodiversity, and provides soft-handoff capability. Each RNC covers several Node Bs. A Node B is a logical entity that is essentially equivalent to a base-station transceiver; it is controlled by the RNC and provides physical radio-link connection between the ME and the RNC. Similarly, the BSS consists of a Base-station Controller that controls one or more Base Transceiver Stations; however, unlike the Node B, each corresponds to one cell. The IuCS and IuPS interfaces connect all mobiles in the access network to the CS and PS domains of the CN respectively.

The CS domain contains the switching centers (the Gateway Mobile Switching Center or GMSC and the Mobile Switching Center or MSC) that connect the mobile network and the fixed-line networks. These are analogous to exchanges in the PSTN, except that the MSC also stores the current location area of the MS within a location register called *visited location register* (VLR). The MSC also implements procedures related to handover between the access networks, that is, when the ME moves from the coverage area of a RNC to another

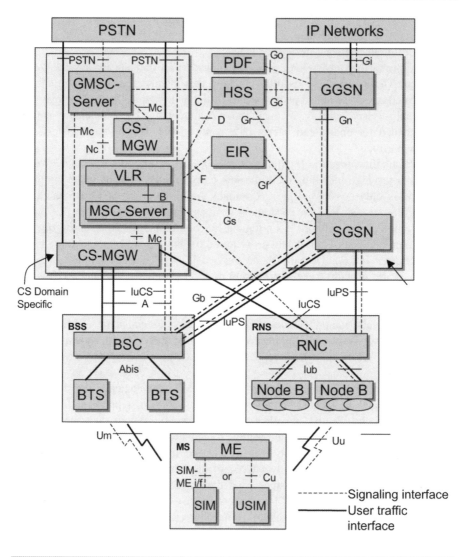

Figure 2.2 Basic PLMN configuration

or from one BSC to another. With the Release 4 and Release 5 networks, the MSC function is split between a circuit-switched media gateway (CS-MGW) and an MSC server, as shown in Figure 2.2. The MGW handles user traffic, whereas the MSC server deals with location and handover signaling. This separation makes the core network somewhat independent of the bearer technology. It is similar to the Next-generation Network (NGN) architecture based on a Softswitch (also known as a Call Agent) developed for fixed networks (3GPP 2000c). The gateway MSC (GMSC) in the core network is similar in function to the MSC, except that it is logically situated at the border between the mobile network and the external networks and acts as a gateway. The GMSC relies on the HLR for location management, whereas the other MSCs are internal to the network and rely on VLRs that are often collocated with the MSC.

The PS domain provides the General Packet Radio Service (GPRS). The PS domain consists of the GPRS support nodes, which are counterparts to the MSC in the CS domain; they maintain the subscription and location information for the mobile stations and handle the user's packet traffic and the PS domain-related signaling. There are two types of GPRS support nodes: the Gateway GPRS Support Node (GGSN) and the Serving GPRS Support Node (SGSN). These are analogous to the GMSC and MSC. The GGSN and the SGSN are sometimes collectively referred to as GSN or xGSN.

For the purposes of location management, the PLMN is divided into several areas of varying scope. The PLMN maintains the location of the mobile node for the purpose of reachability in terms of several location regions (see Figure 2.3). The first of these are the location areas (LA), which are used for locating the user for CS traffic; each is served by a VLR, and a VLR may serve several LA. The routing areas (RA) are used for locating the user for PS traffic; one or more RAs are managed by a SSGN. The UTRAN Registration Area (URA) are smaller than the RA, and cells are the smallest unit of location. Typically, a URA contains the cells controlled by a single RNC. An RA and LA may contain one or more URA. An LA may contain more than one RA, but not vice versa.

The SGSN handles the user's data traffic, including functions such as initial authentication and authorization, admission control, charging and data collection, radio resource management, packet bearer creation and maintenance, address mapping and translation, routing and mobility management (within its serving area), packet compression, and ciphering for transmission over the RAN.

The association information between the PS core network and the MS for an active packet session is encapsulated in a Packet Data Protocol (PDP) context, which contains the information necessary to perform the SGSN functions. It includes information about the type of packet data protocol used, associated addresses, addresses of upstream GGSNs, and the identifiers to lower layer data convergence protocols in the form of access point identifiers, NSAPI and SAPI, to route the packets to and from the access network. Figure 2.4 contains a diagram of the PDP context.

The GGSN is often located at the edge of the PS domain and handles the packet data traffic to the UMTS network from outside the network and vice versa. The GGSN performs an important role in mobility management, packet routing, encapsulation, and address translation. The most visible role for the GGSN is to redirect incoming traffic for a mobile station to its current SGSN.

The GPRS Tunneling Protocol (GTP) is used for carrying traffic between the SGSN and the GGSN. It carries control-plane information, such as commands to create, query, modify, or delete the PDP context, as well as user-plane data.

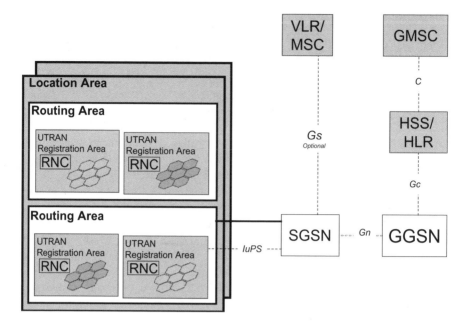

Figure 2.3 A typical PLMN layout

There are a number of other entities and logical functions in the 3GPP architecture that are not shown in Figure 2.2. These include the mobile location centers (MLC), number portability databases, security gateways, signaling gateways, and network management entities and interfaces. While these are important functions, they are not part of the basic transport and service network architecture, and hence are omitted here.

Figure 2.5 shows some components of the IP multimedia subsystem. The IMS provides support for multimedia services, such as voice, video, and messaging over IP networks. IMS uses the Session Initiation Protocol (SIP) for signaling. The Call State Control Function (CSCF) has a role similar to the MSC in the CS domain. It terminates IMS signaling (SIP) and provides call control functions. The IMS also contains media gateways that provide interworking with legacy networks, such as the PSTN, and perform other resource-intensive functions, such as mixing of media streams from multiparty conferences and transcoding. Unlike the CS domain, the signaling entities (CSCF) are completely separate from the media processing. The CSCF communicates with the Media Gateway Control Function (MGCF) using SIP, and MGCF in turn controls the media gateways using ITU-T H.248[3] (ITU-T 2000b). The MGCF also provides necessary interworking with external networks in the signaling plane.

Service Architecture

3GPP has adopted extensive specifications for services and service creation. This section briefly summarizes the main concepts.

---

[3]H.248 is also known as the Media Gateway Control (Megaco) Protocol (Groves et al. 2003)

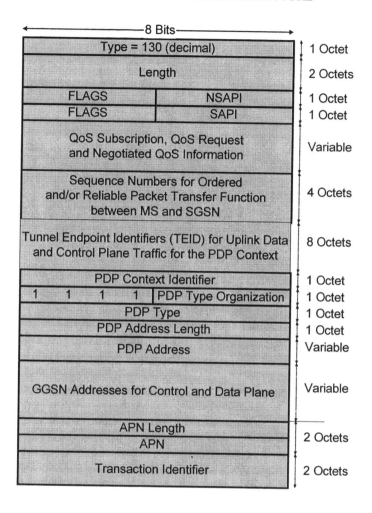

Figure 2.4 PDP context

Services in UMTS are viewed as having a layered structure as shown in Figure 2.6. While several features of this diagram can be debated, the attempt at classifying services is worthwhile. At the lowest level are *bearer services,* such as circuit-switched transport. Short Message Service (SMS) (3GPP 1999c), Unstructured Supplementary Service Data (USSD) (3GPP 1999e, 2000e,f), and User-to-User Signaling (UUS) (3GPP 1999f,g,h) are additional bearer services that can be used by the applications to send different types of content.

Circuit teleservices operate in the CS domain and consist of simple telephone calls, fax, and the like. *Supplementary services* also operate in the CS domain and provide enhancements such as call waiting, call forwarding, and three-way calling. *Non-call-related services* are those that do not directly relate to a call in progress, for example, notification that a voicemail or e-mail message has arrived. *Non-call-related value-added services* are those that do not relate to voice calls but offer, for example, advanced data capabilities such as

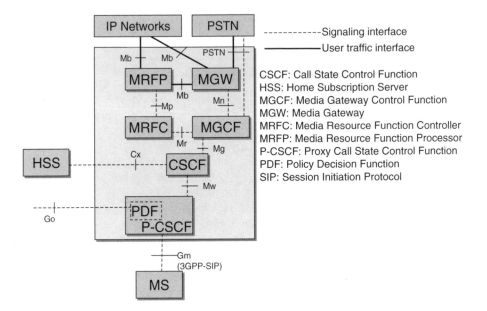

Figure 2.5 IP multimedia subsystem

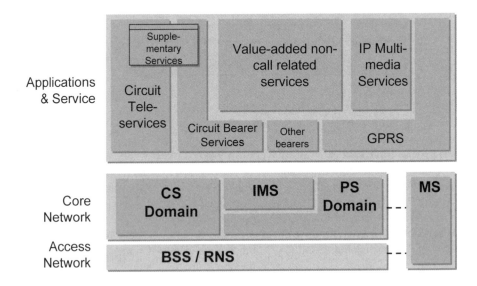

Figure 2.6 UMTS service classification

e-mail access, web browsing, and file transfer. Finally, *IP multimedia services* are those that deal directly with multimedia, for example image and video download and streaming.

It is important to point out that while 3GPP classifies services and provides an architecture and general requirements for services, the actual services themselves are not standardized. Instead, 3GPP standardizes *service capabilities*, which consist of generic bearers – defined by Quality of Service (QoS) parameters such as bandwidth, delay, and symmetry – and the mechanisms needed to realize services, including the functionality provided by various network elements, the communication between them, and the storage of associated data.

An overarching service concept in UMTS is the *virtual home environment* (VHE) (3GPP 2000h, 2002a). The basic idea of the VHE is that, as far as possible, users should have available a consistent and personalized set of services and features as well as a consistent user interface and "look and feel," regardless of which network and which terminal they use. To enable this, the VHE standards aim to provide mechanisms that allow uniform' means for accessing services, as well as a means for creating services.

VHE assumes that a user has a home environment, where one or more user profiles are defined and stored. When a user moves outside this environment, the user profiles can be utilized to provide a "virtual home environment" in the visited network.

Along with the generic bearers (defined by QoS), VHE is enabled by the user profile, referred to as *call control* (CS, PS, or IMS control), and a collection of *service toolkits*. The service toolkits are essentially specifications of protocols, environments, or APIs for developing services of various types. They include the User SIM Application Toolkit (USAT) (3GPP 2000g), the Mobile Execution Environment (MExE) (3GPP 1999a) (3GPP 2000c), Customized Applications for Mobile Network Enhanced Logic (CAMEL) (3GPP 2000b) (3GPP 2000a), and Open Service Access (OSA). We discuss these toolkits briefly in the rest of this section. 3GPP envisions that new toolkits can be added to the 3GPP specifications, and non-3GPP toolkits can be used as required to satisfy the VHE concept.

CAMEL

CAMEL is used to provide network intelligence in the UMTS and is based on the IN conceptual model of separation of high-level services from basic switching and call processing. Unlike fixed networks, for which the IN concept was first developed, in mobile networks, the subscriber can roam between switching centers and foreign networks, and CAMEL allows the switching functions in the foreign network to interact with the service control functions in the user's home network. In a manner analogous to traditional IN services, a CAMEL service residing in the service control function is invoked when a trigger contained in the switching function fires.

CAMEL operates using two protocols: the CAMEL Applications Part (CAP) and the Mobile Applications Part (MAP). The former is similar to the Intelligent Networks Application Part (INAP) protocol in fixed networks and the latter is used for signaling between the mobility service functions, such as the MSC or VLR, and the control function, such as the HLR.

The functional architecture for call control with CAMEL is shown in Figure 2.7 and is summarized briefly here.

There are three logical networks that interact to provide service. The first is the Home Network of the user, which contains two logical entities, as in 2G cellular networks: the

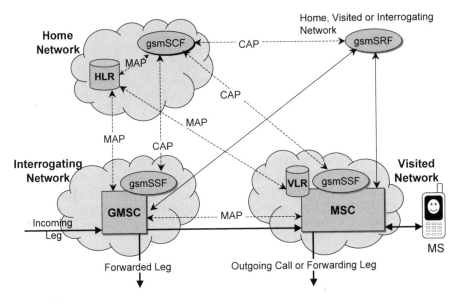

Figure 2.7 CAMEL architecture

HLR, and the execution environment for services, here called the *GSM Service Control Function* (gsmSCF). The latter is analogous to the service control proxy (SCP) in a fixed IN network. The HLR stores the subscriber's location and service profile information, here called the *CAMEL Subscription Information* (CSI), and the gsmSCF stores the service logic.

The second logical network is the Visited Network, where the user is currently located, which contains three logical entities: the MSC, VLR, and a functional entity called the *GSM Service Switching Function* (gsmSSF) that interfaces between the MSC and the gsmSCF. The gsmSSF is analogous to the Service Switching Point (SSP) in a fixed IN network, and the MSC and VLR are essentially as in a 2G cellular network.

The third logical network is the Interrogating Network, that is, the network that needs information in order to provide a service to the user. This contains two logical entities: the GMSC, and a gsmSSF to allow the GMSC to communicate with the gsmSCF. An example of the information the interrogating network needs is the mobile user's location from the HLR in order to deliver a call. Unless the network supports "optimal routing" (3GPP 2000d), that is, routing of calls directly to and from a user's location without first visiting the Home network, the Home Network will always be the interrogating network.

The gsmSRF shown in the CAMEL functional figure corresponds to the IN Intelligent Peripheral, and can play announcements, collect user digits, and the like.

A representative (and highly successful) example of a CAMEL service is wireless pre-paid service, where the subscriber establishes an account with the service provider and pays before use in order to obtain service in home and visited networks. When the mobile user initiates a call, the MSC recognizes the user as a prepaid subscriber; a preprovisioned trigger at the originating MSC (and logically in the gsmSSF function) fires. The gsmSSF queries the gsmSCF to validate the user, for example, to check if the user has sufficient funds. This query consists of a CAP message. If the gsmSSF gives an affirmative response, call handling proceeds as usual, with real-time rating of the call started when the called **party answers**.

MExE

The MExE is the execution environment in the mobile terminal. The MExE client on the terminal interacts with a MExE Service Environment (MSE) in the fixed network to deliver services to the user. Two MExE devices can also, in principle, interact to provide a service.

3GPP assumes that a wide variety of terminal devices will be available and hence defines categories of devices that are assigned into categories, called *classmarks*, based on computational capability.

- Classmark 1 is a device that essentially supports WAP and has limited input and output facilities.

- Classmark 2 is a PersonalJava[4] device with the addition of the JavaPhone API[5]. PersonalJava supports web content access and Java applets, while the JavaPhone API allows telephony control, messaging, and personal information management functions, such as address book and calendar.

- Classmark 3 is based on the J2ME Connected Limited Device Configuration (CLDC)[6] and Mobile Information Device Profile (MIDP) environments[7].

- Classmark 4 is based on the Common Language Infrastructure (CLI) Compact Profile (Ecma 2002).

As an example, Figure 2.8 shows the APIs for Classmark 2 (3GPP 2000c).

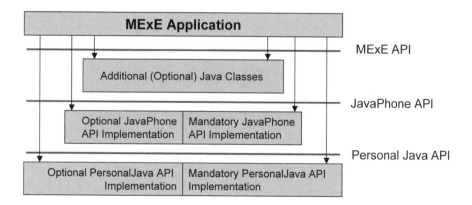

Figure 2.8 MExE API for the Personal Java and JavaPhone environments

---

[4] See http://java.sun.com/products/personaljava for details about the Personal Java Application Environment
[5] See http://java.sun.com/products/javaphone for more information
[6] See http://java.sun.com/products/cldc for more information
[7] See http://java.sun.com/products/midp for more information

Services can be downloaded from the MSE to the terminal's MExE. The MSE may contain proxy servers to translate content to make it suitable for delivery on the mobile terminal. All MExE devices are required to support capability negotiation, which usually takes place before service commences. MExE can thus inform the MSE of the mechanisms, resources, and support it can offer, using Composite Capability/Preference Profiles (CC/PP) (W3C 2004); the MSE may inform the MExE which capabilities it is using or will use. MExE devices may also engage in content negotiation with the MSE, using Hypertext Transport Protocol (HTTP) or the Wireless Session Protocol (WSP), to determine the requested and available form of content. Finally, the MExE can support one or more user profiles, storing them in the ME or USIM.

USAT

In 3GPP terminology, a *universal integrated circuit card* (UICC) is defined as a physically secure device, like an IC card (or "smart card") that can be inserted into terminals. It contains the *subscriber identity module* (SIM) used in 2G systems, or the universal SIM (USIM) used in UMTS systems, for accessing mobile services. The first SIMs, introduced in the late 1980s, had 4-bit CPUs and held about 10 kb of total memory for the OS and user data; now SIMs with 512 kb memory and the compute power of 16-bit processors are being planned (see, for example, http://www.gemplus.com). The USIM thus goes beyond the main initial functions of the GSM SIM, which were to store the mobile user's identity and personal information securely. It can essentially act as an execution environment in its own right, with the ability to utilize certain functions of the ME.

The Universal SIM Application Toolkit (USAT) (3GPP 2000g) is an enhancement of the SAT defined for 2G systems. It provides mechanisms to allow applications on the USIM to interact with the ME, the user, and USAT servers in the fixed network. Among the mechanisms provided are:

Profile download: Allows the ME to tell the USIM what functions it is capable of.

Proactive UICC: Allows the application to initiate actions to be taken by the terminal, such as displaying text or playing tones from the USIM to the terminal, sending an SMS, setting up a data or voice call, retrieving data from the ME, or initiating a dialogue with the user.

Data download: Allows the service provider to download information to the USIM, such as via SMS or cell broadcast.

Call control: Allows the application to intervene when the ME sets up a call, in a manner analogous to how the IN SCP intervenes in PSTN calls. The ME passes all dialed digits and relevant control strings and parameters to the USIM, as well as the current serving ID. The USIM application can allow, bar, or modify the call.

OSA

With OSA, 3GPP has adopted the Parlay service framework (Jain et al. 2004a) as the service framework for 3G networks (3GPP 2000h). OSA enables applications to use network functionality defined in terms of a set of logical Service Capability Features (SCFs) that are

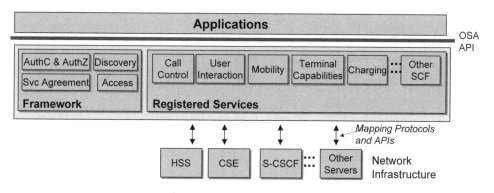

Figure 2.9 OSA architecture

accessed via a language-independent API. The SCFs are encapsulated as Service Capability Servers (SCS). Like Parlay, OSA consists of three parts (see Figure 2.9):

Applications: This is a call control application for call forwarding, a virtual private network (VPN) application, or a location-based service. Applications execute on application servers.

Framework: Before an application can access the SCFs, it must utilize the framework to be authenticated and discover the available SCFs. In Figure 2.9, AuthC and AuthZ stand for authentication and authorization respectively.

SCS: These are abstractions of underlying network functionality. Examples are call control and user location. The MExE and CAMEL are also abstractly regarded as SCS, so that a single functionality, such as call control, may be distributed across multiple SCS.

The basic operations between the application and the framework are authentication, authorization, discovery, service agreement establishment, and access to SCFs. The basic operation between the framework and the SCS is for the latter to register itself with the framework to enable discovery by applications. And the basic operation between the application and the SCS is for the application to issue OSA API commands, including commands to perform service functions as well as to register to be notified of underlying network events, such as call origination. Parlay/OSA is summarized further in Chapter 6 of this book.

2.2.2   CDMA2000

Network Architecture

The UMTS and CDMA2000 systems differ largely in how they handle packet-switched traffic in the core network. The IP multimedia subsystem and the service platform for open services for the CDMA2000 system are similar to the IMS and the service platform for UMTS networks.

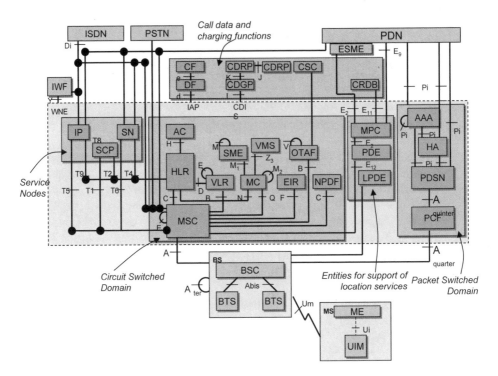

Figure 2.10 CDMA2000 wireless network architecture

For the purposes of present discussion, Figure 2.10 depicts the standard CDMA2000 architecture diagram partitioned into different domains. Although the CDMA2000 architecture specification does not do so, such partitioning makes comparison with UMTS easier.

This chapter does not consider the following domains:

(i) The traditional PSTN and cellular service architecture consisting of SCPs, Intelligent Peripherals, and Service Nodes

(ii) Functions related to call data and charging information collection, such as the Call Data Generation Point

(iii) The ISDN portions of the PSTN network and interfaces to it

(iv) The functions for support of location services

These domains, with the exception of the last, are not specific to the 3G cellular nature of the system, and thus not of interest for architectural discussion; the last is fairly straightforward.

For the comparison with the UMTS network architecture, this discussion considers the following two main domains: (1) the CS domain in the center of the figure that is identical to the 2G circuit-switched cellular architecture, as it is for UMTS; and (2) the PS domain.

From a high-level perspective, the CDMA2000 architecture thus has a CS domain and a PS domain in a manner similar to UMTS. The CDMA2000 network also supports the IMS for Internet multimedia services, and OSA for service creation, in a manner similar to UMTS.

Thus, the main difference between UMTS and CDMA2000 from this architectural perspective lies in the PS Domain. The latter consists of the packet control function (PCF), packet data support node (PDSN), mobile endpoint home agent (HA), and an authentication, authorization, and accounting (AAA) function. (PCF technically is a radio access network function, but shown with the PS domain for convenience, and controls the transmission of packets between the BSC and the PDSN.)

The access network diverts the packet-switched traffic to the PDSN. The PDSN terminates the logical link control layer for all the packet data (refer to Figure 2.14 later in this chapter) and additionally acts as the foreign agent or access router, depending on the network configuration and whether the network uses IPv4 or IPv6 to support IP-based mobility with Mobile IP (Johnson et al. 2004; Perkins 2002a). The PDSN also interfaces to the AAA subsystem for performing AAA for packet access and with the HA and other PDSN to support mobility using Mobile IP.

The services in the CS domain in the CDMA2000 architecture are based on the Wireless Intelligent Network (WIN) (TR-45.2 1997, 2001) standards, which are similar in nature to the GSM MAP and CAMEL architecture explained earlier. Figure 2.11 shows a simplified configuration using WIN. Just as in CAMEL, the high-level services are moved away from the MSC and implemented in a SCP. The MSC consults the HLR and the SCP during the processing of the call, and the SCP or HLR decide what type of service to provide for a particular call. This model simplifies the MSC and makes the service deployment somewhat easier. The Intelligent Peripheral performs simple tasks such as collecting digits or speech-to-text conversion and hands over the results to the SCP for further processing. The Service Transfer Point (STP) is a packet switch (not shown, but resides in the SS7 network) that connects the different components of the network.

Just as in UMTS, 3GPP2 is standardizing OSA and its mapping to the internal network protocols, such as IS-41 and Wireless Intelligent Network Application Part, the 3GPP2 equivalent of MAP.

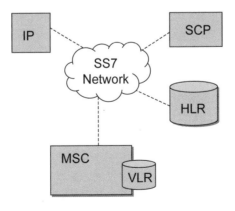

Figure 2.11 WIN components with stand-alone HLR

Service Architecture

The service architecture for CDMA2000 system is similar to that of UMTS, as it relies on the same underlying domains (CS and PS) and adopts the IMS for multimedia services. 3GPP2 also adopts the principles of the VHE, although in a slightly different form.

Thus, the key differences between the UMTS and CDMA2000 architectures are generally found in the network architecture, and particularly in the air interface and RAN design. This situation reflects the main concerns of 3G network designers. As we have mentioned, we believe that, for XG networks, the key features and differentiators in fact will lie at the service architecture levels, as 3G networks do not have such features today.

2.2.3   MWIF

Network Architecture

The MWIF was an industry forum formed in early 1999 by leading 3G operators, telecommunications equipment providers, and IP networking equipment providers to develop all-IP network architectures for the core network and RAN as a counterpoint to the 3GPP R4 architecture. The MWIF core architecture is intended to completely eliminate circuit-switched support except as a compatibility option through a gateway, and the MWIF RAN architecture is intended to support IP to the base station, instead of ATM as in 3GPP R4. In 2002, MWIF dissolved into the Open Mobile Alliance, which took up where MWIF left off, focusing on the service architecture. This section discusses the MWIF core and RAN architectures, with reference to 3GPP R5 for consistency with previous sections.

In comparison with the 3GPP R5 architecture, the MWIF design does not consider the PSTN and focuses only on packet-switched transport. The MWIF architecture is based entirely on Voice over IP after traffic leaves the access network; thus, there is neither an IU-CS interface nor an MSC in the MWIF design. The MWIF core architecture consists of two parts: a layered functional architecture (Barnes, M. 2000) (see Figure 2.12) and a network reference architecture (Wilson, M. 2002) (see Figure 2.13).

The layered functional architecture has four layers and two cross-layer functional areas. The four layers are:

Applications: This layer is specifically for third-party applications available through the mobile operator's network

Services: Applications within the operator's network and such basic networking support services as naming and directory services

Control: Mobility management, authorization, accounting, real-time media management, network resource management, and address allocation

Transport: Basic IP routing, gateway services to access networks

The two cross-functional areas are:

- Operations, administration, management, and provisioning

- Security

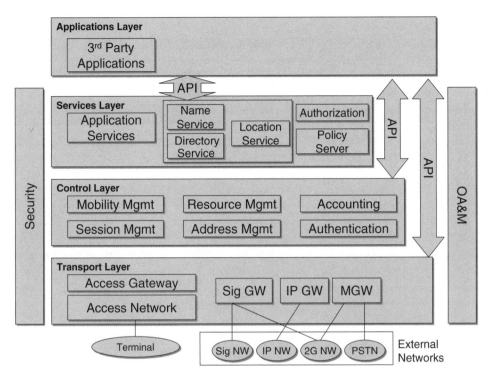

Figure 2.12 MWIF layered functional architecture. Reproduced by permission of OMA

The layered functional architecture was developed into a network reference architecture that assigns particular functional entities to specific network entities. The result is a design for a core network with 68 specific network reference points that act as interfaces between the network entities and outside networks. Where appropriate, MWIF has identified standardized protocols from the suite of Internet standard protocols for each of these interfaces.

A separate effort within MWIF designed a functional architecture for a general radio access network based on open, IP-based protocols, called _OpenRAN_ (Kempf and Yegani 2002). The idea behind OpenRAN was to utilize IP-based signaling and transport for radio access networks where possible instead of ATM and SS7, which are used in the 3G RAN architectures. A feature of the OpenRAN is a separation of the control and data planes to accommodate the expected difference in scalability properties of these two basic functions. However, depending on the radio protocol, the two may merge on the radio layer.

The OpenRAN architecture consists of 14 functional entities separated by 27 reference points. Two reference points connect the OpenRAN control and data planes to the MWIF Access Network Gateway (and thus to the MWIF core network) and two connect the control and data planes to another OpenRAN. Additional reference points are available to connect with legacy circuit-switched networks. The baseline radio protocol for the OpenRAN was CDMA, but other protocols could be accommodated by dropping particular functions specific to CDMA. Like the 3G networks, OpenRAN does not provide voice over IP over the air, but rather terminates voice over IP at a functional entity that adapts the IP traffic to the radio.

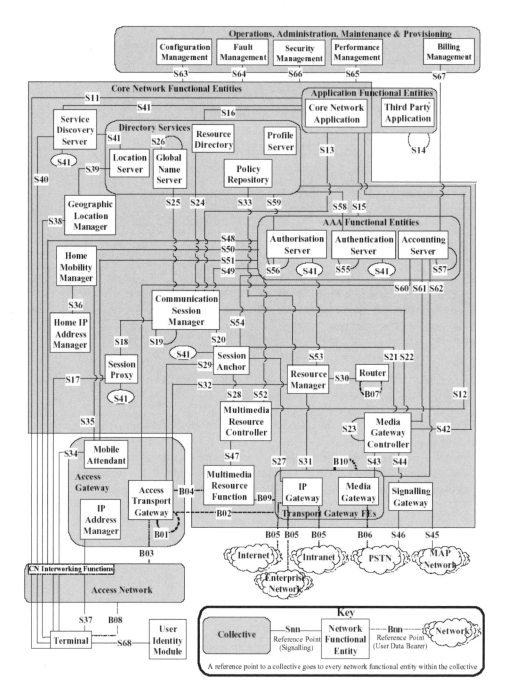

Figure 2.13 MWIF network reference architecture. Reproduced by permission of OMA

Service Architecture

A service architecture is not explicitly defined in MWIF. It is implicitly assumed that telephony-oriented services can be developed with an end-to-end approach using SIP signaling. However, an IMS similar to that for UMTS and CDMA2000 system can also, in principle, be used; it can be connected directly to the IP core rather than to a PS domain.

The design of MWIF reflects the focus of 3G network architects and designers on the RAN and core network. It also reflects a major goal, which is to obtain seamless interoperability with the Internet by using IP as the core base protocol, which is certainly desirable. As we observed in Section 2.2.2, we believe that for XG networks, the key features and differentiators in fact will lie at the service architecture levels.

2.2.4   Limitations of 3G Architectures

The 3G architectures presented above have several architectural limitations. In terms of the network architecture, UMTS in particular duplicates functionality for different traffic types, has a complex protocol stack, and uses a modified SIP protocol that is relatively heavy weight. In terms of service architecture, UMTS and CDMA2000 have a somewhat limited programmability concept, and, while they offer several point solutions, they do not have a coherent programmability solution.

This section briefly discusses the important limitations of the 3G architectures considering the network architecture (including logical separation of networking functions and protocols) and the service architecture (including provision for developing and deploying new services).

Network Architecture

The UMTS, CDMA2000, and MWIF architectures all suffer from various limitations, as summarized here (see Table 2.2).

UMTS. The first fundamental characteristic of the UMTS architecture that becomes obvious is the separation into three domains: CS, PS, and IMS, corresponding roughly to voice, data, and packet-based multimedia services. This separation has the important commercial virtue that it offers a relatively easy migration path from 2G to 3G, preserving infrastructure investments, since the PS domain can be incrementally added to CS, and IMS can be added after that. However, from a network architecture perspective, it has the significant drawback that it entails duplication of functionality. At an abstract level, it is undesirable that new types of network elements (SGSN, GGSN, etc.) be developed to provide the same functionality (e.g., mobility management) for user traffic with different QoS characteristics. It could be argued that the CS and PS would simply "wither away" in time, leaving an IMS-only architecture (and, in principle, a greenfield operator could choose to build an IMS-only network). However, the additional capital expense, as well as the operational expense and complexity in the interim, is significant.

Another issue is the complexity of the transport protocol stack as shown in Figure 2.14 for packet data in the PS domain (Park 2002) in a typical UMTS network that uses an ATM backbone to interconnect access and core network entities.

A user data packet moves up and down the stack several times before it is handed over to an IP native network. This access network architecture not only has several points at

Table 2.2   Summary of UMTS, CDMA2000, and MWIF network architecture limitations

	UMTS	**CDMA2000**	**MWIF**
Integration and interoperability with the Internet	Complex, due to separation of domains, protocol stack and other issues (see rows below)	Complex, due to separation of domains, protocol stack and other issues (see rows below)	Simplest, in principle. However, specification is not comprehensive and system is not deployed (see rows below)
Separate CS/PS domains	Separate PS/CS/IMS domains	Similar to UMTS	Unified handling for all traffic
Protocol stack	Complex stack due to IP over ATM, sequential tunneling, use of GTP	Simpler stack for packet data, removing some stack traversals; Mobile IP is used for mobility management	Simplest, all native IP stack
Routing equipment cost	Use of ATM transport may raise cost compared to native IP	Cost may be lower than UMTS if native IP is used	May be lowest due to economies of scale of standard IP solutions
Real-time packet data services	Problematic due to modified SIP and other latencies	Not clear	Not clear, as native IP-based solutions do not support QoS guarantees well
Coupling of AN and CN specifications	Close dependence	Close dependence	Independence between AN and CN
Service architecture and programmability	VHE concept offers OSA, MExE, USAT etc., but with limited, and not comprehensive, programmability	Similar to UMTS	Not addressed explicitly or in detail
Commercial deployment	Several large-scale deployments	Widely deployed	Not deployed

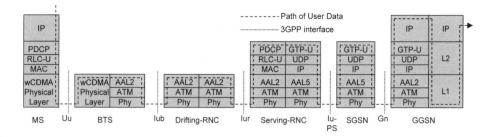

Figure 2.14 Transport protocol stack for typical UMTS network deployment using ATM backbone

which packet segmentation, reassembly, and retransmissions occur, leading to additional delay, but it also adds unjustifiable complexity to the network.

The protocol stack uses GTP to tunnel data in the CN. Closer inspection of the protocol stack shows that in fact there are two GTP tunnels involved: one between the GGSN and the SGSN and another between the latter and the Serving-RNC. Tunneling is a problem shared with the MWIF architecture as well as other all-IP architectures. However, setting up two sequential tunnels is particularly undesirable because of the additional overhead.

The protocol stack further shows the use of ATM transport all the way from the BTS to the GGSN, with IP over ATM AAL2 from the Serving-RNC to the GGSN. IP over ATM has a number of issues, such as fixed ATM cell size leading to packet fragmentation, virtual circuit setup delays, and need for interaction between ATM rate control and higher-layer congestion control mechanisms. Native IP transport over a simple MAC protocol is preferable. From a deployment perspective, market conditions may make the cost of ATM infrastructure greater than using native IP as economies of scale may favor the latter (if they do not do so already).

We observe that GTP is designed to be independent of underlying network protocols and can carry a number of different packet data protocols, including X.25, Frame Relay, and IP, transparently. While such a design is beneficial in principle, since many data protocols could be accommodated easily, in practice, the immense growth and popularity of IP means that GTP is used only to carry a single protocol. Its flexibility thus leads to needlessly adding another protocol to the stack, and one that is not well-known outside the cellular networking community.

The IMS in UMTS can support real-time services, and UMTS has used a significantly modified version of SIP to do so. The modified SIP protocol aims to allow negotiation of communication details (codecs, etc.), ensure network paths of the required QoS are available before the session starts, and provide appropriate charging signaling to prevent service fraud (Kim and Bohm 2003). In addition, the IMS works in conjunction with the PS domain, the application platform, and other infrastructure entities, such as the gateways and subscription servers. In principle, basic SIP session setup can be done using as few as 3 messages and with 1.5 Round-trip Time (RTT) delay. Within the UMTS architecture, the modified protocol can require as much as 30 messages exchanged between different network entities (3GPP 2001a). One important goal of the next-generation network is to eliminate such extraneous interactions, while maintaining the desired security and QoS properties.

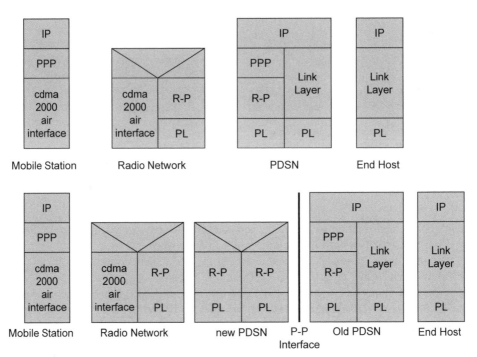

Figure 2.15 Protocol model for IP packet data

CDMA2000 System. The CDMA2000 architecture does not suffer some of the more obvious architectural problems of UMTS. In particular, the protocol stack for CDMA2000 system is shown in Figure 2.15; a data packet need not undergo multiple transformations to reach the Internet. The top portion of the figure shows normal operation while the MS is stationary. The PDSN uses the Point-to-Point Protocol (PPP) to maintain a link with the mobile station and, in effect, this forms a link control layer. This scheme is much simpler than the UMTS approach, because now there is only one logical link control connection between the first hop IP router and the mobile station, and hence the protocol stack is relatively simple.

A major difference in CDMA2000 system architecture is use of Mobile IP combined with AAA functions to support handover, unlike the UMTS network where GTP combined with MAP, used to communicate with the Authentication Center, is used for mobility management.

The CDMA2000 system also introduces an edge-based technique for handoff in PS domain, by stretching the PPP tunnel between the old and the new Packet Control Function. This technique is sometimes referred to as "stretchy PPP" and effectively defers the signaling and end-to-end path update between the MS and its correspondent nodes, handling mobility locally. This scheme is similar in spirit to the fast mobile IP techniques, which are discussed later in this book, and is fairly effective in reducing packet loss during the handover.

MWIF. To a large extent, MWIF is the 3G architecture that comes closest to the basis for an XG architecture. The elimination of a CS domain in the core, with an emphasis on packet switching, and IP in particular, is conceptually highly desirable. The OpenRAN effort takes this to the next logical step, which is IP in the RAN.

A feature of the MWIF core network architecture that differentiates it from the UMTS and earlier CDMA2000 system architectures is that it was designed to be access network independent. Details of how to interface the core network to an access network were confined to a functional element, called the *access network gateway*, at the periphery of the core network. In principle, this should allow an MWIF core to be connected to any of a variety of access networks, including such wired networks as DSL. The later generation CDMA2000 network and the UMTS network have adopted a similar model. (Note the flaw that different access networks may require different custom access gateways.)

The major drawback of the MWIF architecture, of course, is that while laboratory proto-types were developed, it was never fully specified, developed, or deployed in a commercial setting. Thus, it is not clear that the knotty architectural issues of QoS (including handover as well as packet delivery latencies), AAA, and, above all, interoperability with legacy 2G or even 2.5G systems can in fact be solved effectively and gracefully.

Service Architecture

For the purposes of this discussion, UMTS and CDMA2000 system are considered together since their broad features are similar, although, for the sake of concreteness, the focus is on UMTS. This discussion does not include MWIF, since the service architecture of MWIF was not developed in detail prior to disbanding of the group.

The first observation about the limitations of the UMTS service architecture is the VHE concept itself. Allowing services to be portable across networks and across terminals is certainly a worthwhile goal, and providing a common "look and feel" to the user interface is very attractive. This is likely to ease and speed the adoption of services, particularly new services. However, this goal is too narrow. What is required in addition is to ease and speed the development and deployment of large numbers of innovative new services. The VHE concept does not really address this issue, which is vital from the point of view of users as well as service providers and network operators. In fact, one could argue that if the goal of providing portable services and a common look and feel to services hinders their deployment, it should be put aside, at least temporarily.

We believe that deploying new services rapidly and efficiently requires the network to become fundamentally more flexible and programmable. In this regard, the 3GPP APIs and toolkits, namely, OSA, CAMEL, MExE, and USAT, are very relevant. Broadly speaking, OSA and CAMEL address programmability of the fixed network, and MExE and USAT address programmability, that of the terminal. OSA is at a higher layer of abstraction than CAMEL since it can span the CS, PS, and IMS domains, while CAMEL is limited to the CS domain. The definitions of MExE and USAT represent a significant step forward from the traditional notion of programmability in the PSTN, which assumes that terminals have little or no intelligence, and even from the programmability offered by SAT in GSM. The OSA APIs are being used in some form or other by several vendors as well as carriers.

Nonetheless, programmability in UMTS has significant limitations. To begin with, CAMEL essentially offers only the same level of programmability as IN. There is no formal API and programming services requires specialized languages and tools, as well as a degree of access typically only available to network operators. MExE is much more accessible; the MExE Classmark is essentially defined in terms of existing, well-known APIs. One problem here is that the Classmarks tend to form vertical separators; in general, it is hard to see how an application developed for one Classmark could be ported to another without significant effort. (Also, in some cases, the APIs used have not been very successful, such as WAP and CLI, and are not likely to be suitable for next-generation systems.) This essentially reflects the fact that MExE tries to tap into an existing application development community and technology without developing a coherent programmability solution.

The USAT is the most interesting aspect of UMTS programmability. It represents a significant potential opportunity for new services as the capabilities of UICC continue to grow. We believe that UICC should be treated as "first-class citizens" in the XG service architecture, which does not seem to be the case for 3G.

The last aspect of UMTS programmability is the OSA APIs. These are derived largely from the Parlay APIs. Chapter 6 discusses these APIs in detail and further details can also be found elsewhere (Jain et al. 2004a). In general, the APIs are limited as they are network centric, tend to be heavy weight, and do not pay sufficient attention to security for advanced services. Also, a critical failing is that they do not explicitly model users or roles (Jain et al. 2004a).

A significant issue with the OSA APIs, as well as the other UMTS APIs (with the exception of some MExE APIs), is that they provide only a single level of abstraction, and hence access, to the network. Thus, OSA provides a layer of abstraction for developing services that need call control, user-location information, charging services, and the like. However, unlike the JAIN APIs, there is no level of protocol programmability to provide further flexibility and possible performance benefits. As discussed later in this chapter, a system of coherent, well-defined APIs at different levels of abstraction is required. Note in passing that the fledgling service architecture in MWIF does depict multiple levels of APIs (Figure 2.12).

Finally, another limitation in the service architecture is the separation into the CS, PS, and IMS domains. As discussed above, this is undesirable from a network architecture view, but it is also problematic from a service point of view. In particular, it seems a complex application that requires coordinated features from multiple domains and will be more difficult to develop in a coherent manner. Certainly, the system can endeavor to provide an API, like the OSA API, that hides this separation; however, the network support for the API may become complex.

2.3 Approach to a Next-generation Architecture

2.3.1 Rationale and Key Features

The All-IP Network

It is increasingly accepted that the next-generation fixed and mobile telephony architecture will be based on IP (Tachikawa 2003b; Yumiba et al. 2001). However, the motivations for this are typically either that IP networks will be less expensive in some way or are based on somewhat vague generalizations about increased availability of services. The first category of motivations is not sufficiently compelling, and the second needs more critical exposition.

The motivation for an all-IP network is not primarily to reduce cost. It is indeed likely that IP network elements such as routers, firewalls, and proxies will enjoy economies of scale compared to fixed and mobile PSTN components like switches and SS7 elements, and it could be argued that they will be fundamentally less expensive because of simpler design. However, network elements are not necessarily the dominant costs of a network. OA&M and, for a mobile network, the connection costs from base stations to the core fixed network (commonly referred to as *backhaul*), can be dominant. Further, Moore's Law will accelerate this trend.

The first motivation for an all-IP network is that currently it is the best architectural choice to enable a wide variety of innovative and commercially lucrative services. To see this, it is worth considering the development of PSTN services historically. Typically, PSTN services have been offered in the context of an integrated or centralized business model: the same entity (a government body or a company) typically served as the network operator, the service developer, and the service provider to the end user. Service development was done by specialized personnel in closed environments, using specialized languages and tools.

In contrast, the open and rapid proliferation of knowledge about IP networks has tapped into the creativity of legions of programmers and entrepreneurs, and the possibility of developing advanced services at the edge of the network has lowered the barriers to entry into the market. Thus, an important, if nontechnical, reason for choosing an IP-based architecture is that there is much greater availability of personnel, tools, and support for application development in the context of IP networks than for the PSTN. To a large extent, the acceptance of ATM networks has been similarly hindered by the paucity of applications available in a timely manner.

A second reason for an all-IP network is that the relative technical simplicity of the basic IPs, as well as the ability of IP itself to act as a unifying abstraction or *waist* (Deering 2001) that hides and divides the complexity of the protocol stack, also helps to make application development for IP networks easier. This has resulted in the rapid development and deployment of a large number of Internet applications, many of which, like e-mail, messaging, and content distribution, directly compete with the PSTN's current or potential markets.

The rise of these Internet applications means that the telecommunications network must be efficiently integrated with the Internet, either to allow access to Internet applications via the network or to interoperate with them in some way. The use of an IP-based XG architecture, while not strictly essential, makes this integration much easier, and is a third reason stressing services as the key differentiator in the XG architecture.

Having said this, it is worth pointing out some factors that make the choice a little less clear-cut. Firstly, the development of open standard APIs, such as JAIN[8] and Parlay (Jain et al. 2004a), makes the choice of the underlying transport network less important, as the APIs are generally designed to provide programming abstractions that operate across PSTN, ATM, and IP networks. In the JAIN framework, APIs have also been developed for individual protocols so that applications needing improved performance or fine-grained control can obtain it in exchange for increased programming complexity. JAIN APIs have been developed not only for IPs, such as SIP, but also for SS7 protocols, such as ISUP and TCAP.

Secondly, the number, complexity, and size of the Internet-related protocols, continues to grow rapidly and is beginning to rival that of traditional PSTN protocols. In addition, the standardization delay is becoming longer (e.g., standardizing mobile IP took about 10 years) and is approaching that of traditional PSTN protocols. These factors indicate that introducing new applications in an IP-based network could become as slow and difficult as

---

[8]For more information, see http://java.sun.com/products/jain

for the PSTN. However, the open and end-to-end nature of IP networks often means that innovative applications can still be built even if core network support is lacking or delayed, albeit in a less efficient manner.

The Second Waist and Programmability

Given that XG will be defined by the availability of innovative applications, the question arises as to where those applications will come from. It is unlikely that a single company, such as a carrier or a network equipment vendor, can develop a stream of "killer apps" rapidly. Thus, XG must rely on harnessing the creativity and energy of independent software vendors and system integrators to develop new applications and solutions, in a manner similar to the PC industry. In short, the next-generation network must be a *programmable* IP-based network in some sense. This in turn requires that XG provide powerful, convenient, and well-defined APIs for application development.

As mentioned earlier, IP provides an important unifying abstraction that hides the heterogeneity of protocols and networks below it and is particularly important for mobile wireless networks given the proliferation of wireless and access network technologies. Thus, IP, or the TCP/UDP sockets interface, can be regarded as the basic API for IP-based networks. However, it is clear that for developing complex applications, this interface offers too low a level of abstraction in most cases. While HTTP operates at a higher level, and in fact many current web applications are built on it (especially to take advantage of the fact that port 80 is typically left open by firewalls), it is obviously very limited in functionality and again offers little abstraction.

What is required is a *second waist* at the interface between applications and the middleware of the web to hide the heterogeneity of the protocol stack and software layers between the application and the IP waist. This second waist must offer a high level of abstraction, convenience, and flexibility, while still allowing the network as a whole to be secure and its use to be billable. Current efforts toward web services, including the definitions of WSDL, SOAP, and UDDI (or their variants and successors), are steps in the direction of a higher-level API, but more work remains to be done.

Our view of programmability for XG networks is illustrated in a little more detail in Figure 2.16, which shows a set of layered APIs. In fact, the set of APIs can be viewed as an application-oriented abstract specification of the architecture (Brooks 1995). This architecture assumes that voice will remain important but will no longer be a dominant application. In addition, while nonvoice applications created by future service providers (or partners) will be significant, third-party applications will in fact dominate both in terms of traffic and revenues.

The XG architecture offers an API above the core network and also an API above the middleware layer; these correspond to the first and second waist, as discussed above. However, unlike the first waist, the core network API is not limited to being simply the TCP/UDP socket interface. It also provides interfaces to Internet signaling and coordination protocols that can be viewed as being from the network, transport, and session layers (i.e., layers 3–5) of the OSI Reference model, such as SIP, Mobile IP, and so on. Thus, the core network layer API allows applications with the proper security and billing authority to obtain fine-grained control over core network resources and functions.

The middleware API, in contrast, offers applications (again, those with the required security and billing credentials) access to higher-layer functions, such as access to overlay

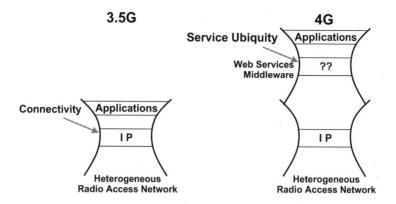

Figure 2.16 The second waist for next-generation networks

network elements for content distribution or multicast, location and other context information servers, security certificate authorities, and the like. Thus, the middleware API offers a higher level of abstraction than the core network API, allowing rapid application development, and independence from changes in networking technology, but does so at the cost of fine-grained control and performance. The middleware API and core network API are roughly analogous to high-level language and assembly language respectively.

Note that in practice, both the middleware API and the core network API may themselves be composed of multiple APIs, at different levels. These details need to be worked out further. Also note that the lowest layer of APIs is required in order to provide security, QoS, and billing functions, especially those involving real-time constraints, as well as service differentiation. Thus, we recommend that any third-party access networks interfaced to the XG architecture also provide appropriate levels of programmability and access.

The idea of APIs is not by itself a radical notion, particularly for a service provider like DoCoMo. The tremendous success of i-mode depends upon leveraging third-party application developers. The issue is whether the APIs are *open* or *closed*, or rather, the degree to which the service provider has control over them. This issue will probably be decided by marketing and strategic concerns rather than technical ones. For the moment, assume that a public API is an API made available to all third parties (in a manner similar to i-mode). It may contain standard open APIs (e.g., 3GPP OSA or Parlay, or their successors) as a subset. On the other hand, a private API is an API controlled by one company and used only internally or by favored partners. Figure 2.17 illustrates, at an extremely high level, an approach to an XG API architecture.

Technology Evolution

APIs enable access to network resources in a well-defined manner. The next question is where the resources are located in the network. Historically, the telecommunications network has consisted, by necessity, of a complex, *smart* core that is tightly integrated in order to meet strict performance and reliability requirements, and simple ("dumb") terminals that are affordable for the mass market and offer a fixed, easy-to-use user interface. This is in

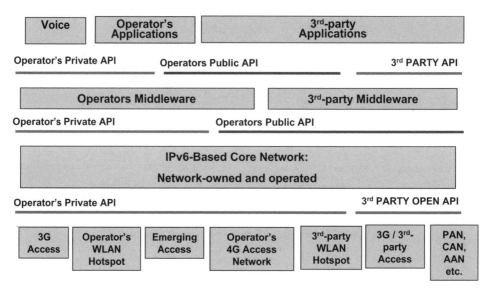

Figure 2.17 XG architecture as defined by layered APIs

contrast to the Internet, which allows for very complex end machines interconnected by a relatively simple core of routers that lead to loose performance and reliability guarantees for the system as a whole. The tension between the smart core and the dumb core has been widely debated (Isenberg 1997) and is one of the key decisions in the architecture. This section examines it a little more closely.

Rather than considering the placement of intelligence only in terms of a core versus edge dichotomy, consider how intelligence has evolved for all parts of the telecommunications system, particularly the cellular network. Figure 2.18 depicts informally how intelligence has evolved in the system components. In general, the intelligence in the core network functions of switching and routing has decreased over the past decades, starting with the introduction of Common Channel Signaling (CCS) in the 1970s and continuing with the Intelligent Network (IN) architecture that removed the service logic from telecommunications switches and placed them in distributed databases.

In fact, SS7 and the IN architecture marked the beginning of a control layer that has gradually increased in sophistication and become more distributed. The so-called next-generation network (NGN) designs of the 1990s (Jain et al. 2004a; Ohrtman 2003), which are perhaps better called *Voice over Packet* (VoP) architectures, utilized packet transport networks (often ATM rather than IP networks) controlled by a higher-layer call agent or Softswitch that remained in charge of all call processing (Jain et al. 2004a). The call agent can also host service logic and provide interfaces to external application servers. Newer architectures call for using Internet protocols like SIP and mobile IP directly, placing functionality in distributed servers like home agents and registrars.

In IP networks, on the other hand, there has been a steady rise in the use of overlay networks as an alternative to placing or using functionality in the core routers. This is driven by standardization delays for new protocols, the increasing complexity and lack of

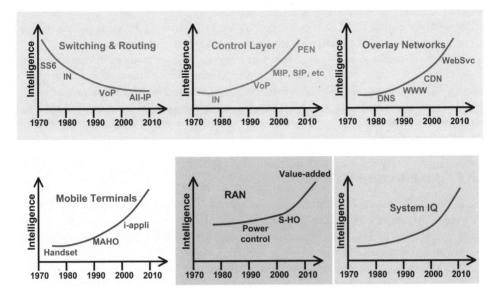

Figure 2.18 Evolution trends for intelligence of system components

openness of the core elements (routers), and the attractive economics of computing at the edge of the network; interestingly, these are all reasons cited as weaknesses of traditional telecommunications networks (Isenberg 1997; Lazar 1997). A degenerate (and arguable) example of an overlay on IP is the DNS; however, application layer multicast, anonymizers, and Content Distribution Networks (CDN) are increasingly prevalent and sophisticated. With VoP designs, these overlay networks can serve both PSTN and IP-based networks.

Similarly, it is clear that mobile terminals and the RAN are increasing in intelligence. For example, the latter has the ability to do more fine-grained power control, soft handoffs, and the like. Recently, there has been a surge of activity in integrating IEEE 802.11b based wireless access networks that are far simpler than traditional cellular base stations. However, it is not clear that this represents a trend; there are numerous efforts in the industry toward providing QoS, security, and power control for WLAN. What is likely is that there will be increasing distribution of control functions, in contrast to the rather centralized designs of base stations and base-station controllers of cellular systems.

We believe that the XG architecture must be consistent with these historical trends. Thus the distribution of intelligence and resources in the architecture can be summarized as follows: Smart terminals, smart RAN, *dumb core*, smart control layer, and smart overlays.

The migration of intelligence outward from the core network also means that increasingly, applications can be provided at the edge of the network. This is a serious potential economic threat to carriers, as it can reduce their role to being simple bit-pipes or, worse, being bypassed altogether.

There are two main responses to this threat. The first is to leverage the information and functionality available in the network to make it indispensable for application developers. Examples of this include the user's profile, location, and context information; the trusted authority and billing relationship that the carrier enjoys with the user; and the ability to integrate diverse aspects of the user's quality of experience to make applications more efficient and attractive. This calls for smart overlay services and programmability.

The second response is to integrate the intelligence and competitive threats at the edge of the network. To some extent, DoCoMo as well as other carriers have started doing this by offering WLAN hotspot services. An extension of this would be to provide not only wide-area connectivity, but also core support functions that can be used by enterprise networks, multihop or ad hoc and peer-to-peer networks, and niche markets, such as automobile area networks, and the like.

2.3.2   Architecture Overview

The design rationale presented in the previous section leads naturally to an XG architecture, depicted schematically in Figure 2.19. It consists of four main abstract layers: overlay, control, core, and access. This section summarizes the main features of this architecture, along with a few details.

The XG architecture contains a core IP network that has relatively little intelligence. Thus, most core network functions, such as routing, are handled by existing and evolving IP technology. Above the core is what we call a high-level control layer. It is important

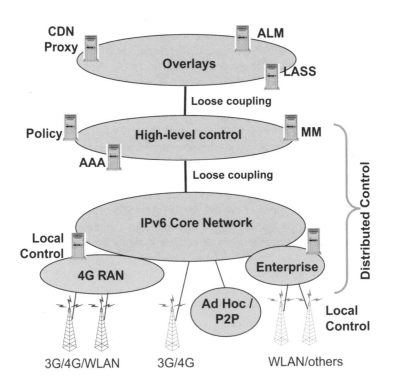

Figure 2.19   Schematic of the XG architecture

to specify not only what this layer does but what it does *not* do. In particular, it does not provide functions for routing or call path setup, unlike the control layer of SS7 in the PSTN, but leaves that to the core. Instead, it focuses on functions that can be made available to applications and overlay network elements, such as access to decision points for AAA, agents for mobility management, and role and rule assignment for policy management. The loose coupling between the control and the core means that the former generally cannot be involved in the fast path of packet forwarding and manipulation.

Below the core is a collection of access networks that serve different market niches and needs. The 4G RAN is the evolution of the current RAN toward higher data rates, support for interactivity and multimedia, and distributed control elements interconnected by an IP network. Since real-time constraints are critical at this layer, relatively tight coordination and coupling between the core and the access networks is required. The core also provides support services and connectivity to specialized networks, such as enterprise networks and multihop/ad hoc or peer-to-peer networks owned and operated by the next-generation network operator or by third parties. These specialized networks are likely to desire local control, especially over key features such as AAA. Providing some autonomy in these areas, while maintaining QoS and reliability, is a challenge that needs to be addressed.

Finally, the XG architecture has an Overlay layer that provides higher-layer functionality and support services for applications, such as Application Layer Multicast (ALM), location services, and content distribution. This overlay can be split into two tiers, with functions that are relatively close to the core (such as ALM) in the lower tier and functions such as location services at a higher tier.

Figure 2.20 shows the functional aspects of XG architecture in more detail. The four horizontal abstract layers discussed above are further subdivided and some of the functions in each specified. The functions are grouped into vertical collections we call "facets" that contain key capabilities that span all or several layers. These facets are security, QoS/resource control, and other similar coordination functions, transport, mobility, interworking, and service control. Note that this diagram shows only one plane of the system. Separate parallel planes deal with OA&M and user equipment; both have a similar layering and facet arrangement, the details of which are not shown here. Each plane and each layer is largely independent of the others, resulting in a highly object-oriented network architecture that is easy to maintain and upgrade.

The lower layer (L1, L2, L2.5) is the access network layer and provides physical and MAC level connectivity, necessary access control and wide-area mobility, and QoS-aware switching capability. This layer is topped with IP-based access network that provides IP connectivity along with necessary access control, integrated QoS management, address assignment, and intersubnet handover capability with fast Mobile IP. These two layers are flexible and are mixed in different combinations depending on the access network technology and particular topology requirements of particular part of the network.

The core network layer consists of a pure IP Diffserv core that provides raw bandwidth to connect different parts of the network. It also contains gateways to connect to external networks, such as the Internet, and employs necessary protection against denial-of-service attacks from outside networks.

The network services that help other layers achieve their tasks are called *support services*. Here, these are divided into two tiers. Tier 1 support services are mostly related to transport functionality of the network, while tier 2 support services provide functionality for end

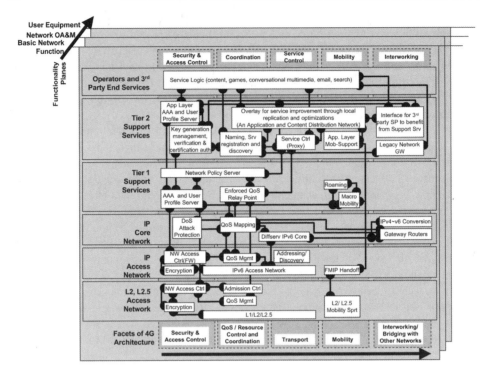

Figure 2.20 Layers and facets of the all-IP XG architecture

services to work properly. Tier 1 services include network-level AAA service, roaming and macromobility management, and a QoS enforcement function that configures different parts of the network to provide QoS in compliance with the network policy and user profile.

Tier 2 support services provide a rich set of services that facilitate end services. These include service registries that allow applications to discover services and interact with them to provide more sophisticated composed services. This layer provides application-layer AAA service for the end services, an overlay network that facilitates application and content distribution and other optimizations, a certification service for the applications, and a set of gateways that provide service level integration with legacy network services such as voice and video in 3G networks or interworking with PSTN.

Clearly there are many possible instantiations of the overview and functional architectures depicted in Figures 2.19 and 2.20. Figure 2.21 shows a view of the architecture with an instantiation in terms of physical network and service elements. Observe that IP is used as the fundamental transport mechanism at all layers, including the RAN and service layers. It is also assumed that appropriate gateways and firewalls connect the IP-based XG network to legacy networks, as well as the Internet at large. Also, observe that there may be functional as well as physical interconnects that jump across layers, for example, from end-user services to the core network, which can be used to facilitate cross-layer adaptation via cross-layer APIs. Needless to say, such cross-layer interaction has to be protected by sophisticated security mechanisms, and is a subject of ongoing research.

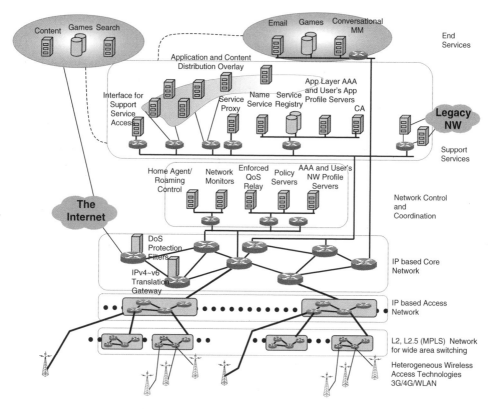

Figure 2.21 An instantiation of the all-IP XG architecture

2.4 Conclusions

Our review of 3G architectures shows significant limitations both at the level of the network architecture and the service architecture. We believe an all-IP architecture is the best choice for developing innovative and lucrative services. We consider the evolution of intelligence in all portions of the cellular architecture. We have developed an approach to an XG architecture, believing that this division of the architecture into four distinct layers, with different degrees of coupling, provides a foundation for providing local control to niche markets while ensuring that the system as a whole retains key control and coordination points for third-party services. Aside from a coherent long-range architectural framework, we have developed research results in the areas of mobility management, security and cryptography, network programmability and support for value-added services that are among the key enablers for the XG architecture; Other articles in this book provide more detail on these areas.

Part II

Overview of Mobile Network Technologies

While the first-, second-, and third-generation cellular networks are based on circuit switching and provide voice-centric services, the next-generation (XG) mobile network with packet switching will support broadband Internet-based multimedia services. The XG mobile network will use a variety of wireless access systems so that a service can be delivered through the system that is most efficient for that service. This part introduces the evolution of the technologies that will support the XG mobile network, in the physical layer, data link layer, and network layer in the ISO-OSI computer network model.

We start from the physical layer. Chapter 3 examines the radio access technologies used in cellular networks. Chapter 4 introduces the evolution of the wireless local area network (WLAN), which is a complementary component of cellular networks, and discusses mobility-related technologies in the data link layer. Chapter 5 discusses mobility-management technologies in the network, or IP, layer.

The development of XG mobile network technology shows trends toward

- broadband radio access that will exceed the performance of any advanced IMT-2000 system, with much higher throughput, much lower latency, and much higher spectrum utilization efficiency;

- seamless integration of different wireless access systems, including cellular networks and WLANs, enabling a mobile device to access the optimal wireless network access system for desired service;

- IP-based networks to support seamless mobility for fast-moving users, based on the IP protocols, using both current and evolving wireless access technologies.

The broadband radio access technologies for the XG mobile network will inherit the technologies of the 3G systems, and introduce new technologies to meet challenges posed by the propagation characteristics of a broader channel bandwidth and by increased performance requirements.

Keeping the leadership in radio access technologies, NTT DoCoMo proposes variable spreading factor orthogonal frequency and code division multiplexing (VSF-OFCDM) and variable spreading and chip repetition factors code division multiple access (VSCRF-CDMA)

Next Generation Mobile Systems. Edited by Dr. M. Etoh
© 2005 John Wiley & Sons, Ltd

for XG broadband radio access. The proposed technologies evolve CDMA-related technologies by introducing new technologies, such as multicarrier and chip repetition, to meet the performance requirements of XG systems.

As laptops, PDAs, and other consumer electronic products with built-in WLAN chips become more and more popular in offices and homes, WLAN will become a necessary component of the XG cellular network. The high penetration decay problem of XG cellular systems that are expected to operate in the over-3-GHz band can be solved by the seamless use of WLAN in indoor environments. Many technologies are needed to evolve WLAN from its enterprise base to a broadly deployed public access network. Fundamental technologies in the evolution of WLAN include those that support mobility (the basic feature of mobile networks), and quality-of-service (QoS) technologies that support multimedia services.

Having proved its ability in the wired network to support and encourage the development of a broad variety of services in a short amount of time and with relatively little effort, the Internet Protocol is a natural choice for the network layer in the XG mobile network. Because the IP network is an open network that is totally different from the traditional closed telecommunication networks, the development of IP-based mobile networks offers considerable challenges. Mobility support in the network layer is a fundamental issue; Mobile IP and its extensions are well-known protocols that support a mobile host moving from one subnet to another. There are many related issues, such as latency, packet loss rate, security, and privacy protection, that offer additional challenges. Finally, in order to provide precise mobility support in the network layer, the system will need to get support from and provide support to lower layers.

3

Radio Access Technologies in Cellular Networks

Gang Wu

3.1 Introduction

The first three generations of mobile communication systems have provided services from voice-only to voice and wideband data. The radio access technologies behind the services have made significant progress in different generations of mobile communication systems: from analog (1G) to digital (2G) and from narrow band (2G) to wideband (3G). This technology is represented in advanced IMT-2000 systems. A next generation (XG) of mobile systems (beyond IMT-2000) requires broader bandwidth to reach the next level of performance. This chapter briefly describes radio access technologies that support the current systems and those that will be needed to support a broadband XG system.

3.1.1 Current Radio Access Technologies

A cellular communication system is an optimized combination of many sophisticated radio access technologies.

- Fundamental technologies for radio communications include channel encoding and decoding, interleaving and de-interleaving, modulation and demodulation.

- Basic technologies for spread-spectrum communications include random and pseudo-random signal generation, synchronization of pseudorandom signals (acquisition and tracking), and modulation and demodulation of spread-spectrum signals.

Next Generation Mobile Systems. Edited by Dr. M. Etoh
© 2005 John Wiley & Sons, Ltd

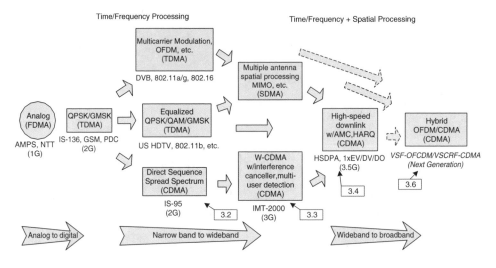

Figure 3.1 Trends of radio access technologies

- System-level technologies include channel configuration, channel estimation, transmission power control, and handover.

For digital cellular systems, radio access technologies in each new-generation system inherit those used in the previous generation and include new technologies. Figure 3.1 shows the history and trends of radio access technologies (Viterbi 1995).

3.1.2 Evolving Radio Access Technologies

This chapter discusses the existing technologies and introduces the new technologies that will be part of XG systems.

- Section 3.2 introduces the concepts that underlie the radio access technologies.

- Section 3.3 discusses *wideband code division multiple access* (W-CDMA), which extends the capabilities of *direct sequence* CDMA (DS-CDMA) with highly accurate transmission power control (TPC), Rake combining, asynchronous cell operation, orthogonal variable spreading factor (OVSF) and code multiplexing, and Turbo coding.

- Section 3.4 discusses *high-speed downlink packet access* (HSDPA), a standard that provides a packet-based data service in a W-CDMA downlink.

- Section 3.5 relates different technologies to the problems they solve and shows which technologies offer potential solutions to uplink and downlink transmission problems.

- Section 3.6 discusses specific broadband radio access schemes that will be applied in XG cellular networks, including *variable spreading factor orthogonal frequency code division multiplexing* (VSF-OFCDM) for downlink transmissions and *variable spreading coding repetition factor* CDMA (VSCRF-CDMA) for uplink transmissions.

3.2 Background of Radio Access Technologies

This section briefly reviews the propagation characteristics in a mobile communication environment, basic multiple access methods, and discusses why CDMA is the best scheme for cellular communications.

3.2.1 Propagation Characteristics in Mobile Environments

In a territory mobile communication system, the path between antennas of base station and mobile station is usually non-line-of-sight. The propagation characteristics vary in time because of the mobility of the station itself, and to changes in the surrounding physical environments. A major objective in mobile communications is to overcome the degradation in communication quality caused by the channel variation.

Figure 3.2 illustrates the multipath propagation in a mobile environment. Radio signals from the same transmitter propagate via different paths, resulting in multipath signals at the receiver with various power level and arrival times.

Figure 3.3 depicts propagation characteristics in a mobile environment. Long-term variation is due to the geographic path-loss law, short-term variation is caused by shadowing fading, and instantaneous variation is due to the change of the surrounding physical environment at the receiver. Raleigh distribution is most often used for describing the instantaneous variation of a multipath channel.

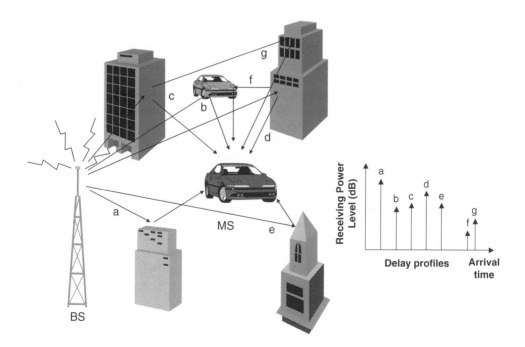

Figure 3.2 Multipath propagation in mobile communication environment

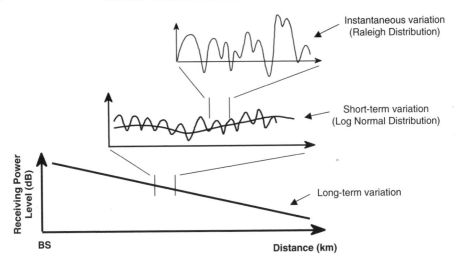

Figure 3.3 Propagation characteristics in a mobile environment

The frequency characteristics of propagation depend on the delay spread of propagation. The longer the delay spread, the larger the impact on the frequency characteristics. Therefore, fading related to frequency characteristics can be classified as frequency-flat fading and frequency-selective fading.

- A frequency-flat fading channel is composed of long-term and short-term variations.

- A frequency-selective fading channel is composed of multipath channels with different time delay spread, each of which is a frequency-flat fading channel.

Theoretical and experimental results of propagation show that a narrowband channel is a frequency-flat fading channel and a wideband channel is a frequency-selective fading channel (Kinoshita 2001). As the channel bandwidth becomes wider, the effect of averaging the total receiving power in the bandwidth gets more significant, and thus the fluctuation of receiving power due to instantaneous variation gets flatter (Kozono 1994). Figure 3.4 shows examples of fluctuation of receiving power in the 1.25-MHz channel for IS-95 and in the 5-MHz channel for W-CDMA.

As indicated in Figure 3.2, there is a time difference among multipath propagations because of the difference in distance among paths. In a narrowband mobile system, this phenomenon brings about intersymbol interference (ISI), because the mixed signal waves caused by multipath propagation cannot be decomposed. However, in a wideband channel, it is possible to decompose these paths by using, for example, a Rake receiver in a CDMA system (Viterbi 1995).

3.2.2 Basic Multiple Access Schemes in Cellular Systems

A cellular system generally consists of base stations (BS) provided by operators and a number of mobile stations (MS) that transmit and receive radio signals to and from a BS.

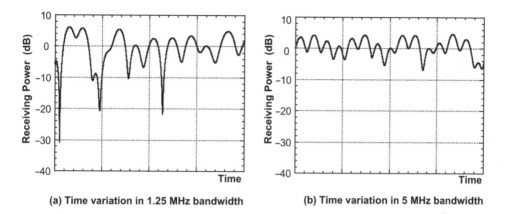

(a) Time variation in 1.25 MHz bandwidth   (b) Time variation in 5 MHz bandwidth

Figure 3.4 Fluctuation of receiving power with different channel bandwidth

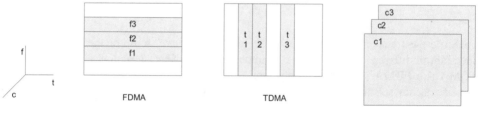

Figure 3.5 Basic multiple access methods

Since there are many MSs in a cell (the coverage area of a BS), multiple access technologies to ensure the transmission of each MS are fundamental for cellular communications.

As shown in Figure 3.5, Frequency Division Multiple Access (FDMA), Time Division Multiple Access (TDMA), and Code Division Multiple Access (CDMA) are three basic multiple access methods that maintain the orthogonality among MSs in frequency, time, and code domains respectively.

- In FDMA systems, each MS tunes its frequency synthesizer to the channel (frequency carrier) assigned by the BS and then transmits signals on this dedicated channel.

- In TDMA systems, a channel with a relatively wide bandwidth is divided into nonoverlapping time slots. All MSs tune their frequency synthesizers to the same frequency carrier, but each MS transmits in a dedicated time slot assigned by the BS.

- In CDMA systems, in contrast, orthogonal spreading codes are assigned to MSs. MSs can transmit in the same frequency and time domains, and their signals are distinguished by these orthogonal spreading codes.

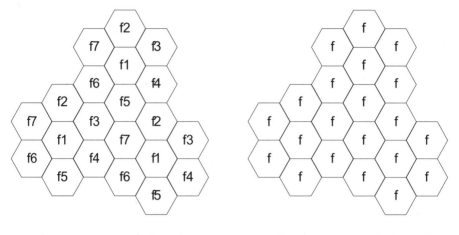

frequency reuse factor of 7   frequency reuse factor of 1

Figure 3.6 Frequency reuse in a cellular system

Since the frequency spectrum is a very limited resource, reuse of the same frequency spectrum in different cells is always an important issue when designing a cellular system. TDMA and FDMA systems can only work if the interference from other cells using the same frequency spectrum is small enough. Figure 3.6 shows an example of frequency reuse with a factor of 7 – that is, with 7 frequencies in use.

The factor can be reduced to 3 in a TDMA or FDMA system using sector antennas. In a CDMA system, however, the frequency reuse factor is always 1 because all cells can use the same frequency spectrum. This gives the following advantages to CDMA systems:

- Larger system capacity. Since the same frequency spectrum can be used in the adjacent cells or sectors, a CDMA system has a larger system capacity than a TDMA or FDMA system in a large scale, multicell environment.

- Soft handover. An MS can communicate with more than one BS at the same time, allowing unbroken soft handover between cells or sectors.

- Easy frequency planning. Frequency planning has been a time-consuming and difficult part of deploying a cellular system. A frequency reuse factor of 1 significantly simplifies the frequency planning task.

Since it was first introduced to the second-generation system IS-95, CDMA has become the fundamental multiple access scheme for the systems of IMT-2000 and beyond.

3.2.3   Principles of DS-CDMA and IS-95

This section reviews the principle of DS-CDMA, on which the 3G systems and beyond are based, and discusses IS-95, the first CDMA commercial system. A basic block diagram of DS-CDMA is shown in Figure 3.7 (Tachikawa 2002b).

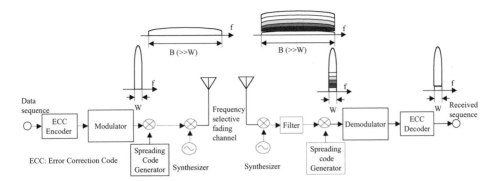

Figure 3.7 Basic block diagram of DS-CDMA. Reproduced by permission of Dr. Sawahashi

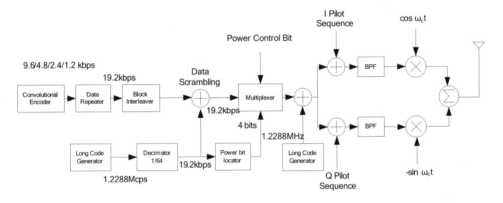

Figure 3.8 Forward transmission structure of IS-95

At the transmitter, the binary data sequence is first encoded and then modulated. The result is a narrowband signal with a bandwidth of W. After the spreading, the narrowband signal becomes a wideband one, occupying the whole channel bandwidth of $B(B \gg W)$. The transmitted signal arrives at the receiver after passing through a Raleigh fading and frequency-selective fading channel due to the propagation via different paths and the physical environment around the receiver. Since all the MSs transmit in an overlapped time period and on the same frequency carrier, the received signal is a mixture of signals sent from multiple MSs. For a signal from a desired MS, the receiver must separate it out of the multiaccess interference (MAI) from other signals. After the filtering and de-spreading, the broadband received signal becomes narrow band again. The information data sequence is finally recovered after demodulation and decoding.

IS-95 (A/B), with the brand name of cdmaOne, was the first frequency division duplex (FDD) DS-CDMA system with a chip rate of 1.2288 Mcps in a 1.25-MHz channel bandwidth. As a principle of IS-95, Figure 3.8 and Figure 3.9 show block diagrams of forward link and reverse link transmissions in an IS-95 system.

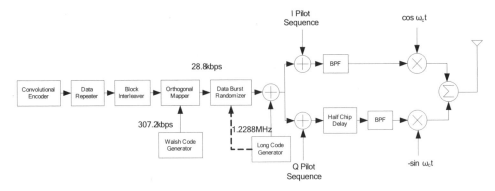

Figure 3.9 Reverse transmission structure of IS-95

In this system, all the BSs are synchronized with the clock on the basis of the time reference from a global positioning system (GPS). Since a DS-CDMA link is interference-limited, technologies to reduce the multipath interference (MPI) and MAI are introduced in the IS-95 system. Walsh codes combined with long pseudorandom code and short pseudo-random code are used for spreading. Pilot-aided coherent demodulation is used for forward link (BS-to-MS) transmission and transmission power control (TPC) is used for reverse link (MS-to-BS) transmission. Convolutional coding is used for error-correction coding (ECC) and quadrature phase shift keying (QPSK) is used for modulation.

For a detailed description of the IS-95 system, see (Garg 2000).

3.3   Radio Access Technologies in Wideband CDMA

Wideband CDMA (W-CDMA) inherits the merits of DS-CDMA technologies that are used in IS-95. The W-CDMA system also includes additional new technologies, such as highly accurate TPC, Rake combining, asynchronous cell operation, OVSF and code multiplexing, and Turbo coding. These technologies are key to the success of the W-CDMA. (Adachi et al. 1998; Dahlman et al. 1998; Tachikawa 2002b)

3.3.1   W-CDMA

W-CDMA introduced intercell asynchronous operation and a pilot channel associated with each data channel. The asynchronous operation brings flexibility to system deployment. The pilot channel enables coherent detection on the uplink and makes it possible to adopt interference cancellation and adaptive antenna array techniques later on. W-CDMA features:

- Fast cell search under intercell asynchronous operation

- Coherent spreading-code tracking

- Fast TPC on both uplink and downlink

- Coherent Rake reception on both links

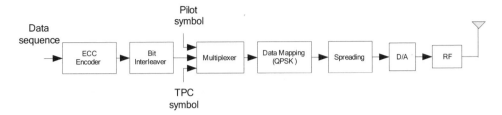

Figure 3.10 Block diagram of transmitting transceiver in W-CDMA system

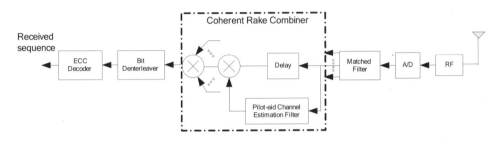

Figure 3.11 Block diagram of receiving transceiver in W-CDMA system

- Orthogonal multiple spreading factors (SFs) in the downlink

- Variable-rate transmission with blind rate detection

In order to explain these techniques, we first give a brief description of the radio access system. Figures 3.10 and 3.11 show the block diagram of transmitter and receiver in the W-CDMA system.

A simplified frame structure used in the system is shown in Figures 3.12 and 3.13. A 10-ms-long frame consists of 15 slots.

Transmission Procedure: At the transmitter, the binary data sequence of the 10-ms frame to be transmitted is fed into channel encoder and bit interleaver. The traditional convolutional coding and new turbo coding schemes (see Section 3.3.4) with a coding rate of 1/3 are used. The output of coding and interleaving is mapped onto the 15 slots. On the downlink, the data sequence of one frame after encoding and interleaving is transformed into a QPSK symbol sequence and is time-multiplexed every 0.667 ms with several pilot symbols as well as a TPC command (Figure 3.12). The QPSK symbol sequence is QPSK-spread – orthogonal binary phase shift keying (BPSK) spreading and QPSK scrambling are applied. On the uplink, BPSK data modulation is applied and the pilot channel is I/Q-multiplexed before QPSK spreading (Figure 3.13). After being power amplified with TPC, the spread signal is transmitted.

Receiving Procedure: The signal sent by the transmitter arrives at the receiver after propagating along different paths. The different distance of each path results in a different arrival time, giving rise to a multipath signal. As soon as it arrives at the receiver, the multipath signal is filtered by a matched filter (MF) that can be implemented using a

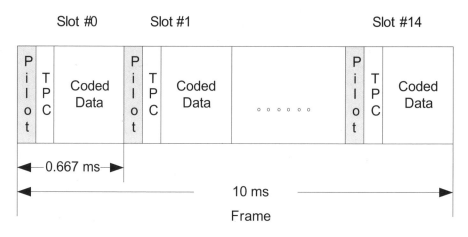

Figure 3.12 Transmission frame structure–downlink frame structure

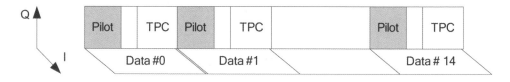

Figure 3.13 Transmission frame structure–uplink frame structure

bank of synchronous correlators. The output is a number of replicas of the transmitted QPSK symbol sequence. Then, a Rake combiner coherently combines these resolved symbol sequences into a soft-decision data sample sequence corresponding to the channel-coded binary data sequence. It is then de-interleaved for a succeeding soft-decision Viterbi decoding process to recover the information data. For fast TPC operations, the Rake combiner output signal-to-interference ratio (SIR) (plus background noise) is measured and compared with the target SIR to generate the TPC command. This TPC command is transmitted every 0.667 ms to the mobile via the downlink, or to the BS via the uplink, to raise or lower the transmit power. At the BS receiver, two spatially separated antennas are used to reduce the mobile transmit power.

3.3.2   Spreading Codes and Asynchronous Operation

Scrambling and channelization codes are two types of spreading codes used in W-CDMA systems.

- A scrambling code is used to distinguish different MSs in an uplink and different cells in a downlink. It is a long spreading code with a length of 38,400 chips in the 10-ms frame period, which guarantees a sufficient number of codes.

- A channelization code is used to identify physical channels. It is a short spreading code with a length from 4 to 512 chips corresponding to the usage.

Both codes are multiplied to spread the encoded and modulated data sequence. It is easier to realize continuous system deployment from outdoors to indoors with an intercell asynchronous system than with an intercell synchronous one. The reason for this is that the asynchronous system does not require any external timing source (such as GPS, used in IS-95). Since a unique scrambling code is assigned to each cell for downlink identification, W-CDMA enables an intercell asynchronous operation. In general, however, the use of different scrambling codes at different BSs increases the cell-search time.

A fast cell-search algorithm involving three steps is described in Higuchi et al. (2000). The downlink control channels of all BSs reuse the same channelization code and the scrambling code sequence is periodically masked over one-symbol duration. It makes the channelization code appear periodically during the scrambling code period. During this masking period, the group identification (GI) code indicating the code group to which the scrambling code of each BS belongs is transmitted in parallel. The GI code can be chosen from the set of orthogonal multi-SF codes to be described in Section 3.3.3.

The three-step cell-search algorithm consists of:

1. Detecting the scrambling code mask timing of the best BS (determined using the least sum of propagation path loss plus shadowing)

2. Identifying the scrambling code group by taking the cross-correlation between the received signal and all GI code candidates

3. Searching for the scrambling code by cross-correlating the received signal with all scrambling code candidates belonging to the identified GI code.

During a soft handoff, an MS must find the best BSs to which it should communicate simultaneously. Since the number of candidate BSs is at most four cells and the MS can be informed of them from the current BS, the cell-search time for soft handoff can be greatly reduced.

3.3.3 Orthogonal Multi-SF Downlink

As the frequency selectivity of the propagation channel strengthens (or the number of resolvable paths increases), the orthogonality among different users tends to diminish because of increasing interpath interference; however, orthogonal spreading always gives a larger link capacity than random spreading. This suggests the advantage of using the orthogonal multi-SF codes in downlink. Multi-SF codes $C_{2^m}^{(j)}$, where m is a positive integer and $j = 1, 2, \ldots, 2^m$, can be generated recursively on the basis of a modified Hadamard transformation. A tree structure of orthogonal multi-SF codes is illustrated in Figure 3.14. For a more detailed description of code generation, refer to Higuchi et al. (2000) or the companion paper by Yang and Hanzo (2003) in the same IEEE issue.

Simplified transmitter and receiver structures of the orthogonal downlink data channel are shown in Figures 3.15 and 3.16.

Data with the symbol rate equal to the *chip rate*$/2^m$ is spread using a single code with the *SF* of 2^m. Since a single spreading code can be used at any data rate, the mobile receiver can be significantly simplified compared to the orthogonal multicode downlink, which simultaneously uses 2^n codes in parallel, each with an *SF* of 2^{m+n}, where $n \leq m$. Noticing that a lower-layer code can be expressed as an alternate combination of the

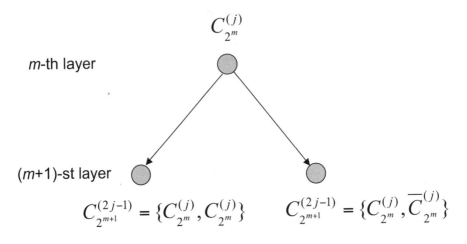

Figure 3.14 Tree-structured orthogonal multi-SF codes

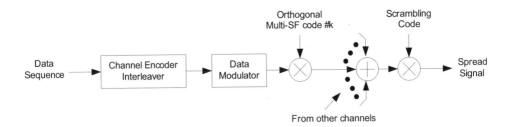

Figure 3.15 Orthogonal downlink transmitter structure

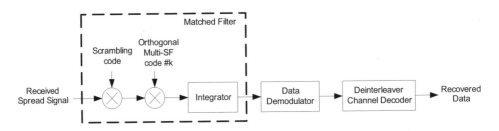

Figure 3.16 Orthogonal downlink receiver structure

sequence of its mother code, further simplification of code usage is possible. Multi-SF codes of the bottom layer (for example, codes of 256 chips/symbol) can always be used irrespective of the data rate, and only the integration time at the receiver needs to be changed. The spreading code does not need to be changed to match the data rate.

3.3.4 Turbo Codes

The W-CDMA system uses ECC to get channel coding gain. In addition to convolutional codes, which are used in IS-95, Turbo codes are introduced to improve system performance. Figure 3.17 illustrates an example of the structure of a Turbo encoder/decoder with a coding rate of 1/3. It is used in W-CDMA to substitute convolutional codes in high-speed transmission.

Actually, convolutional codes are used for low-speed (32 Kbps) voice and data, while Turbo codes are used to encode the high-speed data. A simulation result in Adachi et al. (1998) shows that using Turbo codes when $m = 8$ and coding block length is 3040 bits can achieve about 1-dB coding gain over convolutional codes. As a result, Turbo codes are particularly attractive in data services that permit longer transmission delay.

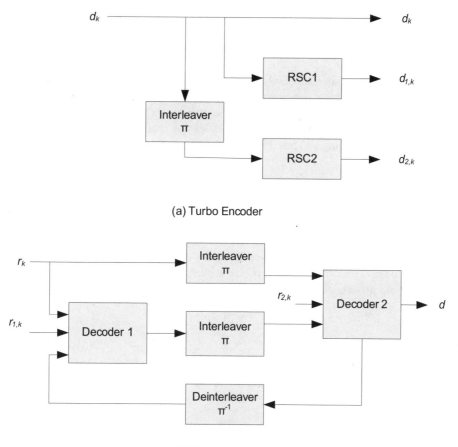

(a) Turbo Encoder

(b) Turbo Decoder

Figure 3.17 Block diagram for Turbo encoding and decoding

3.3.5 Coherent Rake Combining

The wider the channel bandwidth, the higher the decomposition ability of multiple paths. To utilize this behavior, a coherent Rake receiver is used in W-CDMA to obtain the effect of time diversity. Rake reception makes use of all decomposed multipath signals effectively. Figure 3.18 illustrates a coherent Rake receiver structure.

The transmitted spread signal arrives at the receiver after the multipath propagation. After decomposition using a MF, multiple received signals with different delay spreads can be detected. Each of the signals is multiplied by a coefficient weighted by a channel estimator and all the signals are then combined together in a process called *Rake combining.*

In order to estimate channel accurately, the process uses pilot symbols inserted in each time slot, as shown in Figure 3.12. The result of channel estimation is obtained by average weighting the output of channel estimators in several time slots, which are based on the pilot symbols received. A pragmatic approach, 2K-tap weighted multislot averaging (WMSA) (Seo et al. 1998), is used in the channel estimation filter. Instantaneous channel estimation using the pilot symbols belonging to each slot is performed first, and channel estimates of 2K succeeding slots are then weighted and summed to obtain the final channel estimate. A large number of pilot symbols belonging to multiple slots can be used. Then, it is possible to achieve accurate channel estimation, particularly in slow-fading environments, by selecting appropriate weights.

As fading becomes faster, however, tracking ability against fading tends to be lost. This is even true in the case of power-controlled links. Although the received signal amplitude is held almost constant by fast TPC, its phase still varies because of fading. However, this is not a problem because channel coding/interleaving works satisfactorily in fast fading. Channel coding/interleaving and fast TPC complement each other in working against fading.

3.3.6 Transmission Power Control

In DS-CDMA, all MSs with different spreading codes transmit signals to the BS on the same frequency band. If they use the same power to transmit, the BS is likely to receive stronger signals from near mobiles than from far mobiles. This is called the *near–far problem*. To solve this problem, Transmission Power Control (TPC) was proposed to limit an MSIs transmission power, resulting in equal power at the receiver from different mobiles. Since DS-CDMA links are interference limited, the use of TPC can improve the system capacity.

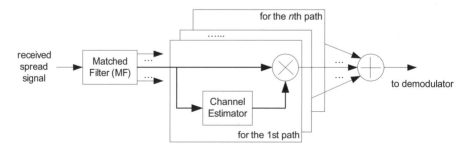

Figure 3.18 Structure of a coherent Rake receiver

In W-CDMA, fast TPC is based on the measurement of SIR of received signals. By minimizing the transmit power according to the traffic load, interference to other users in the other cells is reduced. The pilot symbols used for coherent Rake receiving also play an important role in fast TPC. Both pilot and data symbols are used to measure instantaneous received signal power, but only pilot symbols are used to measure instantaneous interference plus background noise power (followed by averaging using a first-order filter). There are generally two types of TPC, open loop and closed loop. For TPC in the uplink, an MS with open-loop TPC measures downlink SIR and decides the transmission power; an MS with closed loop decides the transmission power on the basis of instruction from the BS. Closed-loop TPC in W-CDMA is based on the outer-loop and inner-loop stages illustrated in Figure 3.19.

In the inner-loop operation, BS measures the SIR of the output from the Rake combiner. If the measured value is larger than the target value, the BS sends a command by setting the TPC command bit to the MS to decrease the transmission power. If the value is lower, it clears the TPC command bit to increase power. Each time the MS receives the TPC command to decrease or increase the transmission power, it reduces or raises power by 1 dB. The time period between two measurements, 0.667 ms, is short enough to realize fast TPC. The outer-loop operation is based on the link quality, as measured by the frame error rate (FER). The comparison between the measured FER value and the target value is transformed into the target SIR to be used in inner-loop operation.

Fast TPC is also applied to the downlink and serves to increase the link capacity. The downlink transmission is based on orthogonal spreading – that is, all downlink channels are synchronous and spread using orthogonal multi-SF codes. With Rake combining, orthogonal spreading provides a larger capacity than random spreading, even in frequency-selective fading channels. However, in the case in which an MS moves away from the BS, the received signal power is reduced because of increasing distance-dependent path loss. Thus, the effects of background noise and other cell MAI become larger. Shadowing also contributes to

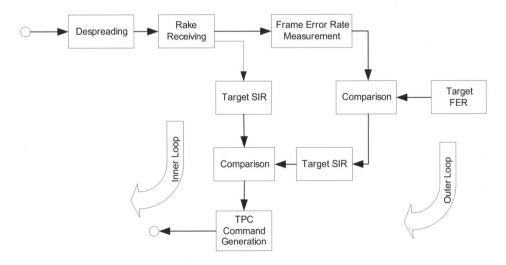

Figure 3.19 Basic multiple access methods

the problem, with the result that the instantaneous SIR on each downlink channel varies randomly, weakening the receiving power and degrading the link quality. When fast TPC is used, the transmitter increases power as the MS moves further away, thus maintaining the link quality.

3.4 High-speed Transmission Technologies in HSDPA

During downlink transmission in a W-CDMA system, user data is carried over dedicated transport channels (DCH), for maximum system performance with continuous user data. The DCHs are code-multiplexed onto one frequency carrier. In the future, user applications are likely to involve the transport of large volumes of packetized data that are bursty in nature and require high bit rates. HSDPA is standardized in 3GPP to provide a packet-based data service in a W-CDMA downlink with a data-transmission rate up to 10 Mbps over a 5-MHz channel bandwidth (Sawahashi et al. 2001). It introduces a new transport channel type, high-speed downlink shared channel (HS-DSCH), that makes efficient use of valuable radio frequency resources and takes into account bursty packet data.

This new transport channel shares multiple access codes, transmission power, and the use of infrastructure hardware between several users. The radio network resources can be used efficiently to serve a large number of users who are accessing bursty data. Several users can be time-multiplexed so that, during silent periods, the resources are available to other users. For example, once a user has sent a data packet over the network, other users can get access to the resources.

To achieve a high-speed downlink transmission, HSDPA implements adaptive modulation and coding (AMC), hybrid automatic repeat request (HARQ), fast cell selection (FCS). These technologies are discussed in more detail below. Similar technologies are also introduced in CDMA2000 systems. Table 3.1 compares HSDPA and CDMA2000 1xEV-DV.

3.4.1 Adaptive Modulation and Coding

Link adaptation in HSDPA is the ability to adapt the modulation scheme and coding rate according to the quality of the radio link (Sawahashi et al. 2001). Figure 3.20 shows the principle of adaptive modulation and coding (AMC).

AMC automatically changes the SF, the number of multiplexing spreading codes, the coding rate of error-correction code, and the level of modulation to achieve high-speed transmission according to user's link condition.

An MS first estimates the propagation status of the downlink and then reports the status periodically to the BS by sending back the channel quality indicator (CQI) signal. Receiving the CQI, the BS chooses a modulation scheme (QPSK or 16-QAM), coding rate (from 1/3 to 1), and the number of multicodes (from 1 to 15) appropriate to the current channel condition. Link adaptation ensures the highest possible data rate is achieved both for users with good signal quality (typically close to the base station) and higher coding rate, and for users with inferior signal quality (typically more distant and close to the cell edge) and with a lower coding rate.

Table 3.1 A comparison between HSPDA and CDMA2000 1xEV-DV

Feature	HSDPA	1xEV-DV
Downlink frame size	2 ms (3 slots)	1.25, 2.5, 5, 10 ms variable frame length (1 slot = 1.25 ms)
Channel feedback	Channel quality reported every 2 ms	C/I feed back every 1.25 ms
Data user multiplexing	TDM/CDM	TDM/CDM (variable frame)
AMC	QPSK & 16-QAM	QPSK, 8-PSK, & 16-QAM
HARQ	Chase combining or incremental redundancy	Asynchronous incremental redundancy
Spreading factor	SF=16 using OVSF channelization codes	Walsh code length 32
Control channel approach	Dedicated channel pointing to shared channel	Common control channel

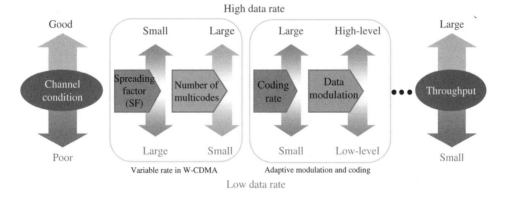

Figure 3.20 Principle of adaptive modulation and channel coding. Reproduced by permission of Dr. Sawahashi

Figure 3.21 shows average BER performance with different modulation schemes. For example, consider the case of QPSK and QAM. It has been observed that to achieve the same BER under these two schemes, E_b/N_0 (signal energy per bit to background noise power spectrum density ratio) for 16-QAM is more than twice that for QPSK. However, the number of information bits in a 16-QAM symbol is twice that in a QPSK symbol. As a result, using 16-QAM can achieve higher spectrum efficiency than QPSK.

A high-level modulation with a large coding rate can reach a high data rate, but it requires an increase of SIR. The effect of MPI can be reduced to $1/SF$ in DS-CDMA.

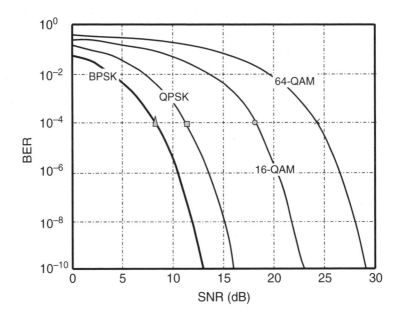

Figure 3.21 Average BER performance with different modulation schemes. Reproduced by permission of Dr. Sawahashi

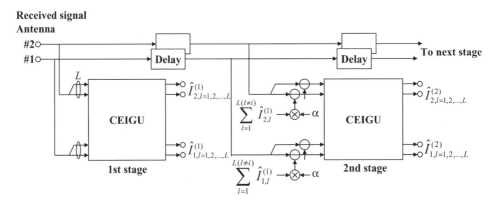

Figure 3.22 Structure of MPIC. Reproduced by permission of Dr. Sawahashi

However, multipath propagation results in a severe frequency-selective fading in W-CDMA using the 5-MHz channel. To achieve the throughput of 2 Mbps under the chip rate of 3.84 Mcps in W-CDMA, *SF* is close to 1, resulting in the significant degradation of SIR because of MPI. This means that high-speed communication is limited to the area near the BS where there is no MPI problem, and the average throughput of the system cannot be improved.

MultiPath Interference Canceller (MPIC) (shown in Figure 3.22) is proposed to solve the problem.

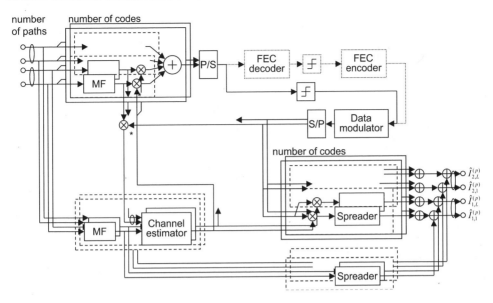

Figure 3.23 Structure of CEIGU. Reproduced by permission of Dr. Sawahashi

MPIC consists of a multistage channel estimation and interference generation unit (CEIGU). The transmitted signals propagate through a multipath-fading channel and are received by a receiver with two-branch antenna diversity reception. During and after the second stage, the MPI replica estimated in the first stage is removed from the received signal for the input signal of CEIGU. Let $_{b,l}^{(s)}$ be the estimated received signal of the l-th ($l = 1, 2, \ldots, L$) path (hereafter called *MPI replica*) at the s-th ($s = 1, 2, \ldots, S$) stage on the b-th ($b = 1, 2$) antenna. The received signal sequence is directly embedded in the first stage, and for the incoming signal sequence at the s-th stage in and after the second stage, the MPI replica of all code channels except for its own path generated in the previous stage, $I_{b,l}^{(s-1)}$, are removed from the received signal sequence. The structure of the CEIGU is illustrated in Figure 3.23.

In each CEIGU, the input sample sequence of each antenna is de-spread by a MF into the resolved multipath components. The channel variation due to fading of each resolved path is estimated by using the common pilot symbols and decision feedback data symbols belonging to the same packet. Then, the phase variation of each path is compensated and coherently Rake combined. The data sequence of the Rake combiner output is de-interleaved and soft-decision Viterbi decoded. The MPIC replica is generated using the decision data sequence, channel estimates, and received power of each path. In this scheme, the accuracy of the MPI replica is improved from the resulting enhancement of channel estimation and decreasing data decision error, because the channel estimation and data decision are updated at each stage. By combining MPIC with orthogonal code multiplexing, when the data decision error occurs on a certain code channel, this decision error can be corrected at the succeeding stage because of the improved SIR. It is proved that the throughput performance is improved significantly by using MPIC (Higuchi et al. 2000).

3.4.2 Hybrid ARQ

When link errors occur (caused by interference, for example), the MS rapidly requests retransmission of the data packets. There are three basic types of ARQ scheme, stop and wait (SW), back-to-N (BTN), and selective repeat (SR). Table 3.2 compares the performance of these schemes.

SR has the best throughput performance, but needs a large buffer and overhead. In contrast, SW has the smallest buffer size and overhead. Combining one of these basic schemes with forward error correction, Type-I and Type-II hybrid ARQ (HARQ) schemes have been developed.

- In Type-I HARQ, a channel-encoded packet is transmitted first. If the packet is received with errors that cannot be corrected completely, the receiver drops the packet and sends a retransmission request to the sender.

- In Type-II HARQ, the sender transmits a packet with an error detection code only the first time. If the packet is received with errors, the receiver stores the packet in the buffer and sends a retransmission request. If the retransmitted packet is received correctly, it is accepted and the erroneous packet in the buffer is dropped. Otherwise, the two erroneous packets are corrected by a decoding algorithm.

Both HARQ schemes repeat the procedure until the packet is accepted. Previous researches show that Type-II HARQ has a better throughput performance but needs a larger buffer.

Figures 3.24 and 3.25 show the principle of Type-I and Type-II HARQ schemes proposed to HSDPA.

The N-channel SW-based Type-I HARQ with packet combining (PC) is accepted as the retransmission scheme for HSDPA. A data packet is divided into N fragments, which are transmitted and processed independently on N channels. This process can improve delay performance while still having the benefit of short overhead. In contrast to the conventional Type-I HARQ (Basic Type-I HARQ) that drops an erroneous packet, the new scheme stores the soft-decision result obtained in decoding of the packet. When the retransmitted packet also contains uncorrectable errors, the receiver combines the two erroneous packets using a Chase combining algorithm (Chase 1985) to improve SIR.

In the proposed Type-II HARQ, an information data sequence is first encoded with a coding rate of R' at the transmitter. Then, a packet with punctured code word with a coding rate of R is transmitted, where $R > R'$. If the packet is not accepted, the packet is retransmitted with another punctured code word. Combining the newly received packet and

Table 3.2 A comparison of three basic types of ARQ scheme

	SW	BTN	SR
Throughput	Good	Better	Best
Delay	Good	Better	Best
Buffer size	Best	Better	Good
Overhead	Best	Better	Good

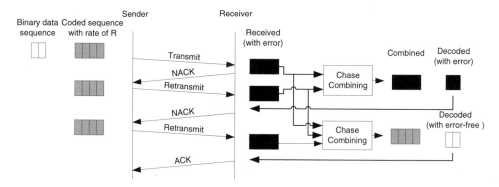

Figure 3.24 HARQ schemes – Type-I HARQ with Chase combining

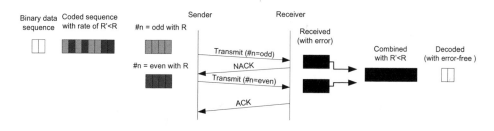

Figure 3.25 HARQ schemes – Type-II HARQ with incremental redundancy

the stored packet, the receiver can decode the combined sequence with a coding rate of R'. As a result, the Type-II HARQ can improve performance with both a time-diversity effect and improved coding gain.

Figure 3.26 shows two examples to compare the throughput performance of these HARQ schemes (Miki et al. 2001).

It is seen that both the proposed HARQ schemes have a significant improvement compared with the basic Type-I HARQ. Also, Type-II HARQ has a better performance than Type-I HARQ with PC at the lower range of E_c/N_0. However, since the HARQ should be used with AMC in HSDPA, a low range of E_c/N_0 for one modulation (64-QAM) is a high range of E_c/N_0 for another modulation (QPSK). The advantage of the Type-II HARQ is not obvious. Moreover, Type-II HARQ must store all the puncture patterns of a de-spread sequence in order to combine code words. This means that the receiver with Type-II HARQ needs a more complex processing mechanism than one with Type-I HARQ with PC. For this reason, the HSDPA system has chosen Type-I HARQ with PC as the fast retransmission scheme.

In current W-CDMA networks, a selective repeat-based Type-I HARQ is used and the retransmission requests are processed by the RNC. A long processing delay results in significant latency to applications. In HSPDA, the above Type-I HARQ with PC is introduced and the request is processed in the BS, providing the fastest possible response.

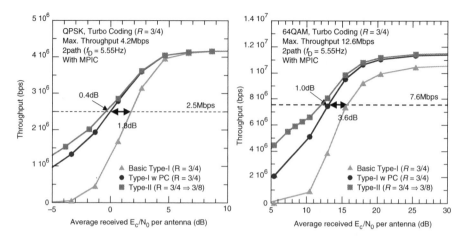

Figure 3.26 Throughput performance of HARQ schemes

3.4.3 Fast Cell Selection

Fast Cell Selection (FCS) (Sawahashi et al. 2001) should be used for intersector diversity, associated with an appropriate scheduling algorithm for decreasing transmission power of HS-DSCH. HSDPA uses FCS for site-selection diversity transmit (SSDT) power control (Furusawa et al. 2000). There are two issues to consider in this choice:

- The effect of FCS depends on the scheduling algorithm that allocates DSCH. There are three major scheduling algorithms, (a) maximum C/I, (b) round robin, and (c) proportional fairness. Each algorithm has advantages and disadvantages.

- The introduction of ARQ enlarges the delay time and increases complexity.

HSDPA offers maximum peak rates of up to 10 Mbps in a 5-MHz channel. However, more important than the peak rate is the packet data throughput capacity, which is improved significantly. This increases the number of users that can be supported at higher data rates on a single radio carrier. HSDPA's high data rates also improve the use of streaming applications on shared packet channels, while the shortened round-trip time benefits web-browsing applications.

Another important characteristic of HSDPA is the reduced variance in downlink transmission delay. A guaranteed short delay time is important for many applications, such as interactive games. In general, HSDPA's enhancements can be used to efficiently implement the *interactive* and *background* QoS classes standardized by 3GPP.

3.5 Radio Access Technologies for Next-generation Systems

This section discusses the radio access challenges facing next-generation systems and the technologies being developed to meet them.

3.5.1   Technical Requirements

A XG broadband mobile communication system will face challenges caused by a channel with a much broader bandwidth; for example, 100 MHz for downlink and 40 MHz for uplink. The propagation characteristics for such a broadband channel cause frequency-selective fading to become more severe, the fluctuation values of receiving power to become smaller, and the ability to decompose delayed waves to become greater.

To reach a level of performance that cannot be achieved by advanced IMT-2000 systems, the XG system will need a much higher throughput (or data rate). The ITU-R SG8 WP8F recommendation document (SG8 2003) requires radio access technologies that support a throughput of greater than 100 Mbps in a high-mobility situation, and 1 Gbps in a nonmobile device.

Furthermore, to support a seamless communication service, the next-generation system will be developed on the basis of one-cell frequency reuse and single-air interface for outdoor and indoor environments (Tachikawa 2003c).

Next-generation radio access will require technologies to support the following:

- Mobile communications with high mobile speed in a microcellular configuration

- Single-air interface in both multicell and isolated-cell environments, with a frequency reuse factor of 1 in the multicell case

- Operation in a high-frequency band with a broad bandwidth

- A much larger system capacity than any 3G system

- A much shorter latency than any 3G system

- A transmission rate higher than 100 Mbps for highly mobile devices and higher than 1 Gbps for nonmobile devices

- A reasonable transmission power for both BS and MS.

Figure 3.27 summarizes the problem space and indicates the corresponding radio access technologies offering solutions.

Note that a combination of technologies is required to solve the problems.

3.5.2   Potential Solutions for Downlink Transmission

In addition to the requirements described in Section 3.5.1, the specific requirements for downlink transmission are summarized as follows and will be explained later in this subsection:

- Multicarrier approach shall be used for the downlink transmission

- CDMA shall be the fundamental scheme.

There are a number of criteria used to determine the basic mechanism of a downlink scheme for broadband cellular systems, such as TDMA or CDMA, single-carrier (SC) or

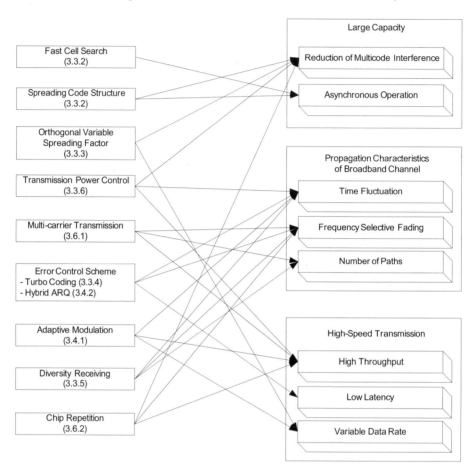

Figure 3.27 Problem space and technologies for solution

multicarrier (MC), and so on (Sawahashi et al. 2003). Table 3.3 compares some of these transmission schemes for a broadband downlink channel.

Single-carrier schemes are no longer available for broadband transmissions. The number of decomposable paths increases as the bandwidth becomes broader. In a W-CDMA system, Rake combining is used to achieve the path (time) diversity. However, too many paths with different delay times generate MI that negates the Rake diversity effect. If the SIR of each de-spread path is very low, the signal after Rake combining cannot achieve the required SIR. A radio access scheme that is robust against MI on a broadband channel is therefore essential to achieve high-quality transmissions.

A multicarrier scheme is also essential for broadband transmissions (Hara and Prasad 1997). In multicarrier transmissions, a broadband channel is first divided into a number of subchannels each with a relatively narrow bandwidth. Correspondingly, a high-speed serial

Table 3.3   Comparison of downlink schemes for wireless access

Access Scheme	Single or Multicarrier DS-CDMA	OFCDM	OFDM	Single or Multicarrier TDMA
Spreading	Yes	Yes	No	No
Number of carriers	1 or several	Many	Many	1 or several
Effects of multipath interference	Degrades signal by negative Rake diversity effect	Robust against multipath interference	Robust against multipath interference	Degradation Degradation due to ISI
Frequency reuse factor	1	1	>1	>1
Spectrum utilization efficiency	Low	High	High	High

data sequence is converted into a number of parallel relatively low-speed data sequences, each of which is transmitted over a subchannel or subcarrier. Such a low-speed data symbol is sufficiently longer than multipath propagation delay that the variation of amplitude and phase within a subcarrier can be treated as a flat fading, and thus the effects of waveform distortion caused by frequency-selective fading can be reduced. For a subcarrier whose received power has dropped because of fading, its decoded error can be compensated by applying ECC across multiple subcarriers whose received power has not dropped, resulting in high-quality reception.

3.5.3   Potential Solutions for Uplink Transmission

In addition to the requirements described in Section 3.5.1, the specific requirements for uplink transmission are summarized as follows and will be explained later in this subsection:

- Single-carrier or multicarrier with a very limited number of subcarriers approach shall be used for the uplink transmission

- CDMA shall be the fundamental scheme.

One of the most significant concerns in the uplink scheme design is the power consumption limit of mobile handsets (Sawahashi et al. 2003). Table 3.4 compares potential broadband radio access schemes for uplink channels. OFDM and OFCDM are not adequate for uplink transmissions. Such schemes need a large number of subcarriers, resulting in a large peak-to-average power ratio (PAPR). Although many methods have been proposed, the problem has not been completely resolved to meet the low power consumption requirement of handsets. In addition, a separate pilot channel is required for each physical channel in an

Table 3.4   Comparison of uplink schemes for broadband radio access

Access Schemes	SC/MC DS-CDMA	OFCDM	OFDM	SC/MC TDMA
Spreading	Yes	Yes	No	No
Number of carriers	1 or several	Many	Many	1 or several
Low power–consuming handset	Yes	No	No	No
Large capacity	Improved by Rake diversity	Accuracy of channel estimation degrades	Accuracy of channel estimation degrades	Degrades due to ISI
Frequency reuse factor	1	1	>1	>1

MS to perform coherent detection and demodulation. For OFDM and OFCDM using many subcarriers, channel estimation must be performed for each subcarrier. If the same amount of power is assigned to pilot channels in both OFCDM and DS-CDMA, the pilot-channel signal power per subcarrier for OFCDM is smaller than that DS-CDMA. Thus, to achieve the same accuracy of channel estimation, a larger E_b/N_0 is needed for OFCDM to meet the requirements for packet error rate (PER), resulting in a lower link capacity.

Multicarrier DS-CDMA is a promising uplink scheme for broadband radio access. DS-CDMA can achieve a better channel estimation accuracy and a higher link capacity than OFDM and OFCDM in uplink transmissions. Moreover, there is an optimal subcarrier band in DS-CDMA that can minimize required transmission power (i.e., received E_b/N_0). This band is determined on the basis of a tradeoff between improved and degraded reception characteristics. The former is achieved by Rake diversity after broadening the frequency band and increasing the number of paths that can be separated. The latter is caused by an increase of MI. Considering the factors, such as the delay profile model, the number of paths, and SF, the received E_b/N_0 required can be greatly reduced by a subcarrier with a bandwidth from 20 to 40 MHz. Accordingly, such a multicarrier DS-CDMA configured on the basis of this optimal subcarrier in accordance with the system band is a promising wireless access scheme from the viewpoint of link capacity.

DS-CDMA also needs to support an isolated-cell environment. In the multicell environment, one-cell frequency reuse in DS-CDMA has the advantage of allowing operators complete flexibility in planning frequency use; however, this advantage diminishes in an isolated-cell environment, where intercell interference can be ignored. The link capacity here turns out to be 20–30% of the SF without voice activity detection. To support a single-air interface for both multicell and isolated-cell environments using DS-CDMA, link capacity needs to be increased for isolated cells. How to increase the link capacity while keeping the orthogonality between different mobile stations is a key issue of the uplink scheme design.

3.6 Broadband Radio Access Schemes for XG Systems

This section introduces the proposal of broadband radio access schemes for XG systems from NTT DoCoMo (Atarashi et al. 2003b; Sawahashi et al. 2003). Such a system inherits the radio access technologies used in 3G systems and introduces new technologies to meet the requirements described above. The proposal is based on an FDD channel configuration with different schemes for uplink and downlink transmissions (Atarashi et al. 2001). The Variable Spreading Factor (VSF) Orthogonal Frequency and Code Division Multiplexing (OFCDM) scheme is proposed for downlink transmission (Atarashi et al. 2001, 2003a). OFCDM is a combination of CDMA and OFDM to support a throughput of 100 Mbps.

OFCDM scheme is proposed for downlink transmission (Atarashi et al. 2001, 2003a). OFCDM is a combination of CDMA and OFDM to support a throughput of 100 Mbps.

The Variable Spreading and Chip Repetition Factors (VSCRF) CDMA scheme is proposed for uplink transmissions (Goto et al. 2003). In VSCRF, spreading factor (SF) and chip (symbol) repetition factor (CRF) are adaptively controlled on the basis of cell structure and number of access users. The downlink and uplink transmission proposals are discussed in detail in the following sections.

3.6.1 VSF-OFCDM for Downlink Transmission

OFCDM is a MC-CDMA scheme (that is a combination of OFDM and CDMA) (Atarashi et al. 2003b; Sawahashi et al. 2003). Generally, it has the following features:

- It reduces the effect of MI by using multicarrier transmission.

- It obtains the effect of frequency diversity by spreading and mapping channel-coded symbols across all subcarriers.

In a cellular environment, OFCDM applies spreading to time domain and/or frequency domain to achieve one-cell frequency reuse. However, in an isolated-cell environment, where intercell interference can be ignored, the spreading in both time and frequency domains is not applied and the scheme becomes pure OFDM. The reason for this is that when spreading in the frequency domain, code orthogonality among multiplexed code channels collapses because of frequency-selectivity fading caused by MI. Similarly, when spreading in the time domain, amplitude fluctuation occurs when Doppler frequency becomes high, giving rise to intercode interference, which in turn causes orthogonality to collapse in the time domain. In either case, the number of code channels corresponding to the band expansion factor cannot be multiplexed. As a result, OFCDM is expected to have a higher system capacity in both cellular and isolated-cell environments.

Principles of VSF-OFCDM

VSF-OFCDM is proposed as the single downlink transmission scheme in both cellular and isolated-cell environments. It adapts the SF in time and frequency domains to cell configuration, propagation conditions (such as delay spread, maximum Doppler frequency, and magnitude of intercell interference), channel load, parameters for radio transmissions (such as modulation scheme and channel coding rate), and so on.

Figure 3.28 shows a conceptual diagram of VSF-OFCDM applied to two-dimensional (2D) spreading (Maeda et al. 2002).

A data symbol is spread into SF_{time} consecutive OFCDM symbols in time domain over SF_{freq} consecutive subcarriers. Here, SF_{time} and SF_{freq} denote SFs in the time and frequency domains respectively. The total spreading factor, SF, can be represented as $SF = SF_{time} * SF_{freq}$. As indicated in Figure 3.29, SF is controlled according to the basic cell configuration. Actually, it is set by an MS according to the downlink control information. SF_{time} and SF_{freq} are adaptively controlled to achieve the maximum channel capacity in either a cellular or an isolated-cell environment on the basis of propagation conditions, channel load, and parameters for radio transmissions. Details of the 2D spreading principle (Atarashi et al. 2001; Maeda et al. 2002) are shown in Figure 3.29, based on an example.

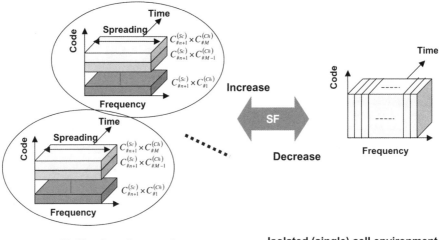

Figure 3.28   Conceptual principles of VSF-OFCDM. Reproduced by permission of Dr. Sawahashi

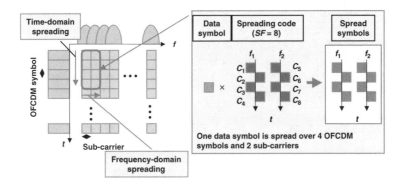

Figure 3.29   Two-dimension spreading principle. Reproduced by permission of Dr. Sawahashi

A data symbol is spread with $SF_{time} = 4$ and $SF_{freq} = 2$ for a total spreading factor of $SF = SF_{time} * SF_{freq} = 8$. It is multiplied by $SF_{time} * SF_{freq}$ orthogonal codes resulting in 4 OFCDM symbols over 2 subcarriers after spreading. By allocating orthogonal codes in 2D in this way, VSF-OFCDM with 2D spreading can multiplex multiple physical channels within a frame and has the following features:

- Physical channels can be flexibly set and released as needed by simply changing the allocation of orthogonal codes

- Physical channels with different symbol rates can be flexibly multiplexed by allocating orthogonal codes with different SFs

- A physical channel with a low data rate can be easily achieved by increasing the SF

- The transmission power of each physical channel multiplexed in the code domain can be flexibly changed

- A code-multiplexed pilot channel can be achieved.

Figure 3.30 shows an example of multiplexing one physical channel having a spreading factor of 16 through time-domain spreading ($SF(16) = SF_{time} * SF_{freq} = 16 * 1$) and another one having a spreading factor of 8 through 2D spreading ($SF(8) = 4 * 2$).

To achieve orthogonality between these two physical channels, a code-allocation scheme called _Orthogonal Variable Spreading Factor (OVSF)_ (Adachi et al. 1998) is used in the multiplexing. Figure 3.30 also shows the corresponding example of the OVSF scheme. Suppose we select $C_{16,1}$ as the orthogonal code for the physical channel $SF(16)$. Then, in order to achieve the orthogonality with SF(16), we need to select the orthogonal code for the physical channel $SF(8)$ by using any of $C_{8,3}, C_{8,4}, \ldots, C_{8,8}$ codes that have been generated from a code other than $C_{4,1}$ lying on upper level of $C_{16,1}$.

Variable-spreading-Factor Control Methods

In VSF-OFCDM with 2D spreading, the spreading in the time domain has a priority over that in the frequency domain. Figure 3.31 shows the general policy used to control the SF in time and frequency domains (Atarashi et al. 2003b; Sawahashi et al. 2003).

In the time domain, SF_{time} is controlled according to the information bit rate and the maximum Doppler frequency. In the frequency domain, SF_{freq} is controlled according to the information bit rate, channel load, and delay spread. However, it is recommended that SF_{freq} be set to 1 when using 16-QAM, and changed for QPSK, for the following reasons:

- Time-domain spreading is more effective with a short frame length. In order to shorten the delay caused by adaptive radio link control, packet retransmission control, and other control processes, a frame with length of $0.5-1.0$ ms is used in the proposed VSF-OFCDM. Within such a short frame length, the time-domain spreading is more suitable than the frequency-domain spreading for minimizing the effects of destroyed orthogonality between code-multiplexed physical channels (that is, intercode interference) for cases other than high mobility.

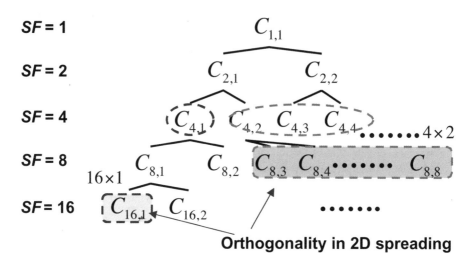

Figure 3.30 Two-dimensional orthogonal-code allocation method. Reproduced by permission of Dr. Sawahashi

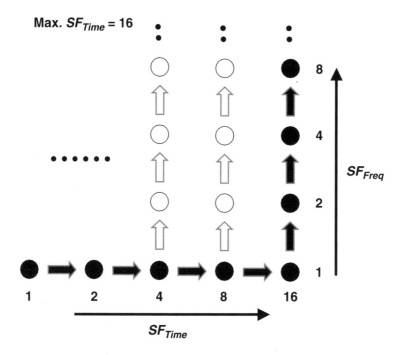

Figure 3.31 Spreading factors in two dimensions. Reproduced by permission of Dr. Sawahashi

- In good channel conditions where multi-ary modulation having little robustness against intercode interference is used, giving priority to time-domain spreading can decrease the required E_b/N_0.

- Low data rate control and data channels with QPSK will set $SF_{freq} > 1$ in addition to controlling SF_{time} so as to decrease the required E_b/N_0 through a frequency diversity effect. This type of control is also effective in a multicell environment, where intercell interference is dominant. Also, the high Doppler frequency caused by the high mobility of a vehicular user will result in the increase of intercode interference in the time-domain spreading. In this case, SF_{time} should be reduced.

Figure 3.32 shows the configuration of the VSF-OFCDM transmitter.

Here, a binary data stream is first applied to channel coding and bit interleaving and then to data-modulation mapping and serial–parallel conversion. These modulated symbols of each parallel stream are then spread in two dimensions to generate physical channels. The bit-interleaving block, serial–parallel converter, and 2D-spreading block adopt the appropriate bit-interleaving pattern, serial-to-parallel conversion and 2D spreading according to the values of SF_{time} and SF_{freq} determined by the variable-spreading-factor controller. The physical channels are allocated different orthogonal codes and code-multiplexed. After converting these streams to OFCDM symbols by using the Inverse Fast Fourier Transform (IFFT) algorithm, a guard interval to reduce ISI is inserted for each symbol.

To demonstrate the effect of VSF control, Figure 3.33 shows average PER versus channel load (number of multiplexed physical channels normalized by the spreading factor: C_{mux}/SF) when varying SF_{time} and SF_{freq}.

The results for QPSK in Figure 3.33 show that we can get a better PER under low channel-load conditions with a larger SF_{freq}. This is because a frequency diversity effect obtained through spreading overcomes the effect of intercode interference caused by the frequency-selective fading. As channel load increases, however, intercode interference becomes more significant. In such a case, PER can be decreased by applying no frequency-domain spreading ($SF_{freq} = 1$). Thus, adaptive control that employs optional SFs in the time and frequency domains according to channel load can achieve high-quality transmission with improved PER.

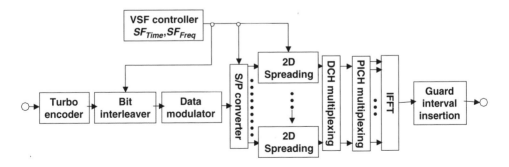

Figure 3.32 Configuration of VSF-OFCDM transmitter. Reproduced by permission of Dr. Sawahashi

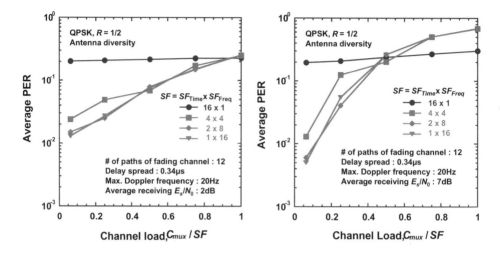

Figure 3.33 Effect of applying VSF control. Reproduced by permission of Dr. Sawahashi

The results for 16-QAM in Figure 3.33 show that a larger SF_{freq} under low channel-load conditions has a better PER because of the frequency diversity effect. This is similar to the case of QPSK. However, it is should be noted that the information bit rate here can be also achieved with a higher quality by simply increasing channel load with QPSK. Meanwhile, for channel load of $C_{mux}/SF > 0.5$, PER for $SF_{freq} > 1$ deteriorates significantly because 16-QAM employing phase and amplitude modulation is subjected to intercode interference. Therefore, setting $SF_{freq} = 1$ in 16-QAM is generally effective under high channel-load conditions.

3.6.2 VSCRF-CDMA for Uplink Transmissions

Variable spreading and chip repetition factors CDMA (VSCRF-CDMA) is proposed as the uplink scheme for both multicell and isolated-cell environments (Atarashi et al. 2003b; Goto et al. 2003; Sawahashi et al. 2003). In a multicell environment, only the spreading in the time domain is used to achieve one-cell frequency reuse. In an isolated-cell environment, on the other hand, a symbol repetition method is introduced to apply repetitive chips to the spread sequence. Let SF denote the spreading factor for total band, $SF_{cellular}$ and $SF_{hotspot}$ the spreading factors for orthogonal spreading code in a multicell and an isolated cell respectively, and CRF, the chip repetition factor. SF is determined by the symbol rate of the physical channel. In a multicell environment, SF is simply equal to $SF_{cellular}$. In an isolated-cell environment, we have $SF = SF_{hotspot} * CRF$, where $CRF \geq 1$.

Principles of VSCRF-CDMA

Figure 3.34 shows a conceptual diagram of the VSCRF-CDMA scheme.

In a multicell environment like a cellular system, VSCRF-CDMA suppresses intercell interference through spreading gain in order to realize one-cell frequency reuse. However, in an isolated-cell environment, such as a hot spot or indoor office, with little intercell

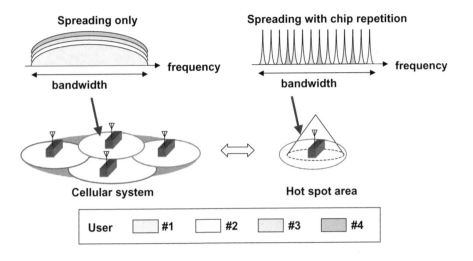

Figure 3.34 Conceptual diagram of VSCRF-CDMA. Reproduced by permission of Dr. Sawahashi

interference, the benefits of one-cell frequency reuse by spreading decrease and the effects of multiple-access interference and multipath interference are significant instead. In such an environment, frequency spectrum efficiency of DS-CDMA deteriorates. Accordingly, VSCRF-CDMA decreases the SF used by DS-CDMA in an isolated-cell environment and applies chip repetition by the amount of that decrease. The application of chip repetition results in the generation of a comb-shaped frequency spectrum, enabling orthogonality in the frequency domain of users simultaneously accessing the system in the uplink (ICC 1998). This approach can decrease multiple-access interference and achieve high-frequency spectrum usage in an isolated-cell environment in contrast to the conventional DS-CDMA with only spreading.

For base-station reception, VSCRF-CDMA applies "loose" transmission timing control to align the reception timing of received signals from each user. Specifically, the received timing for each path of each user is accommodated within the guard interval length. This can achieve complete orthogonality in the frequency domain between signals from different users in accordance with the principle of symbol repetition. Meanwhile, in DS-CDMA that does not perform chip repetition, performing "strict" transmission timing control (so that reception timing for each user's maximum-received-power path is aligned within chip duration) can decrease multiple-access interference.

Control Method for Spreading and Chip Repetition Factors

VSCRF-CDMA uses a spreading and chip repetition factors control block having the configuration shown in Figure 3.35.

For a multicell environment, in which chip repetition is not applied, this block performs two-layered spreading using an orthogonal code corresponding to spreading factor $SF_{cellular}$ and a cell-specific scrambling code (or user-specific scrambling code). For an isolated-cell

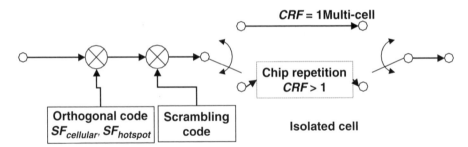

Figure 3.35 Configuration for changing radio parameters in VSCRF-CDMA in the uplink. Reproduced by permission of Dr. Sawahashi

environment like a hot spot or indoor cell, this block performs chip repetition according to chip repetition factor *CRF* after multiplying input data by an orthogonal code corresponding to spreading factor $SF_{hotspot}(< SF_{cellular})$ and a cell-specific scrambling code. To keep the same system bandwidth even when applying chip repetition, the relationship $SF_{cellular} = SF_{hotspot} * CRF$ must hold in this case.

As for the basic principle of orthogonalization in the frequency domain by applying chip repetition, Figure 3.36 shows how a comb-shaped frequency spectrum is generated by compressing Q spreading chips and performing chip repetition *CRF* times.

The stream resulting from this chip repetition can then be multiplied by a user-specific phase vector to generate a comb-shaped frequency spectrum mutually orthogonal with those of other users. In general, applying chip repetition *CRF* times means that signals from *CRF* number of users can be orthogonalized.

Figure 3.37 shows required average received E_s/N_0 (signal energy per bit to background noise power spectrum density ratio) per antenna satisfying an average $PER = 10^{-1}$ versus the number of simultaneous accessing users for both VSCRF-CDMA and DS-CDMA using spreading only (Goto et al. 2003).

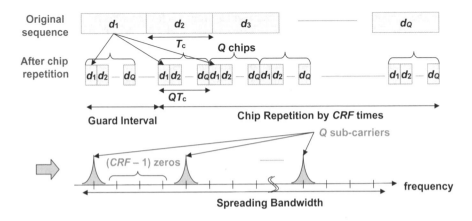

Figure 3.36 Basic principle of chip repetition. Reproduced by permission of Dr. Sawahashi

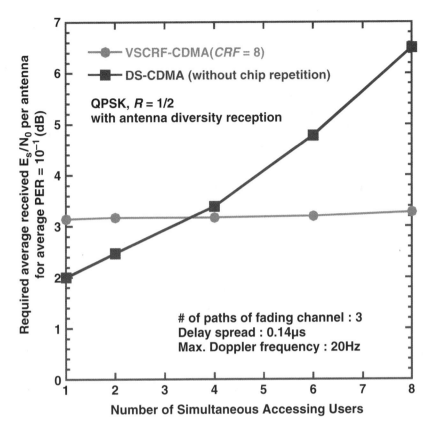

Figure 3.37 Comparison of VSCRF-CDMA and DS-CDMA. Reproduced by permission of Dr. Sawahashi

First, for DS-CDMA that uses spreading only, multiple-access interference increases as the number of simultaneously accessing users increases, resulting in the significant degradation of average PER. Meanwhile, for VSCRF-CDMA, average PER is nearly constant regardless of the number of simultaneously accessing users, since multiple-access interference can be reduced by orthogonalization of the signals of those users in the frequency domain. In short, VSCRF-CDMA can decrease required average received E_s/N_0 compared to DS-CDMA using spreading only.

3.7 Conclusions

The next-generation cellular system needs to achieve the performance that any advanced IMT-2000 system cannot, such as a much higher throughput, a much lower latency, and a much higher spectrum utilization efficiency. A much wider bandwidth channel is expected to be adopted to support the broadband wireless access.

Radio access technologies for the XG cellular system will inherit the technologies of the 3G systems, and introduce new technologies to meet challenges posed by the propagation characteristics of a broader channel bandwidth, and by increased performance requirements. Keeping the leadership in radio access technologies, NTT DoCoMo proposes VSF-OFCDM and VSCRF-CDMA for XG broadband radio access. The proposed technologies evolve CDMA-related technologies by introducing new technologies, such as multicarrier and chip repetition to meet the performance requirements of next-generation cellular systems.

4

Wireless LAN Evolution

Fujio Watanabe, Gang Wu, Moo Ryoung Jeong, and Xiaoning He

4.1 Introduction

The Wireless Local Area Network (WLAN) industry has become one of the fastest growing segments of the communications industry. WLAN equipment shipments grew to almost 12 million units in 2001 and demand is expected to continue growing at a compound annual rate of 23% over the next five years. This growth is due, in large part, to the introduction of standards-based WLAN products. The expectation of the WLAN's continuing growth stems from the promise of new standardized WLAN technologies, from improved cost/performance of WLAN systems, and from the growing availability of WLAN solutions that consolidate voice, data, and mobility functions. This, combined with market forecasts that report that WLAN will experience tremendous growth in the next years, show that WLAN technologies will play a significant role in the future and will have a significant impact on our business and personal life styles (Salkintzis and Passas 2003).

The IEEE 802.11 working group (WG) is already evolving several aspects of the dominant WLAN standard, including security, quality of service (QoS), and coexistence with other unlicensed technologies. Figure 4.1 shows how the WLAN standards have been developed and continue to evolve. There are additional technical challenges to be addressed to resolve problems of high throughput (HT), mobility, interworking among WLAN and external networks, and radio resource management.

Higher data rates requires broad frequency bands, and sufficient broadband can be achieved in higher frequency bands. Since the propagation path loss is in proportion to the carrier frequency, increased propagation loss is unavoidable in the higher frequency band. Accordingly, the service coverage area is shorter than in the current cellular system.

Next Generation Mobile Systems. Edited by Dr. M. Etoh
© 2005 John Wiley & Sons, Ltd

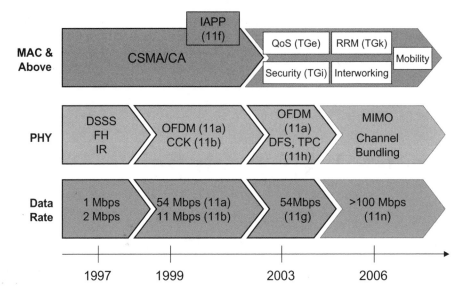

Figure 4.1 Current and future MAC, PHY, and data rate extensions of WLAN standards

Both WLAN and 3G cellular systems can provide higher-speed wireless connections that cannot be offered by earlier cellular technologies. WLANs are more suited to hotspot coverage, that is, broadband wireless access with limited mobility. 3G cellular systems, which have voice support, wide coverage, and high mobility, are more suited to areas with moderate or low-density demand for wireless usage requiring high mobility.

In future, various complementary radio access networks (RANs) will be used in combination with 4G RANs to provide full coverage services. WLAN is expected to be one of these complementary RANs used to achieve broadband wireless service with limited mobility.

This chapter provides an overview of current development in the area of WLANs, with particular focus on the enhancement of Medium Access Control (MAC) layer technologies, instead of the physical layer (PHY). The chapter briefly reviews the current WLAN standards and technologies (802.11-based WLAN), including MAC and PHY, along with an example of the OFDM-based PHY technologies (802.11a 5 GHz). It presents some limitations of current WLAN technologies and discusses some key enabling technologies that are needed to extend WLAN in order to integrate it with next-generation (XG) mobile networks. Finally, we present current research work and ongoing standardization related to these key technologies.

4.1.1 Overview of Current WLAN Standards

The international WLAN specification is developed by the IEEE 802.11 WG, which was established in May 1989 and is composed of volunteers from industry and academia. The first standard, called 802.11, was released to the public after a seven-year development cycle (IEEE 1999a; Salkintzis and Passas 2003). This standard specified three PHY technologies

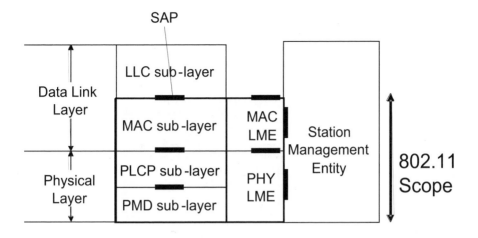

Figure 4.2   IEEE 802.11 scope

and a unified MAC protocol to support 1 and 2 Mbps transmission over wireless media. Figure 4.2 shows the scope and layer configuration of IEEE 802.11 WG.

- The MAC protocol has two functions, the main <u>distributed coordination function (DCF)</u> and the optional <u>point coordination function (PCF)</u>. PCF uses a polling scheme. DCF uses the Carrier Sense Multiple Access with collision avoidance (CSMA/CA) protocol.

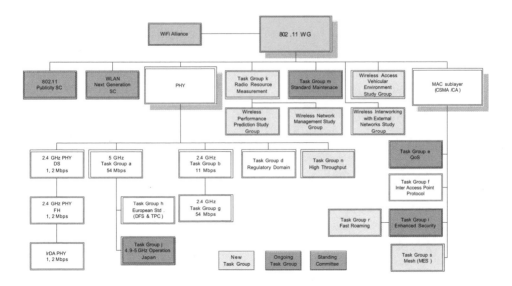

Figure 4.3   IEEE 802.11WG organizational chart

Table 4.1   IEEE 802.11 standards summary

Standard	802.11	802.11a	802.11b	802.11g
Date standard approved	July 1997	September 1999	September 1999	July 2003
Frequency bands and bandwidth	2.4-2.4835 GHz (83.5 MHz) DSSS, FHSS	5.15-5.35 GHz 5.725-5.825 GHz (300 MHz) OFDM	2.4-2.4835 GHz (83.5 MHz) DSSS	2.4-2.4835 GHz (83.5 MHz) DSSS, OFDM
Number of overlapped channels	3 Indoor/ outdoor	4 Indoor (UNII1) 4 Indoor/ outdoor (UNII2) 4 outdoor (UNII3)	3 Indoor/ outdoor	3 Indoor/ outdoor
Data rate per channel	2, 1 Mbps	54, 48, 36, 24, 18, 12, 9, 6 Mbps	11, 5.5, 2, 1 Mbps	54, 36, 33, 24, 22, 12, 11, 9, 6, 5.5, 2, 1 Mbps
Modulation types	DQPSK (2 Mbps DSSS) DBPSK (1 Mbps DSSS) 4GFSK (2 Mbps FHSS) 2GFSK (1 Mbps FHSS)	BPSK (6, 9 Mbps) QPSK (12, 18 Mbps) 16QAM (24, 36 Mbps) 64QAM (48, 54 Mbps)	DQPSK/CCK (11, 5.5 Mbps) DQPSK (2 Mbps DSSS) DBPSK (1 Mbps DSSS)	OFDM/CCK (6, 9, 12, 18, 24, 36, 48, 54 Mbps) OFDM (6, 9, 12, 18, 24, 36, 48, 54 Mbps) DQPSK/CCK (5.5, 11, 22, 33 Mbps) DQPSK (2 Mbps DSSS) DBPSK (1 Mbps DSSS)

- The 802.11 PHY includes Directed Sequence Spread Spectrum (DSSS) and Frequency Hopping Spread Spectrum (FHSS) technologies for radio communications in 2.4-GHz band and infrared communications.

Further improvement of these protocols is described in the following sections.

Since its release, one of the IEEE 802.11 WG's major focuses has been to improve the transmission rate to meet the requirements of broadband wireless access. In September 1999, the 802.11b specification for WLAN in the 2.4-GHz band to support a transmission rate of 11 Mbps was ratified by IEEE (IEEE 1999b). The modulation method selected for 802.11b is known as Complementary Code Keying (CCK) and is based on DSSS. Also in September 1999, the 802.11a specification was ratified to support WLANs in the 5-GHz band, with a maximum transmission rate of 54 Mbps (IEEE 1999c). The European Telecommunication Standards Institute (ETSI) broadband radio access network (BRAN) in Europe and the Association of Radio Industries and Businesses (ARIB) multimedia mobile access communication (MMAC) in Japan also developed WLAN in 5 GHz. These standards are called *HIPERLAN type 2 (HIPERLAN/2) and HiSWANa*, respectively. The modulation scheme adopted in 802.11a, HIPERLAN/2, and HiSWANa is the well-known Orthogonal

Frequency Division Multiplexing (OFDM). OFDM was also adopted by the recently ratified 802.11g specification (IEEE 2003a), which provides a maximum transmission rate of 54 Mbps for WLANs in the 2.4-GHz band. Table 4.1 summaries the 802.11 standards.

Task groups (TG) established within the WG focus on specifying enhancements of MAC and PHY based on a unified MAC and different PHYs. The QoS TG (TGe) and enhanced security TG (TGi) have nearly completed their mandate, and these standards will probably be published by the time this book is released. The current standardization focus for a new generation of WLAN is to enhance the PHY data rate with an efficient throughput performance. TGn is in charge of this work. It is focusing on support for single-link HT rate, support for HT rate in the 20-MHz channel, backward compatibility, spectral efficiency (at least 3 bps/Hz), and support for the TGe QoS efforts (IEEE 2004; Simoens et al. 2003). Many related issues, such as radio resource measurement, fast BSS transition, and interworking are also discussed and standardized in different TG. Figure 4.3 shows the organizational chart of IEEE 802.11 WG.

4.2   Basic Technologies in IEEE 802.11 WLAN

IEEE 802.11 specifies two WLAN system topologies; the infrastructure mode, with at least one central access point (AP) connected to a wired network, and the ad hoc or peer-to-peer mode, in which a set of wireless stations communicate directly with one another without needing a central AP or wired network connection.

Most currently deployed WLANs are based on the infrastructure mode. In a hotspot service, for example, the network service provider builds a multiple-access-point (AP) backbone using fiber or wireless links to the Internet. The WLAN user can get broadband wireless Internet access through the AP coverage area. However, high installation costs or the difficulty of installation in specific places will sometimes make the infrastructure mode impractical. Interest is growing in using the ad hoc and multihop communications mode to extend coverage in wireless networks. In multihop communications, each WLAN user operates not only as a host but also as a wireless router, forwarding packets on behalf of other WLAN nodes that may not be within direct wireless communication range.

The following subsections provide a brief outline of the MAC and PHY technologies in the WLAN standard. See IEEE (1999a) for a more detailed description.

4.2.1   MAC Technologies

The MAC protocol provides two functions, the main distributed coordination function (DCF) and the optional point coordination function(PCF), which can be used to determine when a station has access to the wireless medium.

- Each station can apply the DCF individually to determine when to access the medium, thus distributing the decision-making process among stations.

- The PCF is used during a contention-free period to poll stations. The polled station can access the wireless medium exclusively, without any contention.

The relationship between these two functions is shown in Figure 4.4.

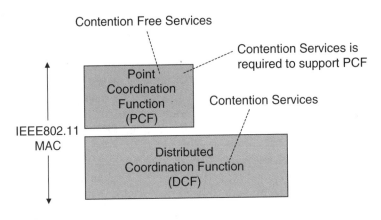

Figure 4.4 DCF and PCF

The two functions are compared and discussed below, and followed by an overview of the CSMA/CA protocol, on which these mechanisms are based.

Distributed Coordination Function (DCF) and Point Coordination Function (PCF)

The distributed coordination function (DCF) is a contention-based function of the IEEE 802.11 MAC. In DCF, when the channel changes from a busy state to an idle state, it does the following:

1. Waits a specific IFS time period.

2. If the channel is still idle, then waits a random backoff time.

3. If the channel is still idle, then it transmits its frame.

Figure 4.5 shows a typical DCF sequence.

In contrast, the point coordination function (PCF) is based on a contention-free mechanism. Figure 4.6 shows the PCF sequence.

The AP polls the STAs to transmit their data frames by using a special poll frame with PIFS. When it receives a poll, the STA immediately sends its data frame with short

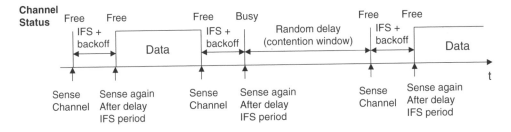

Figure 4.5 DCF in IEEE 802.11 MAC

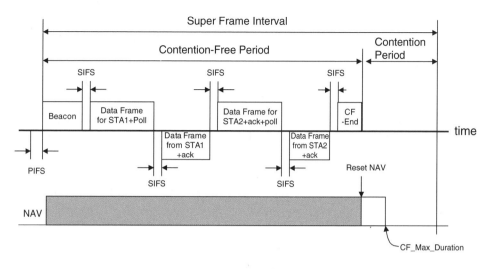

Figure 4.6 PCF in IEEE 802.11 MAC

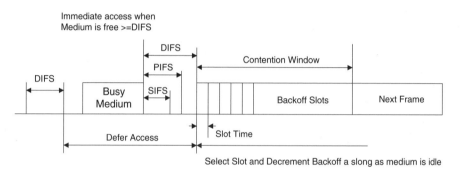

Figure 4.7 STA selection process for APs. Reproduced by permission of Dr. Morikura

interframe space (SIFS). Because PIFS and SIFS are shorter than DIFS, STAs waiting to send frames with DIFS cannot transmit, so there is a collision-free period in PCF.

In addition to the DCF and PCF access functions, the IEEE 802.11 MAC defines a number of management functions. One of these is for an STA to access the specific AP that manages the BSS. Figure 4.7 shows how the STA searches, authenticates, and finally associates to an adequate AP.

The AP periodically broadcasts the operation information of the BSS in what is called a *beacon frame*. The STA listens to all beacons it can hear and then selects an adequate AP to associate. In turn, the AP needs to authenticate the STA sending the association request. The STA finally becomes a member of a BSS after this process is done.

CSMA/CA

With its basic concept of "listen before talk," CSMA is widely used in wired and wireless communication networks as a MAC protocol to coordinate transmissions over a shared communication medium. In contrast to the CSMA/CD protocol used in Ethernet, where collision detection can be easily realized, the CSMA/CA protocol (developed for an 802.11 wireless network) makes an effort to avoid collisions, because the wireless receiver has difficulty with collision detection. The receiver uses the following features and functions:

- Adaptive collision window (CW) based random backoff time to reduce the probability of collisions

- Different interframe space (IFS) to prioritize different types of transmissions

- Acknowledgement frame to realize the stop and wait ARQ

- Request to send (RTS) and clear to send (CTS) handshaking to solve the hidden terminal problem

- Network Allocation Vectors (NAV) to realize virtual carrier sense

As in other random access protocols, the random backoff time in CSMA/CA works to avoid collisions between transmissions from different STAs. The random backoff time can be calculated from this equation:

$$Backoff_{time} = Random() * Slot_{time} \qquad (4.1)$$

In this equation, $Random() = [0, CW]$, $(CW_{min} \leq CW \leq CW_{max})$, and $Slot_{time}$ is the value of the corresponding PHY characteristic. Suggested values are $CW_{min} = 31$ and $CW_{max} = 255$. If it is the current packet's first transmission, CW is set to CW_{min}. After each collision of this packet, the collision avoidance mechanism doubles CW until it reaches CW_{max}.

$$CW_{new} = (CW_{old} + 1) * PF - 1 \qquad (4.2)$$

In this equation, PF is equal to 2. This is referred to as the exponential backoff algorithm. The offered load to the channel is high when experiencing a collision, so increasing the CW to increase the backoff time of each colliding STA helps decrease the collision probability.

The IFS is a time interval after a busy state of the channel. This interval plays an important role in CSMA/CA for collision avoidance and prioritized transmissions. The IFS requires an STA to wait for a period of time after it senses the idle state of the channel. Then, the STA waits for a random backoff time before transmitting its frame. There are four basic types of IFS:

- Short IFS (SIFS)

- Point IFS (PIFS)

- Distributed IFS (DIFS)

- Extended IFS (EIFS)

Each type has a distinct interval time. The four types are designed for transmitting different types of frames. SIFS is used to transmit frames with the highest priority, such as acknowledgment (ACK), CTS, and poll response. PIFS is used in the point coordinate function when an AP issues poll frames. DIFS is used by ordinary asynchronous traffic. EIFS is used when a MAC frame is received with an error. Some examples of IFS relationships are shown in Figure 4.8.

A stop-and-wait ARQ is combined with CSMA/CA. An ACK frame is sent by the STA that successfully receives a data frame. An SIFS is used for sending an ACK frame to guarantee the highest transmission priority.

There is a well-known hidden terminal problem in CSMA-type protocols. RTS-CTS handshaking is used to solve this problem. Accordingly, the concept of network allocation vector (NAV) is introduced. Figure 4.9 shows the time chart of the CSMA/CA with RTS-CTS handshaking.

The source STA sends an RTS to the nearby STAs to make a reservation and start a NAV period. The destination STA sends a CTS to respond to the reservation and start a

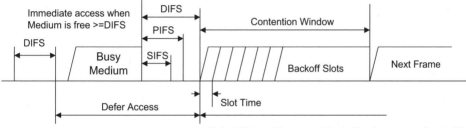

Figure 4.8 Some IFS relationships

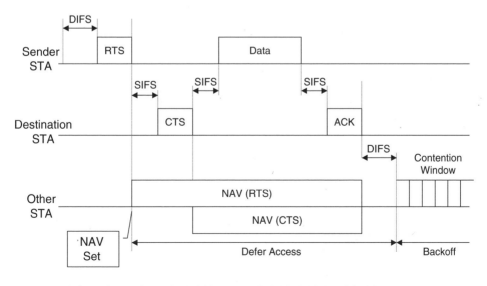

Figure 4.9 IEEE 802.11 MAC RTS-CTS handshaking

NAV period for neighboring STAs. NAV protects the current transmission, thus solving the hidden terminal problem.

4.2.2 PHY Technologies

The four IEEE 802.11 PHY standards are listed in Table 4.2. The fifth is being developed in IEEE 802.11 TGn, targeting a new PHY to support a throughput of more than 100 Mbps. This section briefly introduces the OFDM-based PHY technologies in 802.11, known as 802.11a and 802.11g.

As described in Chapter 3, the multicarrier transmission is an efficient scheme for solving the problem of severe frequency-selective fading in broadband wireless access systems. Figure 4.10 depicts such a mechanism. After experiencing a multipath propagation, an impulse waveform at the transmitter becomes widely spread in the time domain at the receiver. This results in intersymbol interference (ISI) in digital communications. When the symbol rate is low, the problem of ISI can be solved by using an equalizer or canceler at the receiver. The higher the symbol rate, the more complex the equalizer/canceler. This is one of the fundamental problems of broadband wireless access. One solution is a multicarrier transmission that can reduce the symbol rate at each subcarrier, so narrowband solutions can be used in this situation. OFDM is one of the most spectrum-efficient multicarrier transmission methods.

Figure 4.11 shows a block diagram of OFDM transceiver. A channel-encoded data stream is input for the transmitter. The serial data stream is first transformed into parallel and then modulated separately. An Inverse Fast Fourier Transform (IFFT) is used as

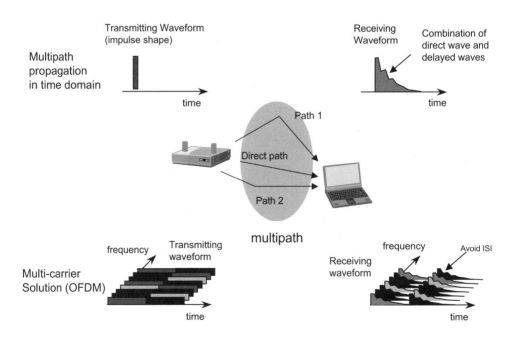

Figure 4.10 Multicarrier transmission in a multipath propagation environment

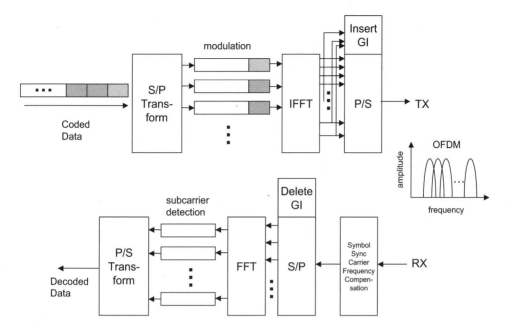

Figure 4.11 OFDM transceiver block diagram

the processing algorithm to create OFDM symbols. To keep the subcarriers orthogonal in a multipath propagation environment, a guard interval (GI) is inserted in each OFDM symbol. After a parallel-to-serial transform, the OFDM symbols are transmitted. The figure shows that the subcarriers overlap each other. These overlapped carriers do not interfere with each other, improving the spectrum utilization efficiency. At the receiver, the GI is deleted and the FFT is used as the algorithm to transform OFDM symbols from a frequency domain into a time domain.

Figure 4.12 shows an important mechanism that uses a GI to reduce multipath effect in OFDM communications. After multipath propagation, the received waveform may involve the direct wave as well as delayed wave components. If there are no means of protection, these components will exist in the results of FFT, that is, each parallel signal stream. The GI is designed to reduce the effect caused by delay spreads. As shown in Figure 4.11, the GI is generated by copying the bottom parts of OFDM symbol and inserting them into the top parts. The multipath effect of the GI is shown in Figure 4.12. The ISI effect can be reduced if the delayed waves arrive at the receiver within the window of GI.

Table 4.2 shows the parameters related to OFDM in 802.11 standards. A 52-subcarrier OFDM symbol consists of 48 subcarriers for information and 4 subcarriers for pilot. Pilot is a known signal sequence to detect and compensate for frequency synchronization errors. The transmission rate varies from 6 to 54 Mbps, according to different modulation schemes and coding rates used. The GI is 800 ns, enabling the WLAN to work in a multipath environment with a root mean square (RMS) delay spread of 100 to 200 ns. Each of the subcarriers is spaced 312.5 kHz apart and the GI is added to each symbol to make the total symbol duration 4 s.

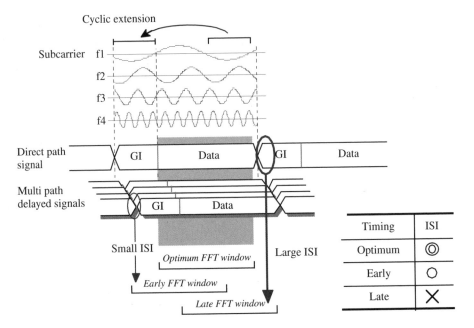

Figure 4.12 Reduction of multipath effect using OFDM

Table 4.2   IEEE 802.11a parameters

Parameter	Value
Data rate	6, 9, 12, 18, 24, 36, 48, 54 Mbps
Modulation	OFDM with BPSK, QPSK, 16-QAM, 64-QAM
Number of subcarriers	52 subcarriers including 4 for pilot
	64 point FFT
FEC	Convolution coding with K=7, R=1/2, 2/3, 3/4
	Viterbi decoding
	Interleaving within an OFDM symbol
OFDM symbol duration	4 s
Guard interval	800 ns
Subcarrier spacing	312.5 kHz
-3 dB bandwidth	16.56 MHz
Channel spacing	20 MHz

Figure 4.13 shows the PHY frame format of 802.11g. For backward compatibility, the Physical Layer Convergence Protocol (PLCP) preamble and header are the same in both 802.11 and 802.11b. The Physical Service Data Unit (PSDU) uses OFDM and has the same structure as in 802.11a. Depending on the 48-subcarrier BPSK, QPSK, 16-QAM, and 64-QAM, the raw data rate can reach to 12–72 Mbps. In order to reduce the effect of fading,

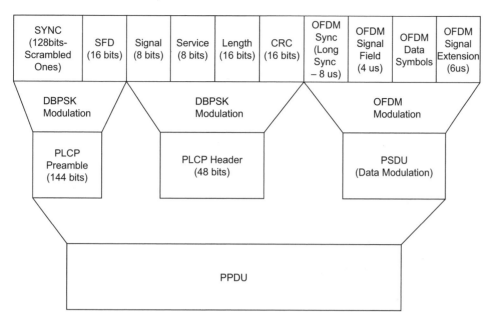

Figure 4.13 IEEE 802.11g PHY frame format

the convolutionary channel coding with a rate of 1/2 and soft-decision Viterbi decoding is specified.

Although 802.11g takes advantage of both 2.4 GHz and OFDM technologies, its performance is not as high as expected. Figure 4.14 shows the upper limits of throughput for 802.11a/b/g (Morikura and Matsue 2001). Note that the throughput of CCK-OFDM does not increase significantly as the PHY layer transmission rate increases. The main reason for this is the relatively long PCLP preamble and header.

4.3 Evolution of WLAN

WLAN has become increasingly popular over the past few years, and customers are demanding additional functionality. To provide high-speed Internet access in a public-access scenario, a WLAN must make an optimal trade-off between bit rates and range. In the home environment, significant challenges include the simultaneous distribution of high-definition video, high -speed Internet, and telephony. Such applications demand efficiency, robustness, and QoS from the WLAN. The forthcoming WLAN system is expected to provide a variety of services not currently available, such as:

- Higher data rates (more than 100 Mbps) and low power consumption

- Extended coverage areas and scalability using the multihop/mesh network

- Coexistence of heterogeneous access devices in the same environment

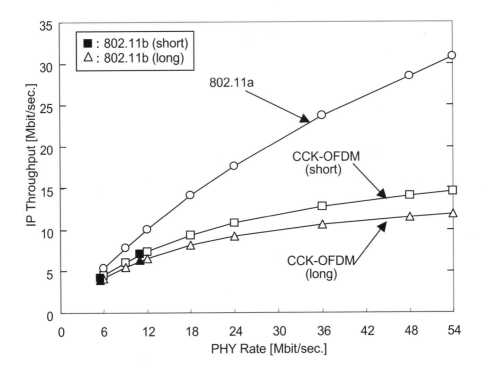

Figure 4.14 The maximum IP throughput. Reproduced by permission of Dr. Morikura

- Seamless mobility support:

    - Handoff mechanism and seamless AAA during handoff

    - Interworking with other systems, seamless mobility between various access technologies, allowing continuity of existing sessions

- Differentiated service support for differing reliability needs

- Indoor location estimation

- Quality of service assurance, including support of real-time applications

- Enhanced security features, including authentication/authorization and data cipher

A number of the issues that limit current WLAN services can be addressed through new technologies. This chapter focuses on the WLAN issues that will be most urgently needed to create solutions complementary to XG mobile networks. The following sections discuss in more detail the technologies related to mobility support, QoS, and enhanced security.

4.3.1   Higher Data Rates and Low Power Consumption

Typical office applications, such as the downloading of large e-mail attachments, are data intensive. In a public hotspot, such as a hotel or airport, the time available for download is likely to be limited. A public wireless access solution should ideally be able to offer very fast transmission capacity.

Both simulation and experience have shown that the throughput in an 802.11a network is actually limited to a point significantly below the 54 Mbps theoretically achievable by the PHY layer. There is also a theoretical maximum throughput for 802.11 MAC (Xia and Rosdahl 2002; Xiao and Rosdahl 2002). However, a WLAN that uses the CSMA/CA mechanism employs four different interframe spaces (IFSs) to control access to the wireless medium. These IFSs act as overhead, which limits the improvement of throughput performance. To reduce this MAC overhead, new systems may use multiple antennas solutions, bandwidth increment, turbo codes, and higher-order constellations, all of which can help to increase the theoretically achievable capacity (Simoens et al. 2003).

The TGn of IEEE 802.11WG is now working on improving the current MAC and PHY throughput. The next generation of WLAN should be able to improve throughput performance significantly, with data rates of more than 100 Mbps.

However, much of the research that targets maximum throughput does not consider increased power consumption. Energy efficiency is becoming crucial to the design of next-generation wireless systems, especially for WLAN that is used by mobile devices with limited battery life. Although WLAN does include a power-management scheme, further power efficiency from both PHY and MAC solutions will be needed.

4.3.2   Extended Coverage Areas and Scalability

Multihop mesh network communication is gaining popularity, both for pure ad hoc communication networks and for coverage extension in wireless networks. A mesh network differs from an ad hoc network in that each WLAN node operates not only as a host but also as a router. User packets are forwarded to and from an Internet-connected gateway in multihop fashion. The network is dynamically self-organizing and self-configuring; the nodes in the network automatically establish and maintain routes among themselves. This makes the meshed topology reliable and it provides good area coverage. Systems are scalable and initial investment can be minimal because the technology can be installed incrementally, one node at a time, as needed. As more nodes are installed, both reliability and network coverage increase (Fitzek et al. 2003; Jun and Sichitiu 2003). This option would decrease installation costs for WLAN hotspots of the next generation.

A mesh network's traffic pattern is different from that of an ad hoc network. In the mesh network, most traffic is either to or from a gateway, while in ad hoc networks, the traffic flows between arbitrary pairs of nodes. Because of poor support for multihop operations in the current IEEE 802.11 standard, current WLAN systems show poor performance for such multihop/mesh networks. To improve this, we need to find more-efficient MAC schemes that make it possible to operate these devices in multihop mode without excessive performance degradation. In the IEEE 802.11 WG, a Mesh Network Study Group was approved to be a TG in March 2004 to create a new standard for mesh networks over WLAN.

4.3.3 Coexistence of Access Devices

The WLAN operates in the 2.4-GHz industrial, scientific, and medical (ISM) unlicensed band. In the unlicensed ISM band, frequencies must be shared and potential interference tolerated as defined in Federal Regulations Part 15 of Federal Communications Commission (FCC). Spread spectrum and power rules are fairly effective in dealing with multiple users in the band as long as the radios are physically separated, but not when the radios are in close proximity. This would be a problem for IEEE 802.11 WLAN and Bluetooth that, for example, come together in a laptop or desktop.

To operate in the 5-GHz range, WLAN must share with other systems, such as military, aeronautical, naval RADARs, and satellite systems. In Europe, for example, WLAN operating on the 5 GHz band is required to implement dynamic frequency selection (DFS) and transmit power control (TPC) in order to share with radar systems.

Current research is focused on the coexistence of wireless devices in the 2.4-GHz band and other bands.

- The IEEE 802.15.2 standard specifically addresses coexistence between WLAN and Bluetooth systems. This standard has adopted an adaptive frequency hopping (AFH) mechanism, which modifies the Bluetooth frequency hopping sequence in the presence of WLAN direct sequence spectrum devices (Golmie 2003; Golmie et al. 2003).

- The TGh standard in the IEEE 802.11 WG met the European regulatory requirement for coexistence with radar systems.

- The IEEE 802.19 Coexistence Technical Advisory Group (WG19) is working on policies that define the responsibilities of 802 standards developers to address issues of coexistence with existing standards and other standards under development.

4.3.4 Seamless Mobility Support

Smooth on-line access to corporate data services in hot spots should allow users to move freely from a private, microcell network to a wide-area cellular (3G) network. In the next generation, various complementary RANs, including WLAN, will be used in combination with 4G RANs to provide full coverage services. Seamless communications over these heterogeneous environments will require effective vertical handoff support.

Current applications primarily move data through the WLAN. In future, users expect to use VoIP over WLAN through the corridor or public space. With VoIP, a user requires handoff support to keep voice connection when moving from one AP to another. In other applications, such as video streaming, users want a seamless connection while roaming through different rooms and corridors.

Mobility support and security are not currently sufficient to support a seamless connection over WLAN. Currently, WLAN does not have any coordination when the station (STA) moves from one AP to another, which causes connections to break during the handoff. Fast-scanning and fast-authentication technologies will be key factors in reducing the handoff blackout time.

To create solutions for these needs, the research community is studying authentication, authorization, and accounting (AAA) and QoS mapping between different access networks (Koin and Haslestad 2003). Standards work in this area is being done by the 3rd Generation Partnership Project (3GPP). WGs are currently developing technical requirements for UMTS-WLAN interworking systems, reference architecture models, network interfaces, and AAA. The IEEE 802.11 WG has also formed a Study Group on Wireless Interworking with External Network, which will soon become a TG, working to standardize an interworking interface between WLAN and other wireless networks.

There are two interworking solutions, tight coupling and loose coupling, based on the type of integration formation. The two solutions have different pros and cons:

- Tight coupling uses the WLAN as a part of 3G RAN in which all necessary functions are located in the core network. This solution has the advantage of fully integrated mobility management (handover) and possible QoS mapping by the 3G core network. The 3G core network also provides sufficient AAA functionality. However, deployment is time consuming, and significant standards work will be needed.

- Loose coupling considers WLAN as equivalent to the 3G networks. It adapts the IP protocol architecture and requires few changes to the WLAN standard. It has a low deployment cost and fast time to market. However, it is not easy to achieve QoS mapping or mobility support, and there is a possible risk of AAA compromise to 3G mobile networks.

4.3.5 Location Estimation by WLAN

The recent growth of interest in pervasive computing and location-aware systems and services provides a strong motivation to develop techniques for estimating the location of devices in both outdoor and indoor environments. Indoor location estimation is particularly challenging because of the poor coverage of global positioning systems (GPS). There are several approaches that use existing wireless LAN infrastructures.

Early work in this area included the RADAR system (Bahl and Padmannabhan 2000), which showed that accurate indoor location estimation could be achieved without deploying separate sensor network infrastructures. Their idea is to infer the location of a IEEE 802.11b wireless LAN user by leveraging received signal strength information available from multiple WLAN beacons.

In following work (Bahl et al. 2000), RADAR was enhanced by a Viterbi-like algorithm that specifically addresses issues, such as continuous tracking and signal aliasing. The Nibble system (Castro 2001) took a probabilistic approach in a similar WLAN environment.

The MultiLoc system (Pandya et al. 2003), which utilizes information from multiple wireless (or wired) technologies, was proposed. The MultiLoc system employs two simple sensor fusion techniques to illustrate the benefit of combining heterogeneous information sources in location estimation.

DoCoMo USA Labs proposes two location-estimation algorithms (Gwon et al. 2004), Selective Fusion Location Estimation (SELFLOC) and Region of Confidence (RoC), which can perform estimation and tracking of the location of stationary and mobile users. More research is still needed for practical deployment. For details of the research, see (Gwon et al. 2004).

4.3.6   Differentiated Services Support

The current service provided by WLAN is a best-effort data service; that is, all customers have the same priority to access a WLAN access point (AP).

Different usages, however, should be able to demand different levels of reliability. A user who is browsing the Internet, for example, might be tolerant of delays and occasional connection failures. However, a user who is accessing an FTP server using the WLAN might want a constant and reliable connection. The new WLAN system must be able to differentiate services on the basis of each user's needs.

In the current IEEE 802.11 standard, all stations have the same distributed interframe space (DIFS) value and perform the backoff window calculation scheme in the same way. As a result, the current IEEE 802.11 standard can provide only a best-effort service, as all stations have the same priority.

4.3.7   Quality of Service Assurance for Real-time Applications

Traditionally, real-time multimedia applications, such as voice service, have been the most basic and important features offered by service providers. The most important quality measures for real-time applications are jitter (the time between two sequential frames), and the end-to-end delay (the time for transmitting a packet from one end to the other) due to the unknown transmission time of a polled station in PCF (Mangold et al. 2003).

- In DCF mode, the timing of a station accessing a channel is unpredictable, so DCF mode is not suitable for real-time applications with stringent delay and jitter requirements.

- Even though PCF mode supports real-time applications, there are very few equipment manufacturers that have implemented PCF in their product because of its high protocol overhead. A new QoS enhancement of the IEEE 802.11 WLAN standard includes three features that support real-time applications:

- Transmission opportunity (TXOP) is defined as the starting time and duration of a transmission.

- The TXOP gives a backoff entity the right to deliver a MAC service data unit (MSDU), and thus provides an important means to control MSDU delivery delay. No backoff entity transmits during the target beacon transmission time (TBTT). This rule reduces the expected beacon delay.

- Direct communication between two WLANs is allowed without involving communication with AP. Further details of QoS enhancement mechanisms are given in Section 4.5.

4.3.8   Enhanced Security

Current security vulnerabilities in IEEE 802.11 WLAN are introduced briefly here and discussed in more detail in Section 4.6 and in Chapter 11.

The necessary level of privacy and authentication can depend on the application, or on the location in which a WLAN is deployed. Enterprise applications, for example, have security needs that are different from those of public space applications. A particular residential application might need the same security level of an enterprise, while another might not. The security technology solutions, therefore, need to be broad enough to support a variety of application spaces. The solutions must be easy to use because the same laptops and devices will be used for Internet access for all types of applications and in all locations (Park 2003).

Current WLAN security, especially wired equivalent privacy (WEP), is known to be a problem area. One major problem with WEP is that its secret keys (shared by wireless devices and the APs) are relatively shorter than those of other security protocols. Secret keys are typically, 40 bits in WEP, although the standard allows up to 104 bits. WEP security also suffers from poor key management, which can leave the keys in a device unchanged for long periods of time. If the device were lost or stolen, an attacker could use the key to compromise that device, and any other devices sharing the same key (Bing 2002). Dynamic key management based on the 802.1X could help mitigate the threat of WEP keys falling into the wrong hands, as well as increase complexity. Because of the current WEP vulnerabilities, TGi (the Security Task Group) is developing a new security standard for IEEE 802.11 WLAN as an amendment. See Section 4.6 for further details.

4.4   Mobility Support

This section explains the channel scanning and the authentication methods that support WLAN mobility.

The fast roaming/fast handoff Task Group (TGr) was newly formed within the IEEE 802.11 WG in March 2004 and is investigating further improvement of the fast-handoff capability.

4.4.1   Fast Channel Scanning

The scanning process – when mobile stations scan for available networks to determine which network to join – is one of the most time-consuming processes in the handoff (Mishra et al. 2002b). 802.11 Wireless LAN has two ways of scanning: passive and active. Passive scanning listens for beacon frames from access points (APs). Active scanning involves a transmission of probe request frames for soliciting a probe response frame from APs. When it receives beacon frames or probe response frames from an AP, the mobile station gathers information about the reachability and the characteristics (such as capability, supported rates, and timing information) of the AP. Two new fast channel scanning technologies, adaptive beaconing and fast active scanning, have recently been proposed in the TG.

Adaptive Beaconing for Fast Passive Scanning

Passive scanning has high latency. In this type of scan, the mobile station must stay on each channel for at least one beacon interval. The value of this interval is usually set to a large number (on the order of 100 msec) to reduce the beacon transmission overhead and the power consumption of mobile stations in power-save mode.

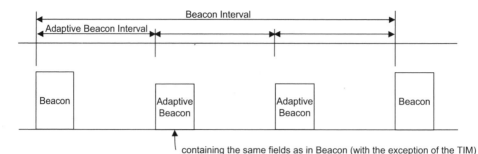

Figure 4.15 Passive scanning improvement

In adaptive beaconing (Orava et al. 2003), adaptive beacons are transmitted with the frequency based on the network load (see Figure 4.15). Adaptive beacons contain the same fields as those in a beacon frame but do not have a traffic indication map (TIM) indicating traffic buffered for specific mobile stations in power-save mode. Mobile stations doing passive scanning quickly gather information about the reachability and the characteristics of the AP by receiving either beacons or adaptive beacons. Mobile stations in power-save mode save power by waking up only during beacon transmissions.

Fast Active Scanning

Active scanning also has high latency. In this type of scan, the mobile station must stay on each channel long enough (up to 50 msec (Mishra et al. 2002b)) to receive probe responses from as many APs as possible (Figure 4.16). Probe requests are broadcast using the DCF, so there is contention among the probe responses from APs and data frames from mobile stations. This contention is resolved using random backoff after a DCF interframe space (DIFS).

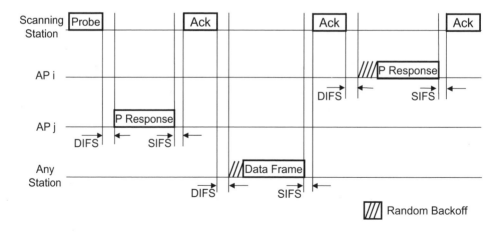

Figure 4.16 Current active scanning scheme

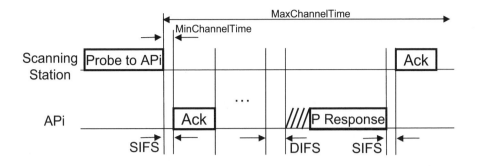

Figure 4.17 Proposed active scanning scheme (option 1)

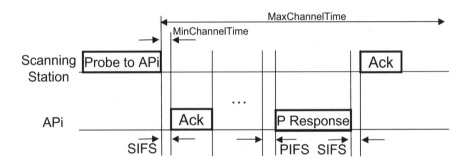

Figure 4.18 Proposed active scanning scheme (option 2)

In fast active scanning (Jeong et al. 2003a,b), a mobile station is allowed to send a directed probe request to APs. These APs are selected using site reports from a current AP with neighbor AP information (IEEE 2003e). When it receives a directed probe request, the neighbor AP acknowledges the request and then sends a probe response (Figures 4.17 and 4.18). Alternately, the neighbor AP replies with a probe response within a short interframe space (SIFS) (Figure 4.19). If the AP opts to respond to the probe response later, it sends the probe response after the medium is idle for a PCF interframe space (PIFS) (Figure 4.18).

When the selected AP is reachable, the mobile station receives the probe response more quickly, because unnecessary probe responses from other APs are eliminated, and the desired probe response transmission is sent with high priority using SIFS or PIFS (Figures 4.18 and 4.19). When the selected AP is not reachable, the mobile station learns this more quickly by receiving either an acknowledgement or a probe response within SIFS.

Performance of Fast Scanning

With a low network load, fast active scanning is flexible and is completed in less than 1 msec (Jeong et al. 2003a,b). With a high network load, fast active scanning takes more time and is costly in terms of bandwidth consumption, as in conventional active scanning. This

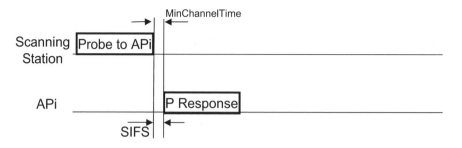

Figure 4.19 Proposed active scanning scheme (option 3)

is more bandwidth consuming because each mobile station performs scanning with separate exchanges for probe requests and probe response frames.

Adaptive beaconing has a longer scanning time but consumes less bandwidth by finding the right trade-off between the scanning time and bandwidth consumption, depending on the network load. An appropriate combination of adaptive beaconing and fast active scanning is required for further study.

4.4.2 Fast Authentication

A couple of authentication solutions for WLAN (Ala-Laurila et al. 2001a; Bostršm et al. 2002a) have been studied. These solutions are based on a single subscriber identity (SIM), which is used in the GSM/GPRS. The main benefit of this method is that it combines different accounts for WLAN and GSM into a single account using GSM and WLAN. Another benefit is easy roaming. Unlike most Internet service providers, mobile operators have the infrastructure and support roaming between different operator networks. So these solutions focus on single bill and roaming rather than supporting authentication method during handoff. The main design challenge for these solutions was transporting standard GSM subscriber authentication signaling from the terminal to the authentication center using the IP protocol framework (Ala-Laurila et al. 2001a).

Unlike the solutions described above, DoCoMo USA Labs has focused on the fast authentication mechanism for supporting mobile users moving from one AP to another within the coverage area of a WLAN system. Mobile communication systems, such as 2G and 3G do not require authentication during handoff because their security and encryption features guarantee that the user is valid. WLAN currently defines three mobility types that do not include seamless handoff (IEEE 1999a):

No-transition: There are two subclasses that are usually indistinguishable:

> **Static:** No motion

> **Local movement:** Movement within the PHY range of the communicating stations (STAs), that is, movement within a Basic Service Set (BSS)

BSS-transition: A station movement from one AP to another within the same Extended Service Set (ESS)

ESS-transition: Station movement from an AP in one ESS to an AP in a different ESS. This is supported only in the sense that the STA can move. Maintenance of upper layer connections cannot be guaranteed by IEEE 802.11; in fact, disruption of service is likely to occur.

The definition of handoff in this discussion includes some features of the first two mobility types described above, but other functions, such as seamless connection, are still missing. When an STA moves from one AP to another, there is no coordination on the network side. Therefore, an authentication for the STA is required whenever the STA moves. Although the IEEE 802.1X authentication method (IEEE 2001b) is widely used to access WLAN networks to carry (extensible authentication protocol) EAP, the communication time between the AP and the Authentication Server (AS) in this method is time consuming. In IEEE 802.1X, the AP is called the *Supplicant* and the AP is called the *Authenticator*. The processing time of IEEE 802.1X probably will not meet the latency of a real-time application connection. Figure 4.20 illustrates the IEEE 802.1X procedure.

IEEE 802.1X and EAP for authentication are executed whenever the WLAN terminal tries to associate the APs. This means that these processes will run whenever the handoff occurs. This is a long process, and the real-time application packet cannot be transmitted while processing is taking place, so many packets will be dropped or discarded at the AP or STA. Eventually the real-time application will be dropped, too. Therefore, it is necessary to reduce the authentication processing time in order to keep the real-time application connection.

The original IEEE 802.11 standard (IEEE 1999a) uses preauthentication to reduce the authentication processing time. This method was not defined in the corresponding clauses (IEEE 1999a), but is defined in the new security enhancement draft (IEEE 2003c) that

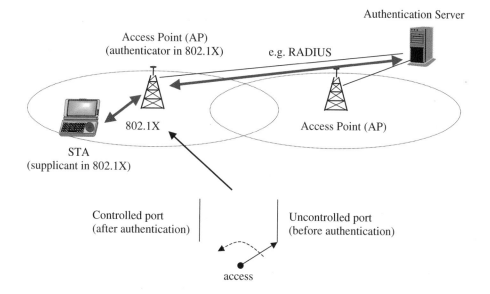

Figure 4.20 802.1X process between the supplicant and the authentication server

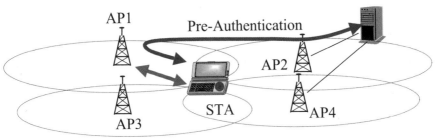

Figure 4.21 Preauthentication

is currently being standardized. The scheme of preauthentication uses the IEEE 802.1X protocol. The IEEE 802.1X Supplicant of a roaming STA can initiate preauthentication by sending an EAP over LAN (EAPOL)-Start message via its old AP, through the distribution system (DS), to a new AP. The current associated AP must forward the data frame to the basic service set ID (BSSID) of the targeted AP via the DS. The preauthentication acquires the Pairwise Master Key Security Association (PMKSA), which is the resulting context from a successful IEEE 802.1X authentication exchange between the Supplicant and Authenticator. In other words, the STA gets an authentication for the target AP (AP2) based on IEEE 802.1X through the current associating AP (AP1) as shown in Figure 4.21.

However, this preauthentication scheme is fully dependant on mobility prediction. This means that the prediction of target AP must be correct unless all possible APs need to be authenticated. In addition, each AP has to store each STA's PMK for a time. This storage process is called *PMK caching*.

Mishra and coworkers (Mishra et al. 2002c, 2003c) have proposed other similar approaches. One of these is to use the interaccess point protocol (IAPP) (IEEE 2003d) that was recently standardized in IEEE 802.11.

As shown in Figure 4.22, AP1 transfers the STA's security context information to the AP2. The AP2 has the security context in cache, so once the STA moves to AP2, the STA can do a fast reassociation. This scheme relies on mobility prediction. It also relies on knowing which APs are neighboring the current APs, because the target AP's coverage area should be overlapped with the current AP's. Otherwise, no handoff occurs.

To make this easier, the research group of the University of Maryland has proposed the use of the AP's neighborhood graph map (Arbaugh n.d.; Mishra et al. 2003a,c). The neighborhood graph is an approximate AP location graph map representing a mobility path between APs. This map is constructed on the basis of the AP MAC address that is sent from the new AP when the STA moves from a current associated AP to a new associated AP. Mishra (Mishra et al. 2003c) proposes three methods for key distribution to authenticate STA. On the basis of the AP's neighborhood graph, these methods are improved and the target APs are clearer. These methods are:

Static roam keys: The AS pushes a unique seed for encryption key derivation (such as a pairwise master key or PMK) to each AP. The encryption key is then derived via some form of handshake. One disadvantage of this scheme is that the past communication

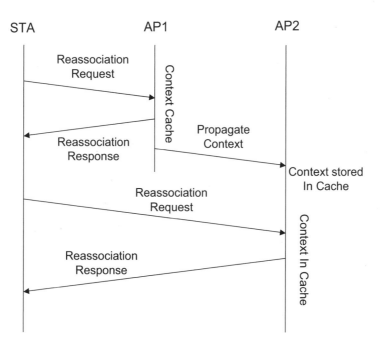

Figure 4.22 Context (authentication message) transfer by IAPP (IEEE 2003d)

is subject to compromise if the AP is compromised. Also, there is a large memory requirement for the AP unless it is combined with a means of proactive distribution.

IAPP with proactive caching: The current AP creates the next PMK for the target AP and these keys are distributed by IAPP (Figure 4.23). The next PMK derived by the current AP can be different for each STA. One advantage of this is the mobility prediction, for which it is necessary to have information about the AP neighborhood. Another advantage is that the compromised AP only compromises the current and the next encryption keys, not future encryption keys.

Proactive key distribution: This method relies on AP neighboring graphs, and the PMK is distributed on the basis of these graphs. Therefore, this method can eliminate problems with sharing key material among multiple APs. Other disadvantages are that it increases network traffic load and that the AP neighboring graphs are unclear.

DoCoMo USA Labs also proposed the handoff key method (Watanabe et al. 2003), which gives the STA temporary access until IEEE 802.1X authentication is completed. This scheme uses a shared key called a *handoff key*, which is distributed to all active STA and APs. With this proposal, our intent is to allow immediate data transmission and data encryption by the handoff key during the handoff process. To meet this goal, we propose a new key method that achieves authentication of the STA much faster.

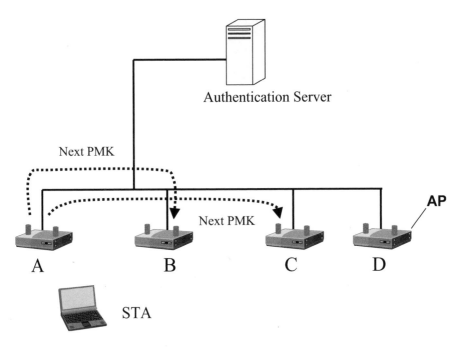

Figure 4.23 IAPP caching of next PMK to neighbors (Mishra et al. 2003c)

Figure 4.24 illustrates a WLAN network configuration. In this figure, the STA associates with AP1. The access router 1 (AR1) has two APs. AR1 and AR2 belong to the authentication, authorization, accounting foreign server (AAAF1). The STA originally belongs to authentication, authorization, accounting home server (AAAH). Whenever authentication is needed for the STA, the authentication request is sent to the AAAH through the AAAFs.

If there is a handoff, the STA that is currently associated with the AP moves to AP2. IEEE 802.1X authentication is required before any access, so the STA must wait until the IEEE 802.1X authentication is approved to receive transmissions. Our method focuses on the real-time application running on the STA, so very fast authentications are necessary when the STA moves from one AP to another. In order to avoid disconnection during the IEEE 802.1X authentication time, we propose a secure temporary access key scheme using the handoff key. This handoff key would only be used during the handoff process to encrypt the data transmission.

The creation of the handoff key is illustrated in Figure 4.25. Once IEEE 802.1X authentication is done, it is necessary to create an encryption key (e.g., PMK) to encrypt the data transmission more securely. For example, all APs under the AAAF1 know the method of key generation for creating a handoff key for the STA. The key-generation process shown in Figure 4.25 is transferred to AP1, AP2, and AP3 by the AAAF1. It is important to note that the secret parameter consisting of various parameters (e.g., AAAF_ID identity and the common parameter of AAAF) is shared by the APs belonging to the AAAF1. The secret parameter is only known to the related APs, in this figure AP1, AP2, and AP3. The secret parameter is transferred to each AP in a secure manner. For example, this parameter could

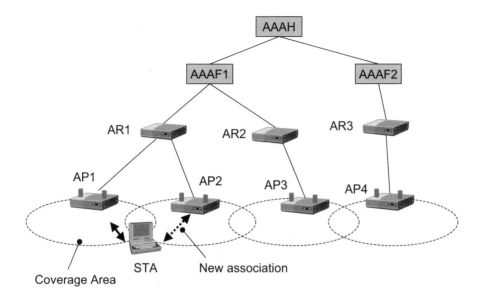

Figure 4.24 Basic WLAN network configuration

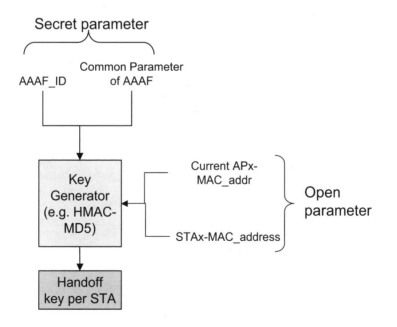

Figure 4.25 An example of handoff key creation

be included in the RADIUS attribute. Note that the STA never knows the common parameter of AAAF, so this scheme is securely protected from DoS attack.

An open parameter is also necessary to create the handoff key. In this case, the open parameter is known by all APs. This open parameter might consist of the current APx-MAC_address and STAx-MAC_address. Both the secret parameter and the open parameter are put into the key generator. As output, a handoff key is created for each STA. These handoff keys can be used to encrypt data during handoff.

Figure 4.26 shows the decoding process when the new AP1 receives a data frame encrypted by the handoff key. When the STA sends the reassociation request frame to AP1, this frame includes the source STA MAC address (= STA MAC address) in the frame header as well as the current AP MAC address (= AP2 MAC address) in the frame body. These two addresses are easily accessed, so this information is not secure. However, the STA does not know how to create the handoff key process with secret information. It is not easily vulnerable to DoS attack. After receiving the reassociation request frame, AP1 creates the handoff key for this particular STA on the basis of the algorithm illustrated in Figure 4.25. Whenever a data frame without IEEE 802.1X is received by AP1 during handoff, the source STA MAC address is verified, and the data frame is decoded. Therefore, the real-time application data frame can be transmitted without waiting for STA authentication between STA and the AAAH.

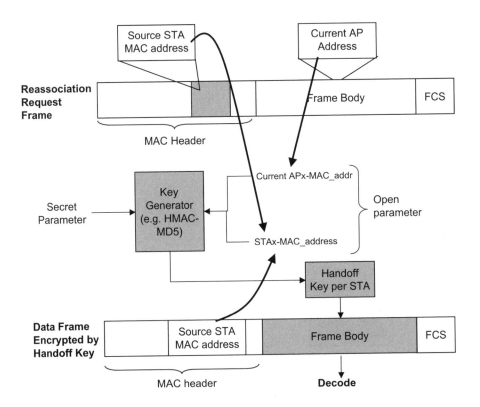

Figure 4.26 A new AP decodes the data frame encrypted by the handoff key

4.5 Quality of Service

IEEE 802.11e TG was established to accommodate QoS support as a new functionality. This section provides an overview of the latest information on the IEEE 802.11e standard. This standard is still evolving, so this section also includes some basic information on the mechanisms used, per the IEEE 802.11e draft version 6.0 published in November 2003.

4.5.1 EDCA and HCCA

The IEEE 802.11e standard introduces the term TXOP, which is a period time during which two stations can communicate with each other in a contention-free manner. In the legacy IEEE 802.11 standard, after the STA or AP obtain the wireless channel, the system only allows the AP and the STA to exchange one pair of frames at a time. Then the STA has to compete for the channel again. The 802.11e standard allows the AP and STA to exchange multiple frames when the STA obtains the channel. This multiple frame exchange support is possible because of the TXOP. The TXOP is defined by a starting time and a maximum duration. Within the TXOP, the STA and AP can exchange frames without having to compete for the channel again.

The IEEE 802.11e standard provides two mechanisms, Enhanced Distributed Channel Access (EDCA) and HCF Controlled Channel Access (HCCA), to support applications with QoS requirements. As shown in Figure 4.27, both mechanisms are based on the DCF mode.

The EDCA mechanism (previously known as EDCF) delivers traffic by differentiating user priorities. This differentiation is based on how long a station senses the channel to be idle before backoff or transmission, the length of the contention window used for the backoff, or how long a station may transmit after it acquires channel access.

The HCCA mechanism (previously known as HCF) allows the reservation of transmission opportunities with the HCF Controller (HC) located in the AP. On the basis of its requirements, a non-AP STA requests the HC for TXOPs, for its own transmission to the

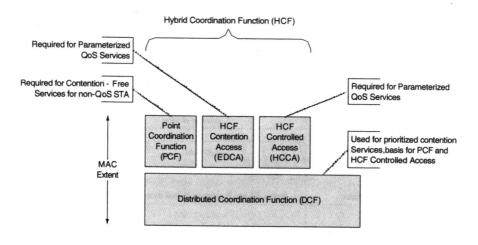

Figure 4.27 EDCA and HCCA architectural diagram

QAP and transmissions from the QAP to itself. The TXOP request is initiated by the Station Management Entity of the non-AP QSTA. On the basis of the admission control policy, the HC either accepts or rejects the request. If the request is accepted, the HC schedules TXOPs for both the QAP and STA. For transmissions from the non-AP QSTA, the HC polls the non-AP STA on the basis of the parameters supplied by the non-AP QSTA at the time of the request. For transmissions to the non-AP QSTA, the QAP obtains TXOPs from the collocated HC directly, and delivers the queued frames to the non-AP QSTA, again based on the parameters supplied by the non-AP QSTA. These two mechanisms are illustrated in Figure 4.27 and discussed in more detail in the following sections.

EDCA

The EDCA mechanism allows each station to vary the amount of idle time it has to sense a channel before backoff or transmission and the maximal length of the contention window used for the backoff.

The 802.11e standard (shown in Figure 4.28) defines a new IFS parameter called *arbitration IFS (AIFS)*. Each 802.11e station can have its own AIFS values based on the station's priority. As discussed in the previous section, stations can be prioritized by using different AIFS values. The shorter AIFS value a station has, the higher priority it has to obtain the channel.

The IEEE 802.11e standard also allows different stations to have different maximal contention window limits. The smaller limit a station has, the more likely it is that it can access the channel because it is more likely for a station to retry the transmission. Figure 4.29 uses an eight-queue example to illustrate how the 802.11e AP works. In this example, frames are pushed into different queues according to their priorities. Each queue is configured with different access parameters, that is, different AIFS and different CW values.

With these two basic schemes, the 802.11e standard allows service differentiation between different stations.

The EDCA method is still a contention-based service, so stations must compete for the wireless channel. As a result, the exact time that a station can obtain the channel is still unpredictable as it is unpredictable in the legacy 802.11 standard. Again, this makes the

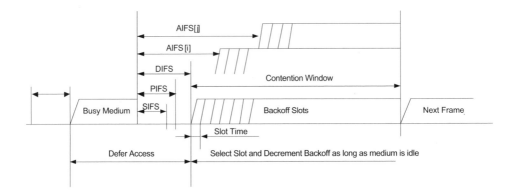

Figure 4.28 EDCA time diagram

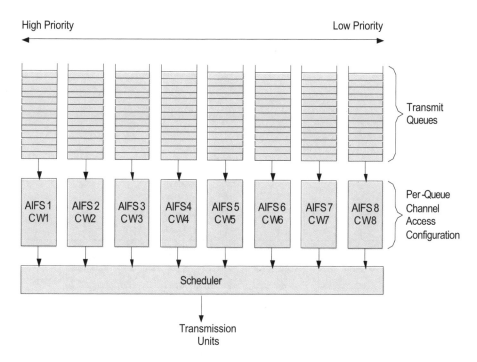

Figure 4.29 EDCA service priorities

delay and jitter between two consecutive frames difficult to predict, so the ECDA is not suitable for real-time applications with strict delay and jitter requirements. The HCCA mode design helps solve this problem.

HCCA

In the legacy IEEE 802.11 standard, the contention-free period can only occur by using PCF model periodically and has certain fixed limits. The new IEEE 802.11e standard removes this limitation. The HCCA mechanism allows each station to negotiate with the AP and generate a CFP period during both CFP and CP periods. The station can first initiate a TXOP reservation request to the AP. When the AP receives the TXOP reservation request, the admission control unit at the QAP decides whether to admit the TXOP from the station. As shown in Figure 4.30, when the AP admits the TXOP reservation request, the HCCA polls the admitted station periodically.

Figure 4.31 illustrates the detailed transmission sequence. In the HCCA mode, the AP uses PIFS, which is shorter than both DIFS and AIFS. As a result, the HCCA has the highest priority to access the channel.

DLP

Another new feature of the IEEE 802.11e standard is the Direct Link Protocol (DLP). In the legacy IEEE 802.11 standard, when the station is operating in the infrastructure mode,

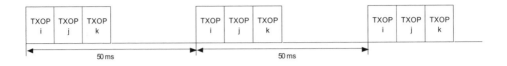

Figure 4.30 TXOP reservation process

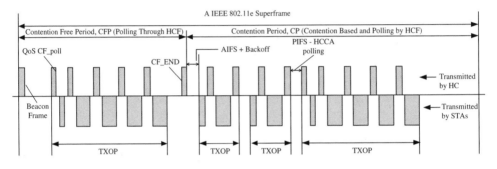

Figure 4.31 Transmission sequence

all the frames are sent to the AP and then forwarded to the proper destination station by the AP even if both stations are within the transmission range of each other can.

In the IEEE 802.11e standard, two IEEE 802.11e stations can communicate directly with each other when they are operating in the infrastructure mode with help from the AP. As shown in Figure 4.32, the DLP setup process includes five steps (In this figure, assume that both the stations are 802.11e DLP-capable and that DLP is allowed.)

1. STA-1 wants to communicate directly with STA-2. STA-1 sends a DLP request frame to the AP. The DLP request frame includes information about the data rate, capabilities of STA-1, and MAC addresses of both STA-1 and STA-2.

2. The AP forwards this DLP request to STA-2 if this DLP function is supported and allowed.

3. STA-2 sends a DLP response frame to the AP, which contains information about the data rate, capabilities of STA-1, and the MAC addresses of both STA-1 and STA-2.

4. The AP forwards the DLP response frame to STA-1.

5. STA-1 is allowed to directly communicate with STA-2.

To terminate the communication, STA-1 sends a DLP-teardown frame to the AP. The AP then forwards the DLP-teardown frame to STA-2, which terminates the DLP session between STA-1 and STA-2.

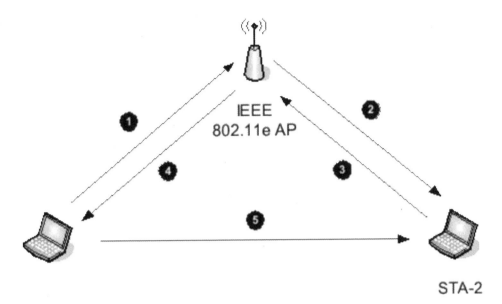

Figure 4.32 DLP setup process

4.6 Security

After a WEP weakness was revealed by U.C. Berkeley and the University of Mary-land (Arbaugh et al. 2002b; Borisov et al. 2002), the IEEE 802.11 WG and the Wireless Fidelity (WiFi) Alliance that provides WLAN interoperability needed to come up with an immediate security solution. In October 2002, before the 802.11 TG I standardized secu-rity enhancements, the WiFi Alliance announced a new security improvement for WLAN, called WiFi Protected Access (WPA) (*Wi-Fi Protected Access* n.d.). The WPA document and protocol are based on a temporal key integrity protocol (TKIP), which was described in an unapproved IEEE 802.11i/D3.0 draft (IEEE 2002). WPA was intended to be forward compatible with the IEEE 802.11i specification, but the current draft D7.0 is not backward compatible with WPA, because the group transient key (GTK) creation is slightly different from that in D3.0.

The new features of WPA provide automatic key management, so the cryptographic key is provided to the user and is changed frequently. IEEE 802.1X (IEEE 2001b) and the EAP are used for user authentication. Figure 4.33 shows the current WLAN security migration. The WiFi Alliance is now trying to replace the current WEP standard with WPA, and they have started a WPA certification program.

Concurrently with the WPA certification program, the IEEE 802.11 TG I is finalizing its security enhancements specification (IEE 2003). This specification supports a new security network association, called *Robust Security Network Association (RSNA)*. In this specifica-tion, authentication and association between a pair of STAs is established in a four-way handshake.

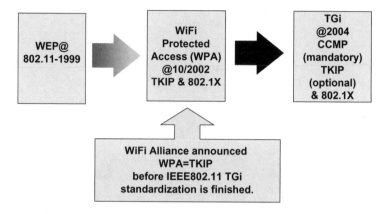

Figure 4.33 Current WLAN security migration

Two additional components are implemented to have further security enhancements. IEEE 802.1X (IEEE 2001b) provides a port-based security system and AS. Pre-RSNA is also defined for backward compatibility and therefore needs to include WEP and IEEE 802.11 entity authentication (Open or WEP authentication). Unfortunately, WPA that is compliant to TGi D3.0 is not interoperable with TGi security (current draft D7.0). The cipher suite selectors of the RSN Information Elements (IE) is used to identify different algorithms in order to recognize the WPA or TKIP.

The IEEE 802.1X standard consists of an IEEE 802.1X controlled port and an IEEE 802.1X uncontrolled port. IEEE 802.1X ports on top of the MAC layer are present on all STAs in RSNA; its reference model is illustrated in Figure 4.34 (IEE 2003). The IEEE 802.1X port implemented at STA is called the *Supplicant*, and the port implemented at the AP is called the *Authenticator*. The controlled port can prevent any data traffic between the AP and the STA until the IEEE 802.1X authentication procedure is complete and the STA is authorized. Although the IEEE 802.1X specification requires the EAP authentication protocol, the IEEE 802.11i standard does not require an EAP implementation.

The second component is the AS. The AS communicates with each STA implementing the IEEE 802.1X Supplicant through the Authenticator for mutual authentication of the AS and the STA. The AS can be located in the DS or in the AP.

IEEE 802.11i provides three cryptographic algorithms to protect data traffic: WEP, TKIP, and Counter mode with the CBC-MAC protocol (CCMP). WEP and TKIP are based on the RC4 algorithm and CCMP (*Counter with CBC-MAC (CCM)* 2003) is based on the advanced encryption standard (AES) (FIPS 2001). In this new security standard, CCMP is a mandatory implementation as long as IEEE 802.11 devices are RSNA compliant. Implementation of TKIP is optional for the RSNA because devices supporting only WEP can be upgraded by the supplier to support TKIP. IEEE 802.11 recommends not using TKIP except as a patch to pre-RSNA devices called *WEP* devices because the encryption algorithm and integrity mechanisms are not as strong as those of CCMP (IEE 2003).

The process of RSNA is as follows. IEEE 802.1X authentication is used, and the EAP authentication (IETF 1998) process starts. Mutual authentication occurs between the Supplicant and the AS, most likely using EAP-TLS (IETF 1999). Both the Supplicant and the AS

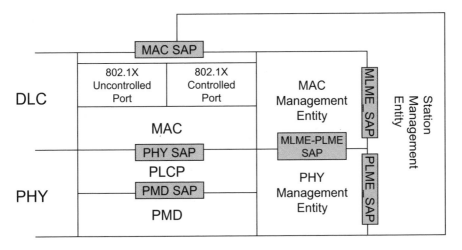

Figure 4.34 IEEE 802.11 with 802.1X reference model

generate a PMK as a shared key. Then, the PMK is transferred from the AS to the Authenticator over a secure channel, such as RADIUS. Upon successful IEEE 802.1X authentication, the Supplicant (STA) and Authenticator (AP) have mutual authentication based on a four-way handshake, which is used to derive a fresh pairwise transient key (PTK). Once the Supplicant and Authenticator have authenticated each other, the IEEE 802.1X controlled port is unblocked to pass any general data traffic. This general data traffic between the STA and AP is encrypted by the TKIP or CCMP protocol. An example of this authentication and key management is shown in Figure 4.35. In this example, the RADIUS protocol is used as a secure channel between the AP and the AS.

The sequence of events is as follows (refer to Figure 4.35):

1. The STA and the AP identify the RSNA capability on the basis of the RSN IE in the Beacon, Probe Response, and (Re)Association Requests.

2. The STA associates with the AP through open authentication.

3. The AP blocks all data frames from the STA until the STA is authenticated by the AS or the AP.

4. The AP requests that the EAP identity carried by IEEE 802.1X have mutual authentication. In this example, the EAP/TLS is used to have mutual authentication of the STA and the AS.

5. During the EAP/TLS process, a shared key called the *PMK* is derived at the STA and the AP.

6. When mutual authentication is successfully completed, the AS sends the PMK to the AP over the secure channel. In the figure, a RADIUS server sends the PMK to the AP.

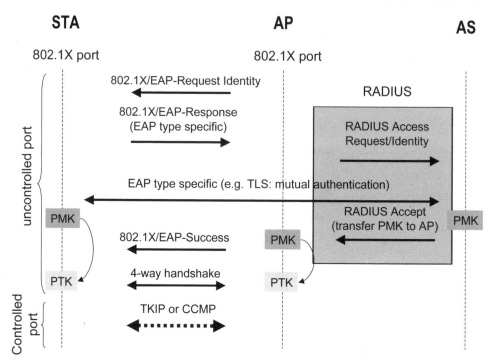

Figure 4.35 An example of authentication and key management

7. On the basis of the PMK, the STA and the AP create a PTK after a four-way hand-shake. The PTK is used between a single STA and a single AP.

8. Upon completion of a four-way handshake, the IEEE 802.1X controlled port can pass all data frames between the AP and the STA. The data frames are encrypted by the PTK or a GTK if the data is broadcast or multicast.

With the improvements from WEP to TKIP encryption, the following additional steps are added to this sequence of events:

1. A transmitter calculates a keyed cryptographic message integrity code (MIC) to prevent a forgery attack.

2. In TKIP, the message integrity can be compromised, so the TKIP implements countermeasures to prevent forgery.

3. A frame sequence counter is used to prevent replay protection.

4. A new cryptographic mixing function creates a temporal key. This key mixing function is designed to defeat weak key attacks.

Details of the TKIP and CCMP algorithms can be found in the IEEE 802.11i standard, which is expected to be published in mid 2004.

5

IP Mobility

James Kempf

5.1 Introduction

To those familiar with 3G mobile standards, the Internet looks like a completely chaotic collection of randomly connected routers and hosts. At best, a local order can be seen, imposed on the fundamental anarchy by the packet-routing topology. 3G network architectures, such as the GPRS IM subsystem, are typically illustrated by a well-organized diagram containing boxes and arrows. The boxes represent network elements that vendors sell as products; the arrows represent open interfaces between the boxes. Standardized protocols defined on the open interfaces allow network elements from different vendors to interoperate. The standardized protocols themselves are usually relatively complex and cover many different functions. In addition, the 3G architectures have been developed with a particular service provider business model in mind: the vertically integrated telecommunications provider offering circuit-switched voice and packet-switched data service.

The Internet architecture is nothing like that. The Internet architecture can be seen as a collection of principles that are used to guide the development of open, interoperable protocols. The principles are defined in RFC 1958 (Carpenter 1996), and are periodically updated as new applications, services, business models, or technologies arise and push the boundaries of the old principles (Kempf and Alstein 2004). The need for new protocols arises in response to particular problems involved in providing new network services and in the introduction of new technology. The protocols themselves are typically designed to be small and to perform one function well, rather than covering a broad range of functions. The formally defined interactions between these functions are usually kept to a minimum, in order to limit the complexity of the protocols. This approach keeps the Internet flexible and highly adaptable to fast changes in technology (driven by Moore's Law). If a change in

Next Generation Mobile Systems. Edited by Dr. M. Etoh
© 2005 John Wiley & Sons, Ltd

an application or technology arises that makes an older protocol obsolete, a new protocol can be developed to replace the older protocol, while leaving other protocols untouched. No particular service provider business model underlies the Internet architecture. Indeed, when the US government ran the Internet, they forbade commercial use. Rather, these architectural principles were originally developed through consideration of the engineering properties of packet data networks. Much later, after the Internet was privatized, a few successful business models were developed.

In the next generation of wireless networks (XG), all-IP wireless networks are much more likely to look like the chaotic Internet than like the neat 3G architectures. The primary function of the Internet is to interconnect networks of different link types, and the number of available wireless link options is growing as discussed in Chapters 3 and 4. As a result, the Internet will, in all likelihood, grow to encompass both 3G networks and new networks based on new wireless link protocols. The resulting combination, together with new services available on both as discussed in Chapter 2, will foster new business models for XG all-IP networks. Historically, the processes and technologies involved in the Internet have proven well suited to supporting this type of situation.

This chapter discusses one of the fundamental issues involved in providing next-generation wireless networks: network layer mobility in all-IP wireless networks. Because the Internet was originally developed on the basis of a model of fixed hosts and routers, mobility does not fall out naturally from basic Internet Protocol (IP) packet routing. IP mobility management requires the addition of a new protocol, Mobile IP, to handle the routing changes caused by moving hosts. But mobility management involves more than simple routing changes. In public access wireless networks, users show up in the network without a fixed connection, often as roamers from other service providers. Authenticating a newly arrived user, determining whether a particular user is authorized for network service, and generating accounting records so that the service provider can get paid are required. In Chapter 11, authentication, authorization, and accounting (AAA) technologies for XG all-IP wireless are explored in more detail; this chapter covers the basic interaction between Mobile IP and AAA. The security of signaling and data traffic is also a heightened concern in wireless networks, because basic security cannot be enforced by physically limiting access to network links. Chapter 10 discusses the basic cryptographic algorithms for security in mobile networks in more detail; this chapter reviews network security solutions for the problem of securing signaling and data traffic. Finally, most retail wireless network users do not want to be bothered by the administrative and technological details caused by their motion. They would prefer seamless mobility so that they can obtain access to their services any time and anywhere. This chapter discusses protocols that support seamless mobility that allows the illusion of a fixed network attachment even though the wireless device is moving.

Most of the protocols discussed in this chapter are either Internet Engineering Task Force (IETF) standards, are still being developed as standards, or are still experimental and are being discussed in the Internet Research Task Force (IRTF) prior to standardization. Discussion of standards is appropriate because, at this time, most of the standards have not yet been deployed or, in some cases, have not even been implemented as products. The large cellular standards organizations (3GPP and 3GPP2) are still discussing how these protocols might play a role in their next generation, all-IP networks. Consequently, the exact mix of protocols that goes into the next-generation wireless network is still experimental and under development in the IETF.

5.2 The Internet Architecture

As mentioned in the introduction, the Internet architecture is a collection of principles for designing interoperable protocols. The most important of these is the end-to-end principle. The end-to-end principle leads to a packet-switched network technology in which endpoints address each other through fixed addresses. Though not an intrinsic consequence of the end-to-end principle, the original designers of the Internet chose to use the endpoint addresses for routing packets through the network as well. This design decision leads to a fundamental problem for wireless hosts, because a wireless host changes its topological location and its routing address as it moves through the network.

This section explores the end-to-end principle more deeply, discusses its impact on the addressing architecture of the Internet, and describes how Internet naming works, all as background for how the Internet addressing and naming architecture leads to a dilemma for mobile hosts.

5.2.1 The End-to-end Principle

Perhaps none of the Internet architectural principles is as misunderstood in the 3G mobile telephony community as the end-to-end principle. Unfortunately, some in the Internet community interpret "end-to-end" to imply that there is literally nothing in the middle, and this is not exactly what the original Internet architecture was about. Earlier descriptions of the end-to-end principle refer to it as fate sharing (Clark 1988), and this term is more descriptive of what this important architectural principle is trying to achieve when applied to protocol design. In essence, fate sharing means that the protocols at the two endpoints in a network conversation contain within themselves the state that is required to maintain that conversation. As a result, the state for the conversation is shared between the endpoints rather than distributed throughout the network, as is typically the case for circuit-switched telephony networks. The motivation for the development of fate sharing came from a basic engineering goal in the original design: the failure of a particular path between two endpoints in a network conversation should not result in the conversation being terminated. The only failure that is acceptable for disrupting network service in the Internet is complete and total partitioning of the network.

As an alternative to fate sharing, the designers considered an architecture in which switching elements within the network contain the state associated with a conversation. In this design, reliability depends on redundancy. The state for the conversation is replicated, usually among distributed switching elements. If one element fails, another takes its place. However, replication has numerous problems. One problem is that the algorithms for achieving reliable replication are difficult to build. In addition, replication only protects against failure to the extent that the state and switching elements providing the backup are not destroyed. Finally, replicated systems tend to be more expensive because they require multiple switches providing the same function, some of which are not put to daily use if they act as hot standbys.

Fate sharing has two important advantages over redundancy. First, the intermediate packet switching nodes, or routers, contain no state with the details of the traffic flows running through them. This means that the routers can be very simple, and therefore are easier to design and build, cheaper to manufacture, and more reliable to operate. Second, hosts are trusted more than the network because they contain the state associated with their own ongoing conversations. The network is assumed to be unreliable and liable to fail at any time. In today's commercial Internet, the end-to-end principle applies to any function that cannot be correctly and reliably implemented without the involvement of the endpoints.

5.2.2 Internet Architectural Elements

From the discussion above, it should be clear that there are really only two important elements in the Internet architecture:

Hosts are the endpoints of communications. Network applications run on hosts. Hosts are responsible for maintaining the state involved in their communications with other hosts.

Routers are packet switches that route packets between hosts. Routers exchange information about the reachability of hosts but do not maintain any state associated with a specific host communication.

The routers that perform packet forwarding constitute the routing fabric of the Internet.

The original IP assumed that the connections between routers and hosts were always available and that the connections had a certain capacity or bandwidth. The bandwidth of the connections was assumed not to be of particular concern to hosts, just to routers that wanted to optimize their packet-forwarding performance. Also, the character of the communication between hosts and routers was assumed to be independent of the underlying link characteristics. As will become apparent later, these assumptions turn out to be largely incorrect for wireless networking.

5.2.3 IP Addresses and Routing Topologies

Application of the end-to-end principle requires that hosts have some method for addressing other hosts. The IP uses fixed-size addresses for this purpose. Every host in the Internet has an address. A source host addresses a destination host by including the destination host's address in the destination address field of the IP packet header. The source host also inserts its address in the source field of the packet header, so the destination host knows who originally sent the packet. The routing fabric between the hosts uses the destination address on the packet to route the packets sent by the source to the host having that destination address. Thus, IP addresses have a dual use in the Internet:

- As the endpoint identifier for a host

- As a location identifier for determining where in the routing fabric a host is located.

The original IP, called *IPv4*, was defined with addresses having 32 bits (4 bytes). However, in the early 1990s, it became apparent that the growth in the number of hosts on the Internet would eventually result in the exhaustion of the IPv4 address space. In addition, because of historical circumstances, the actual distribution of the right to use large chunks of the IPv4 address space has been somewhat uneven worldwide. Certain geographic areas likely to have large growth in the need for IP addresses have rights to IPv4 address space chunks that were projected to be inadequate for even near-term growth. As a result, the body responsible for overseeing Internet standards, the IETF), developed IPv6. IPv6 addresses have 128 bits (16 bytes). IPv6 has many other improvements over IPv4, but a larger address space is the primary one.

An IP address for a host can be further subdivided into two parts:

- A network or subnet identifier that identifies a fixed-end network in which the host is located

- An interface or host identifier that identifies to which host on that fixed-end network the address belongs.

In IPv6, the addressing architecture (Hinden and Deering 1995) divides the 128-bit address into two 64-bit (8-byte) fields. The most-significant 64 bits contain the subnet identifier; the least-significant 64 bits contain the host identifier. In IPv4, the number of bits devoted to the subnet and host identifier can vary on 8-bit (1-byte) boundaries (Fuller et al. 1993). A subnet mask, allocated by the local network administrator and distributed to hosts and routers when they obtain their IP address, determines how many bytes in an address are to be considered as the subnet identifier and how many are to be considered as the host identifier. The collection of routers that make up the subnets in the local administrative domain and out into the Internet is called the *routing topology*. A host's IP address assigns the host a fixed spot in the routing topology through the subnet identifier. In essence, the host is a leaf in the branching tree of the Internet.

5.2.4 Fully Qualified Domain Names and DNS

IP addresses are used by the routing fabric and the host's network layer, but application layer protocols and, in particular, users rarely deal with IP addresses. Although people are accustomed to remembering telephone numbers, they typically find it easier to remember names. To make the identification of hosts easier for users and application programmers, the Internet maintains a parallel set of host identifiers. These identifiers are text-based names and are called *Fully Qualified Domain Names* (FQDNs). FQDNs consist of a series of ASCII character strings with a maximum of 63 characters per string, separated by a period ("."). character. Only the alphanumeric characters, "+", and "−" are allowed in the names, and characters are not distinguished by case. The first string in the name identifies the host or end node. The series of period-separated strings after the host name identify a hierarchy of administrative domains, increasing in scope from an immediately local domain to a top-level domain, for example, "suntana.ebay.scorp.com". In this FQDN, "suntana" is the name of the host, and "ebay", "scorp", and "com" are the names of the administrative domains. The string "com" identifies the top-level domain. The final domain can also be a country domain, such as "uk" for Great Britain or "de" for Germany. A top-level domain name denotes an administrative domain that covers the entire Internet, while a country domain covers a particular country or geographical area. Domain names below the top level cover smaller areas, typically a commercial enterprise, university, city, or other group. FQDNs are typically embedded in URLs and other locators that people use when interacting with Internet browsers or other applications.

FQDNs can be bound to information about the host they identify. In particular, an FQDN can be bound to a fixed IP address for a host. A software application that has obtained an FQDN as part of a URL or through some other means uses a protocol called the *Domain Name Service* (DNS) (Mockapetris 1987) to translate an FQDN into a routable IP address. DNS obtains the name translation from a tree-structured collection of servers, rooted in a few servers that provide name translation for an entire top-level domain or country domain. Local servers mirror some part of the global name space. The information about a particular FQDN is installed in the local DNS server by a network administrator or through other means and is propagated out into the global name space according to the rules for publishing FQDNs established by the local administrative domain. Changing the FQDN to IP address mapping typically requires a considerable amount of time and effort, though there are some protocols that can accomplish this task more quickly. A host can find the IP address of another host elsewhere in the Internet by translating the FQDN through DNS. In this manner, a software application that provides users with some Internet service, such as web browsing, translates the easy-to-remember but unroutable names into hard-to-remember but routable IP addresses.

5.3   Network Layer Mobility in the Internet

The addressing and naming architecture described in the previous section can be problematic for mobile hosts in wireless networks. When an Internet host becomes mobile, the routing of packets to the host can be disrupted. Suppose a host is assigned an address in a particular subnet and that address is bound to a particular FQDN in the DNS by the network administrator. As the host moves between wireless access points, it may, at some point, move from one subnet into a new subnet. When this happens, the subnet identifier portion of its IP address is no longer valid. In addition, the host's address to FQDN binding in the DNS is no longer valid. Users who obtain a URL and use DNS to look up the address will obtain an IP address that points to the original subnet, not to the subnet where the mobile host is currently located.

The basic problem with mobility in the Internet arises because an IP address has the dual role of routing identifier and endpoint identifier. Movement of a mobile host to a new subnet invalidates the original IP address as a routing identifier, but the fact that the original IP address is still available through the DNS means that it is still valid as an endpoint identifier. In addition, applications that are conducting a conversation with the mobile host continue to use the old address as an endpoint identifier for the conversation, but the address can no longer function as a routing identifier. As a result, packet routing to the mobile host fails and end hosts desiring to initiate a new conversation with the mobile host cannot succeed.

5.3.1   Basic Mobile IP

The solution chosen by the IETF for managing mobility in the Internet is the Mobile IP protocol. There are separate versions of Mobile IP for IPv4 (Perkins 2002b) and IPv6 (Johnson et al. 2004). Mobile IP effectively separates the routing identifier and endpoint identifier for the mobile host. The endpoint identifier is the original IP address, assigned to the mobile host when it is in its home network. This address is called the *mobile host's home address*. The home address is propagated into the DNS and is used by other hosts in the Internet to establish initial contact with the mobile host. This address remains valid for software in conversation with the mobile host even when the mobile host moves to a different subnet, possibly distant in the routing fabric from the original home subnet.

When the mobile host moves into a new subnet, it obtains a new IP address as a routing identifier for the new subnet. The new IP address is called the *care-of address*. This address remains valid for the mobile host while it is visiting the subnet. When it leaves the subnet, the mobile host must obtain a new care-of address. However, if a host in conversation with the mobile host uses the mobile host's home address, the packets are routed to the home network, not to the care-of address. In order for packets to be able to reach the mobile host at its new care-of address, a router in the home network must be aware of the care-of address and ready to forward packets to the mobile host. This router is called the *home agent*.

On the surface, the home agent–based architecture of Mobile IP may look like the gateway approach used by the 3G protocols as discussed in Chapter 2. In the gateway approach, a routing gateway sits between the (typically) cellular network and the Internet. This gateway often performs more than simple routing, including complex traffic analysis for the purposes of charging, QoS, and service provisioning. While these functions can be performed at the Mobile IP home agent, they are not part of the Mobile IP protocol, and not part of the basic Internet wireless/mobile network architecture. In addition, the 3G gateways also contain an interface between legacy, SS7-based cellular control protocols, and the IP network. Since Mobile IP was originally developed as an IP protocol, there are no legacy protocols with which to interface.

The mobile host is responsible for maintaining the binding between the home address and the care-of address at the home agent. When the mobile host moves to a new subnet and receives a new care-of address, it sends a message to its home agent containing the binding between the new care-of address and the home address. In Mobile IPv4, this process is called *home-agent registration*. In Mobile IPv6, it is called *home-agent binding update*. In both cases, the end result is that at any point in time, the home agent has a mapping between the home address and the care-of address that it can use on packets sent to the home address to forward them to the care-of address.

The home agent uses an IP tunnel to forward packets to the mobile host. An IP tunnel consists conceptually of a full IP packet, containing a header, inside another IP packet. In practice, IPv6 provides support for reducing the overhead of having two full IP headers in a packet; however, the principle is the same. The IP tunnel allows the packet to traverse the routing fabric between the home agent and mobile host as if it had originally been sent directly to the care-of address. Once the packet gets onto the subnet where the mobile host is located, the tunnel header is stripped and the inner IP packet is delivered to the application layer at the home address.

Exactly how the mobile host obtains a care-of address and how the IP tunnel from the home agent is terminated is different in Mobile IPv4 and Mobile IPv6. The next two sections describe tunnel termination and last hop routing in Mobile IPv4 and Mobile IPv6.

Home-agent Tunnel Termination in Mobile IPv4

In Mobile IPv4, the last hop router for the mobile host's current subnet typically manages the assignment of a care-of address for a newly arriving mobile host and takes care of terminating the tunnel from the home agent. Packets are tunneled from the home agent to the care-of address on the last hop router, with the mobile host's home address as the source address on the tunnel header and the care-of address as the destination address. The last hop router strips off the tunnel header and delivers the packet to the mobile host. The special last hop router that supports Mobile IP functions is called a *foreign agent*. The care-of address in this case is typically located at the foreign agent.

It is important to note here that the IP packet header on the inner packet delivered to the mobile host on the last hop subnet contains a topologically incorrect address for the mobile host. The source address of the packet is the correspondent host address, but the destination address is actually the address of the mobile host in the home network. If this packet were to be released anywhere else in the network, it would end up back at the home network and not at the mobile host. The foreign agent maintains a mapping between the care-of address and home address for the mobile host, and, more importantly, the foreign agent also maintains a mapping between the home address and the link layer address of the host. As a result, the foreign agent can use on-link forwarding, which is below the IP layer and therefore does not require routing, to send the packet directly to the mobile host.

Home-agent Tunnel Termination in Mobile IPv6

In Mobile IPv6, the last hop router is a standard IPv6 router with no modifications. A mobile host obtains its care-of address in one of two possible ways:

- IPv6 defines a protocol called *stateless address autoconfiguration* (Thomson and Narten 1998) by which a host can construct an address from its interface identifier and the subnet identifier advertised by the last hop router. Stateless address autoconfiguration allows a host to automatically configure an address valid on the local subnet without any additional support. The mobile host uses stateless address autoconfiguration to configure a new care-of address, after it discovers the last hop router and the subnet identifiers advertised by the router.

- The host can use the IPv6 version of the Dynamic Host Configuration Protocol (DHCP) (Droms et al. 2003) to obtain an address from a server.

For Mobile IPv6, the care-of address is colocated on the mobile host with the home address, and is called a *colocated care-of address.* Colocated care-of addresses are also possible in Mobile IPv4, but because there is no protocol in IPv4 allowing a host to automatically configure its IP address, a mobile host that wants to use a colocated care-of address must use DHCP for care-of address configuration.

Foreign agents were removed from Mobile IPv6 because they added an additional component that must be deployed for a wireless network. With Mobile IPv6, any IPv6 router can act as a last hop router for a wireless network, considerably reducing the cost and complexity of wireless network deployment.

Figure 5.1 illustrates the general Mobile IPv6 design.

5.3.2 Routing Inefficiencies

Normally, packets exchanged between two hosts in the Internet are routed by the most direct path available between them. The basic Mobile IP routing algorithm described above, on the

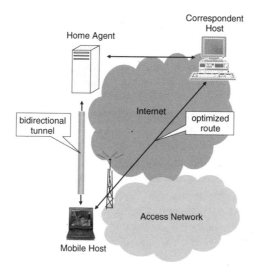

Figure 5.1 Basic IP mobility

other hand, involves a routing indirection. This indirection occurs because the correspondent host sends packets to the mobile host's home address, where they are intercepted and tunneled to the mobile host's care-of address. These two routes form a triangle, with the correspondent host to home agent route forming one side, and the home agent to mobile host route forming another. The third side is the direct route between the correspondent host and the mobile host, which would normally be taken if Mobile IP were not in use. Unless the home agent happens to be on the direct route, the direct route is always quicker than the sum of the other two. This routing indirection is known as triangle routing. Figure 5.1 shows the triangle route through the bidirectional tunnel.

Another source of routing inefficiency in Mobile IP is reverse tunneling. Packets from the mobile host to the correspondent host must arrive at the correspondent host with the mobile host's home address as the source address of the packet. This is necessary because the Mobile IP stack and application software on the correspondent host has identified the mobile host by the home address, not the care-of address. However, in many networks, network operators place an egress filter on last hop routers that causes a router to drop packets if the source address of the packet is not topologically correct when the router tries to forward the packet off the local link. The egress filter is a basic security mechanism that prevents a host from masking its identity by using an invalid address. If the mobile host releases packets with its home address into the foreign subnet, they will be dropped by the filter. As a consequence, the mobile host must tunnel packets through the home agent to the correspondent host. The tunnel header contains the home agent's address as the destination address and the mobile host's care-of address as the source address. The inner packet header contains the correspondent host's address as the destination address and the mobile host's home address as the source address. The home agent strips the tunnel off the outbound packets and forwards them on to the correspondent host, just as the foreign agent does for inbound packets in Mobile IPv4. Reverse tunneling adds triangle routing in the mobile host to correspondent host direction.

Triangle routing and reverse tunneling can become a source of considerable inefficiency, especially for real-time voice and video traffic. An extra amount of latency is added by the longer route, and if the packets must traverse continents or oceans in their route, the latency might exceed what is acceptable for a real-time conversation between people. Unacceptable real-time latency can occur if any of the links have high latency, for example, a high-latency radio access network.

A solution to the problem of routing inefficiencies is route optimization. Route optimization allows the correspondent host to send packets directly to the mobile host's care-of address and for the mobile host to reply using its care-of address as the source address, bypassing the home agent and removing two sides of the triangle. Route optimization requires the mobile host to send the correspondent host a binding update whenever the mobile host changes its care-of address. The correspondent host's routing can thereby quickly track the mobile host's movement. Exactly as for the binding update to the home agent, the binding update causes the correspondent host's Mobile IP stack to change the binding between the mobile host's care-of address and home address. In Mobile IPv6, a special routing header is inserted into the packet by the correspondent with the home address for the mobile host, while the destination address is set to the care-of address. Similarly, the mobile host sets the source address to the care-of address and includes the routing header. The routing header allows both sides of the conversation to identify the packet as for the mobile host, because the home address acts as the node identifier.

In Mobile IPv4, route optimization was designed but was never accepted as a proposed standard by the IETF Mobile IP working group (Perkins 1998), and was subsequently dropped by the working group. In Mobile IPv6 (Johnson et al. 2004), route optimization is built into the base Mobile IP protocol. In addition, all IPv6 correspondents are encouraged to support binding update if they want to support route optimization, including fixed hosts, though the Mobile IP specification cannot require it for fixed hosts. Additional security considerations involving binding updates in Mobile IPv6 are discussed later in this chapter. IPv6 is not yet fully deployed, so there is some hope that route optimization will become widely deployed with the IPv6 Internet, especially for hosts that support real-time voice and video traffic.

5.3.3   Mobile IP Handover

In any wireless network, movement invariably causes the mobile host to move outside the range of the base station or wireless access point currently handling its traffic. Radio-opaque obstacles can also appear, cutting off access to the mobile host's currently active wireless link. Wireless link layer protocols typically support a protocol for allowing the mobile host's network interface card and driver software to reestablish a new link with a new base station to which the network interface card's radio has better reception. Each wireless link layer has its own particular design philosophy and realization of how to do link handover. The basic Mobile IP handover protocol is not affected by how link layer handover is performed. In particular, changes between two different base stations or access points on the same subnet are not important to a Mobile IP host, because the care-of address must change only if the mobile host moves to a new subnet. A mobile host's Mobile IP stack performs movement detection to determine whether the host has changed to a new subnet.

To enable movement detection, the mobile host must be able to determine whether it has left its current subnet. The foreign agent or last hop router periodically sends out a router advertisement message beacon containing the identity of the subnet. The mobile host compares the subnet identifier to its currently identified subnet, and if the subnet identifier has changed, the mobile host proceeds to obtain a new care-of address. The mobile host can also use other methods. One method is the reception of an ICMP Host Unreachable message. This message can be sent by the access router when the mobile host attempts to send traffic using a care-of address with the wrong subnet identifier. This can occur if the mobile host moves into a new subnet during the time between router advertisement beacons. The mobile host then solicits a router advertisement from the access router, and changes its care-of address.

Standard Mobile IP handover is adequate for traffic, such as web browsing, which has lax latency constraints and handles dropped packets by having the correspondent retransmit them. For latency-sensitive traffic, such as real-time voice and video, however, the period of the router advertisement beacons and time during which the mobile host is changing its care-of address can result in packet losses that cause a human perceptible change in the communication. Mobile IP fast handover handles these cases. There are separate and somewhat different fast handover protocols for Mobile IPv4 (El Malki 2004) and Mobile IPv6 (Koodli 2003a). This chapter describes the overall architecture of fast handover, with a slight emphasis on the Mobile IPv6 mechanism, as it is the most highly developed. Figure 5.2 shows how packets get dropped prior to update of the home agent and correspondent host.

From an architectural standpoint, the fast handover protocols provide a mechanism to handle four different functions:

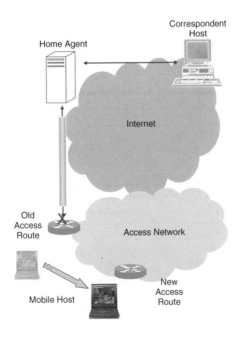

Figure 5.2 Dropped packets due to handover

- Fast movement detection

- New care-of address preconfiguration

- Prehandover subnet change signaling

- Localized routing failure repair

The fast handover protocols require an alternative movement detection algorithm, typically a link layer hint that provides the mobile host an indication by which it can deduce whether it has changed subnet as soon as it is up on the new access point. This hint is typically the link layer identifier of the new access point, which the mobile host can map into the identity of the corresponding access router, using information obtained from the old access router prior to handover. This hint allows the mobile host to start changing global routing immediately, rather than waiting for a router advertisement beacon. In Figure 5.3, the general scheme of fast handover is shown for the Fast Mobile IPv6 protocol.

Care-of address preconfiguration requires that the same information as fast movement detection–about the access routers and their subnets in geographically adjacent cells–be available prior to handover. Using this information, the mobile host can configure a new care-of address, which it can then use as soon as it has detected a move into the new subnet. If the adjacent subnet information is not available, the mobile host cannot preconfigure its care-of address and, therefore, it must wait until it is physically connected to the new subnet before it can determine its new care-of address. Localized routing failure repair can still be performed from the new subnet after the new care-of address is determined; however, many dropped packets can occur during the link switch and care-of address configuration.

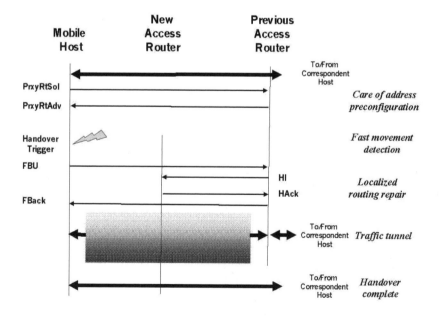

Figure 5.3 Fast handover scheme

Prehandover subnet change signaling is an optional part of the protocol and is only possible if the mobile host receives a timely indication, prior to actual link movement, as to which neighboring subnet the mobile host will move. This indication can come in the form of a hint from the link layer or a message at the IP layer from the current access router. The message sent from the current access router is called a *Proxy Router Advertisement*, and it contains the same information as would be obtained from a Router Advertisement message on the new subnet. When this information is available, the mobile host or access router can initiate localized routing failure repair prior to the link switch, thereby removing any source of lost packets during handover. However, because timely indication of movement is not possible on all wireless link layers (in particular, it is not possible on the popular 802.11 wireless LAN protocol (IEEE 1999e)), dropped packets may be inevitable during the actual link switch and for a short period thereafter until localized routing failure repair can be accomplished from the new link.

Routing failure repair allows packets to continue being routed to the mobile host on the new link while the mobile host is changing the care-of address to home-address mapping at the home agent and, if route optimization is in effect, any correspondent hosts. Packet delivery is accomplished through a bidirectional tunnel between the old router and the mobile host at its new care-of address. The tunnel header in both directions locates the mobile host using its new care-of address. Packets destined for the mobile host arriving at the old router are tunneled to the mobile host. Packets from the mobile host destined to the correspondent are tunneled to the old router. The tunnel is maintained until the mobile host has finished changing the global routing, at which point the home agent and correspondent hosts are delivering packets directly to the new care-of address. If the signaling for performing routing failure repair can be accomplished prior to handover, as discussed in the previous paragraph, packet delivery can be almost seamless. If that is not possible, the signaling must be accomplished as soon as the mobile host arrives on the new link. In either case, the new access router must confirm that the new care-of address is unique on the link. This is accomplished in one of two ways: inter-router signaling if prehandover link change information is available or a specialized neighbor advertisement option, if not, as follows:

- The inter-router signaling prior to mobile host movement is straightforward: the old access router reports the proposed care-of address to the new router and the new router confirms the address (or not, if the proposed new care-of address is not unique). The old router then informs the mobile host.

- The specified neighbor advertisement option is sent to the new router, either as part of the routing failure repair signaling or as a separate message, when the mobile host arrives on the new link. The new router checks the new care-of address and responds if it is not unique. If the neighbor advertisement option is sent together with the routing failure repair signaling, the new router strips it out and sends the routing failure repair signaling on to the old router to initiate the routing repair.

Currently, fast handover for Mobile IP is deemed experimental because of a lack of clear understanding about security. Research work in the next few years is expected to result in a better understanding of security requirements and mechanisms for satisfying them.

5.3.4   AAA and Security

Access authentication and security are two important requirements for mobile networks. Network access authentication is important for controlling which hosts are allowed to enter a public access network. Support for network-access authentication is provided by Mobile IPv4, but not by Mobile IPv6. This is a specific architectural choice. Mobile hosts are expected to use standard network-access authentication in IPv6, in order to avoid requiring special network-access mechanisms for wireless networks. However, in Mobile IPv6, route optimization presents a special security problem. Because binding updates cause routing changes for hosts, they require proper authentication. Mobile IPv6 provides a special protocol for security on binding updates to correspondent hosts. In addition, Mobile IPv6 requires additional security on signaling message exchanges between the mobile host and home agent.

Another issue in wireless link security is the security of the local link. In IPv4, address resolution and router discovery (RD) on the local link are unsecured, but in IPv6, the SEND protocol provides security on address resolution and RD. The last section of this chapter discusses the issue of location privacy, that is, how to prevent unauthorized agents from obtaining information on the geographical location of a mobile host (and thus, its user). This topic is not unique to wireless networks, as unauthorized collection of location information on fixed hosts can also occur. Chapter 11 discusses AAA in more detail, and cryptographic algorithms for security are discussed in Chapter 10.

AAA for Mobile IPv4

In a public access network, a host must be authenticated to make sure that it is authorized to enter the network. The ISP running the network requires some type of accounting information so that the customer can be billed at the end of the month. This process is called *authentication, authorization, and accounting* (AAA).

A dial-up network requires a fixed host to dial in via a modem, which provides a point-to-point connection between the host and the network. The first network element with which the host comes in contact is the Network Access Server (NAS). The host and NAS exchange configuration and AAA information via the Point-to-point Protocol (PPP). PPP is known as a Layer 2.5 protocol because it runs below the IP layer but above any link layer protocol, and it runs before the host's IP service is set up. The PPP exchange configures the host with an IP address and last hop router address, allows the host to exchange authentication information with the NAS, and sets up the accounting. The NAS, in turn, consults the local AAA server to authorize the host, and, if the host is a roamer, the local AAA server consults the host's home AAA server. When the host has been authorized to receive IP service, packets can start flowing at the IP layer. A similar procedure is used in multiaccess networks, such as Local Area Networks (LANs) or with DSL, if PPP is run.

Mobile IPv4 does network admittance differently. Because address and last hop router configuration is done by Mobile IP, the address and last hop router configuration part of PPP is not needed. Instead, Mobile IPv4 uses an extension to the home-agent registration message for AAA initialization with the home agent. When the mobile host registers a binding between the care-of address and home address, it includes the registration extension with the mobile host's authentication information. The home agent then performs the functions of the NAS. The extension includes a security parameter index that identifies the context of the authentication and Message Authentication Code (MAC) calculated over the message. The MAC is calculated using the HMAC-MD5 algorithm (Perkins 2002b). The mobile host may optionally be required to authenticate itself with a foreign agent before registering, by including an extension on the foreign-agent registration. Additionally, the Mobile IPv4 specification includes an authentication extension that the foreign agent may include when performing a registration.

Security for Mobile IPv6

As mentioned above, Mobile IPv6 uses standard IPv6 network access authentication methods for authenticating and authorizing the entry of a mobile host to the network, which is discussed in Chapter 11. These methods may include a link layer authentication procedure (such as is used in dial-up networks), a procedure specifically for the wireless link protocol (such as 802.1x in 802.11 networks (IEEE 2001c)), or a procedure that runs over IP itself (such as PANA (Forsberg 2004)). However, Mobile IPv6 requires additional security for route optimization and uses a different approach for security between the home agent and mobile host compared to that used by Mobile IPv4.

Binding Update Security in Mobile IPv6

In principle, a Mobile IPv6 binding update can be sent to any node on the Internet. This prospect makes security for binding updates a daunting challenge. Public key techniques requiring certificates, such as those associated with IPsec (Kent and Atkinson 1998), are excluded because they would require deployment of a global public key infrastructure. Cryptographic techniques with lesser infrastructure requirements for key exchange (for example, AAA) are potential candidates, but they would also be restricted to those correspondents that support the requisite infrastructure. So a method is required that does not need any more infrastructure than is available with the base Mobile IPv6 protocol.

The protocol used by Mobile IPv6 to secure binding updates, called *return routability*, leverages the presumed security of the routing infrastructure. The mobile host and correspondent host establish a shared key between them immediately after the subnet change and before the binding update, using the return routability protocol. The mobile host calculates a MAC on the binding update sent to the correspondent host, with the shared key. The key is valid for only a limited time (approximately 7 min), and the mobile host must refresh the key by performing the return routability procedure on each binding update, unless the binding updates are closely spaced in time.

Figure 5.4 illustrates the protocol. The protocol is initialized by having the mobile host send off two messages to the correspondent host, the home-address initiation test (Home Address Test Init) and care-of address initiation test (Care-of Address Test Init). The home-address initiation test is reverse tunneled through the home agent and is protected by an AH or ESP digital signature that the home agent verifies before it strips off the tunnel header. The home agent forwards the home-address test initialization to the correspondent host. The care-of address test initialization is sent directly to the correspondent host without any cryptographic verification, because the mobile host and correspondent host do not have any security association.

When the correspondent host receives the home-address test initialization, it returns a home-address test message (Home Address Test) to the mobile host through the home

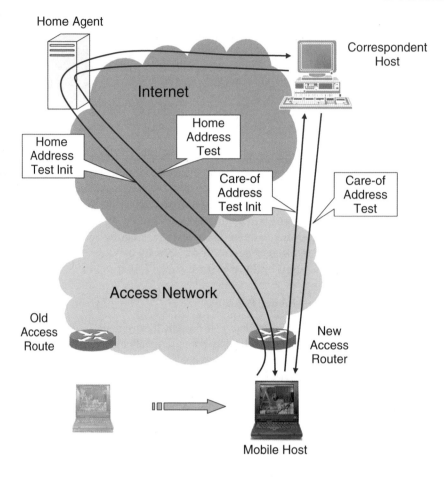

Figure 5.4 Return routability protocol

agent. The home-address test message contains part of the shared content needed to calculate the shared key and an index identifying the content. Similarly, the care-of address
test initialization message triggers a care-of address test message (care-of Address Test)
containing another part of the shared content and the index. When the mobile host receives
both home-address test and care-of address test messages, it combines the shared content
along with other identifying information to construct the shared key.

The correspondent host does not keep track of content necessary to generate the shared
key, however. Doing so would lead to a potential attack, in which an attacker could send
repeated home test initialization messages from different addresses to cause the correspondent host to run out of memory. Instead, the correspondent host keeps track of the key index
that can be used to retrieve the content needed to regenerate the key when the binding update
message is sent. The binding update message contains the index, and the index is changed
from time to time in order to foil eavesdroppers.

The purpose of the home-address test message is to verify that the mobile host is, in fact, at the home address. The purpose of the care-of address test message is to verify that the mobile host is, in fact, located at the care-of address where it claims to be located. If the home-address test message were omitted, an attacker could claim to be at a particular home address and divert traffic for the mobile host. If the care-of address test is omitted, an attacker could claim to be at a particular care-of address where another victim host is located, and thereby launch a denial-of-service attack by causing the victim to be bombarded with traffic.

This procedure is not completely invulnerable to attack. If an attacker can snoop both the home-address and care-of address test messages, the attacker can obtain both halves of the shared content for key construction. Using this content and the other parameters necessary to construct the key, the attacker could fabricate a binding update and sign it with the key. This type of attack is considered unlikely, because it would require the routing infrastructure between the home agent, mobile host, and correspondent host to be subverted. It is in this sense that the return routability procedure depends on the security of the routing infrastructure.

Mobile Host/Home Agent Security in Mobile IPv6

For a mobile host to be able to perform return routability, it must have a security association with its home agent that allows it to tunnel packets having digital signature contained in Encapsulating Security Payload (ESP) authentication header (AH) (Kent and Atkinson 1998). This prevents intermediaries from altering the home-address test initiation message en route. In addition, the same security association can provide confidentiality for binding update packets and ICMP messages sent between the mobile host and home agent. The mobile host may optionally protect payload traffic through the home agent using the security association, including encryption, if desired. Payload data protection is required if multicast group membership or stateful address configuration protocols are run between the home network and mobile host.

IPsec was not designed with mobility in mind, so some special measures are needed to use IPsec with Mobile IP. An Internet specification on using IPsec to protect signaling describes these (Johnson et al. 2004). The precise ordering of headers in the IPsec-protected packets must be specified, because certain headers need to be outside of the IPsec encapsulation, while others do not. The specification also restricts exactly how IKE (Harkins and Carrel 1998b) is used if dynamic keying is desired. If preshared secrets are used for the IKE main mode transaction, the home agent cannot identify which mobile host is performing the transaction because the signaling only contains the care-of address. In this case, only IKE aggressive mode can be used. In addition, the home agent requires some method of updating the IPsec security association database with the new care-of address when a binding update is sent. Otherwise, the security association must be renegotiated from the beginning. The binding update itself contains a Home Address option, and therefore the home address is used as the source address for matching the IPsec security association database entry rather than the new care-of address, which is the source address on the binding update packet. These steps avoid the circular dependency problem, in which a binding update triggers an IKE transaction that cannot complete until the binding update does.

Local Link Security

As described above, forwarding on the basis of the IP address subnet prefix only allows a packet to be routed as far as the last hop router. A key step in delivering a packet to a host on an IP subnet is mapping the host's IP address to a link layer address. The packet is then delivered by the link layer transmission mechanism, which differs depending on the link layer. In order for this step to occur, the router must maintain a cache containing a mapping between the IP address and link layer address. This cache is built up using signaling between the router and hosts on the last hop subnet. In IPv4, the signaling protocol used to determine a mapping between the IP address and link layer address on multiaccess links is the Address Resolution Protocol (ARP) (Plumber 1982). ARP is a separate protocol that runs directly on the link layer and does not run on IP.

Unfortunately, ARP was designed long before concern for security was as high as it is today. ARP is broadcast, so any node on the local subnet can hear a request for an address resolution from the access router. Consequently, an attacker could respond with its address and thereby steal traffic from the legitimate owner of the address. On wired networks, this has traditionally not been a serious problem. Multiaccess links, such as Ethernet, have traditionally been used in enterprise and other private networks where physical access to the premises has been considered sufficient to deter attack, whereas most public access networks have been dial-up, point-to-point links. Point-to-point links do not use ARP, because the access router can obtain a mapping between the link address and the IP address from the NAS. However, on publicaccess wireless networks, such as 802.11 wireless LAN, ARP is used just as for any other multiaccess Ethernet link, and this kind of ARP spoofing is easy to do and occasionally does occur.

In IPv6, local link address resolution was completely redesigned to run on IP, and is called Neighbor Discovery (ND) (Narten et al. 1998). An IPv6 node (including a router) that wants to discover the mapping between an unknown link layer address and a known IPv6 address, multicasts a Neighbor Solicitation message on the local link. The node owning the IPv6 address responds with a Neighbor Advertisement that contains the mapping. The problem with this process is that there is typically no security on the ND protocol packets, so any node on the local link can claim to have the right to use the address, and thereby spoof the victim host into sending traffic to the attacker. A host discovers its last hop router in IPv6 in a similar manner. The host multicasts a Router Solicitation message and the router replies with a Router Advertisement. In addition, a router may multicast an unsolicited Router Advertisement beacon periodically, to inform standard Mobile IPv6 hosts that are newly arrived on the link about the router. But, as in ND, the RD packets typically are not secured, so any node can claim to be a router. The ND specification recommends using IPsec AH on the packets, but IPsec will not work unless manual key distribution is used. Manual key distribution is too cumbersome for mobile networks, because it requires manual configuration of all hosts when they enter the network, including roaming mobile hosts from other access providers. More information about attacks on ND can be found in IPv6 Neighbor Discovery trust models and threats (Nikander et al. 2004). SEcuring Neighbor Discovery (SEND) (Arkko et al. 2004) provides security on ND and RD using two different techniques: Cryptographically Generated Addresses (CGAs) (Aurea 2004) and router certificates. These techniques secure address resolution at the IP layer, but the local link remains vulnerable to attacks at the link layer if the link layer is not secure.

To secure ND, a node sending a Neighbor Advertisement uses IPv6 address autoconfiguration to generate a special kind of address, known as a CGA. The node first generates a public key, and then takes a hash of the public key and a few other pieces of information to form the interface identifier field (last 64 bits) in its IPv6 address. This technique ties the host's address to its public key, and thereby to a signature on the Neighbor Advertisement message. If the signature validates, a recipient of the Neighbor Advertisement with a CGA address and a signature knows that the sender has the right to claim the address. SEND is also used to secure IPv6 duplicate address detection (Thomson and Narten 1998).

To secure RD, the last hop routers in the access network are configured with digital certificates signed by the ISP. The certificates contain the router's public key, and an extension indicating which subnet prefixes the router is allowed to route. A host moving into the subnet through handover or bootup obtains a Router Advertisement as usual, but the Router Advertisement contains a digital signature. If the host already has the router's public key and certificate, it can validate the signature using the key. If not, the host obtains the certificate by sending a Delegation Chain Solicitation message to solicit part of the certificate chain back to a commonly held root certificate, typically the certificate of the ISP, which is preconfigured on the host. The router replies with the Delegation Chain Advertisement containing the certificate chain part. The host validates the chain and uses the public key to validate the signature.

In Figure 5.5, SEND is shown. A SEND-secured router advertisement received by the newly arrived SEND host triggers certificate chain solicitation through the Delegate Chain Solicitation (DCS)/Delegate Chain Advertisement (DCA) message exchange. The SEND host then generates an RSA key, and from that a CGA, and performs duplicate address detection to make sure the address is unique on the link. Duplicate address detection is

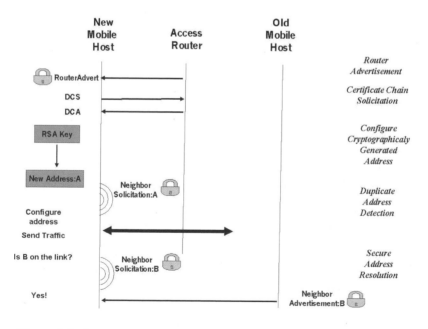

Figure 5.5 Secure router discovery and neighbor discovery using SEND

secured by signing the Neighbor Solicitation. Later, when the host wants to find the address of another host on the link, it performs address resolution by soliciting the address with a secured Neighbor Solicitation, and the solicited node replies with a Neighbor Reply secured with a digital signature.

Location Privacy and Localized Mobility Management

The care-of address in Mobile IP identifies the subnet in which the mobile host is located. As the mobile host moves about the Internet, a stream of binding updates containing the binding between the home address and care-of address are issued from the mobile host to the home agent and correspondent hosts. These binding updates contain precise information about the topological location of the mobile host in the routing infrastructure. In addition, if the interface identifier portion of an IPv6 address can be tied somehow to the owner of the mobile host (for example, through a telephone number), the identity of the user could be determined. With a moderate amount of additional information on the mapping between topological addresses and geographical location, an unauthorized agent monitoring these messages could obtain a trace of the geographical location of the mobile host, and thus the geographical location of its user. Since the user and location of the mobile host are exposed by these updates, this problem is known as location privacy.

While the IETF has not yet issued a standard in this area, work is in progress to address location privacy. To prevent anyone from learning the exact identity of the mobile host, the mobile host can use randomly generated interface identifiers (Narten and Draves 2001) rather than interface identifiers that can be somehow tied back to the owner (randomly generated interface identifiers are also possible with CGAs). The interface identifiers can be changed periodically and a new care-of address obtained. To prevent anyone from learning about the binding changes, the home-agent registrations or binding updates between the mobile host and the home agent can be encrypted and sent with IPsec ESP (Kent and Atkinson 1998), because the mobile host and home agent can easily set up a security association. Such a security association could be set up with a correspondent host as well, but setting up a security association between two random hosts in the Internet is difficult, as discussed previously in this chapter. However, if route optimization is used, an eavesdropper could still obtain information about the host's geographic location because the source address on the packets changes. If route optimization is not used, the source address only changes on the packets tunneled from the home agent.

The exact location of the mobile host can be obscured, but not completely eliminated, by interposing a routing proxy between the mobile host and correspondent host. A routing proxy is a network element that intercepts packets for a host, encapsulates them in a tunnel packet, and tunnels them on to the host without performing any further processing. A routing proxy differs from a router in that it uses tunneling to forward all packets, and it does not participate in the routing information protocol used to exchange routing information between routers. Instead, the host itself, or an intermediate routing proxy, changes the routing at the routing proxy. The foreign agent in Mobile IPv4 and home agent in Mobile IPv4 and Mobile IPv6 are examples of routing proxies.

A routing proxy obscures the mobile host's location by allowing the mobile host to use a globally visible care-of address that can only be mapped to a certain topological region of the network covering a large geographic or organizational domain, such as a country or an ISP. This is sometimes called a *regional care-of address*. The mobile host obtains a regional care-of address from the routing proxy and obtains a local care-of address from its foreign agent or local subnet. Initially, the mobile host issues a binding update to the home agent, binding the home address to the regional care-of address. Each time the mobile host moves to a new subnet within the region covered by the routing proxy, it obtains a new regional care-of address and issues a binding update to the routing proxy, binding the regional care-of address to the local care-of address. The global binding at the home agent is not changed until the mobile host moves outside the coverage area of its current routing proxy. The mobile host can perform route optimization at the correspondent host, but only needs to do so once, to establish a binding between the regional care-of address and the home address. Correspondent hosts see only the home address and regional care-of address; they do not see the local care-of address. The mobile host's location is not exposed except on the link between the routing proxy and the mobile host itself, where the local care-of address appears in the tunnel header. This technique of managing addresses is also called *localized mobility management*, because mobility is managed strictly within the domain of the routing proxy. HMIPv6 (Soliman et al. 2004) is an example of a localized mobility management protocol for Mobile IPv6. In HMIPv6, the routing proxy is called a *Mobility Anchor Point* (MAP); Figure 5.6 illustrates how HMIPv6 hides the location of a mobile host.

Another side benefit of routing proxies is that they provide additional efficiency for managing binding update times and signaling loads. The mobile host must send only one binding update to the routing proxy when it enters the coverage domain for the routing proxy, whereas binding updates would have to be sent to the home agent and also to all correspondent hosts if route optimization were in effect, every time the mobile host moved to a new subnet. The overhead of sending binding updates can be considerable, particularly when the return routability security protocol is required or the recipient is in another continent.

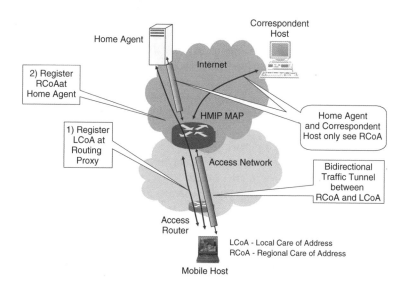

Figure 5.6 HMIPv6 for location privacy

A major drawback of routing proxies is that they introduce a single point of failure into the routing infrastructure. The routing proxy contains bindings for all mobile hosts across a wide geographical area or large organization. If the routing proxy fails, these hosts are suddenly left without service. The techniques for introducing reliability at single points of failure, such as replication and super-reliable systems, tend to be expensive. The routing infrastructure itself achieves reliability by using redundancy, so that if any single router fails others can take up the load. No routers need be dedicated as hot spares; they can all be put to daily use. Localized mobility management also adds an additional layer of tunneling between the routing proxy and the mobile host. This may be an issue for tightly bandwidth-constrained wireless links or when the frame size on the wireless link is small.

5.4   Achieving Seamless Mobility

While Mobile IP can achieve seamless packet forwarding for mobile hosts, moving a host's network layer point of attachment from one subnet to another may require additional measures to make the transition in routing seamless. In addition to forwarding on the basis of topology, routing may have associated with it certain treatments that modify forwarding behaviors. For example, on low-bandwidth links, header compression may be performed between the router and the host. Establishing the state or context associated with the header compressor on a new router typically requires several packets before full compression is achieved. During that time, the mobile host's application protocols are not obtaining the full bandwidth of the link. The compressor can be hot started by transferring the context from the old router when the routing changes, avoiding the need for sending uncompressed or partially compressed packets over the link. Other examples of such treatments are the quality of service (QoS) requested by the host and the authorization credentials of the host.

With choices in wireless media expanding, future mobile hosts may provide more than one wireless interface. WAN media such as GPRS are typically more expensive and have limited bandwidth but have broad geographical availability. Wireless LAN media are cheaper and have higher bandwidth but have geographical availability limited to hot spots. The ability to move a Mobile IP home-address binding from one wireless interface to another allows a wireless service customer to choose which wireless medium is most appropriate for the current traffic pattern. Movement of a Mobile IP binding from one wireless interface to another (or between a wireless interface and a wired interface) is known as intertechnology or vertical handover.

When a mobile host has multiple wireless interfaces, figuring out which wireless link types are available in a particular access network may prove to be a problem. Theoretically, a mobile host could keep all wireless interface cards active, scanning for wireless access points all the time. As a practical matter, however, wireless interfaces tend to consume power. Limiting the number of active wireless interfaces to just the one required for connectivity provides better power utilization. Candidate Access Router Discovery (CARD) provides a means whereby a mobile host can learn which handover candidate access routers are available in a network. CARD may also be useful for moving a mobile host to a new interface on a single wireless interface, if the wireless link layer technology allows triggering handover from the IP stack, and for mapping access point and access router link layer identifiers to IP addresses for fast handover protocols.

5.4.1   Header Compression

2G and 3G cellular wireless links tend to be relatively low bandwidth and have high latency. For such links, limiting the amount of nonessential data sent over the wireless link is important. IP packets have large headers, but for data packets, the header as a fraction of the overall packet size is relatively small, about 2% for a 1500-byte packet, which is the most common packet size seen in the Internet. For real-time voice packets, however, the header size as a fraction of the overall packet size is considerably larger. For example, in voice packets with the IP + UDP + RTP protocols used to transmit voice over IP, the header doubles the packet size for IPv4 and triples it for IPv6. IP headers are used to direct packet forwarding through the routing fabric, but are not necessary on the last hop, because the last hop router uses the link layer address of the host to deliver the packet. The packet must arrive on the host at the application layer of the stack with a correctly formed IP header, but it does not need to have an IP header when sent over the wireless link. Clearly, removing the header can result in greatly increased wireless link utilization efficiency.

IP packet headers have a significant amount of redundancy, both within the header itself and between headers on different packets. For example, the source and destination address on an IP packet flow does not change between packets. Header compression works by recording header information in a context on the last hop router and in the host. If no change occurs in the header, the header can be compressed prior to sending over the wireless link and reconstituted on the other side. The packet only requires a small context identifier on the air so that the right header compression context for the packet can be found when it reaches the other side. A header change requires the context to be updated and the new information to be sent over the air.

A variety of header compression schemes have been proposed for IP, but the RObust Header Compression (ROHC) (Bormann et al. 2001) scheme has the best performance for cellular links. The ROHC scheme classifies the IP + UDP + RTP header fields according to their predictability. Most fields do not change between packets and can be eliminated. Only five fields require a more complex mechanism:

- IP addresses

- UDP checksum

- RTP marker bit

- RTP sequence number

- RTP timestamp

The IP addresses and RTP timestamp can be predicted from the RTP sequence number. The RTP marker bit only needs to be sent occasionally, and the UDP checksum must be sent when present. ROHC operates by establishing a function between the timestamp and the other fields and then reliably communicating the timestamp. When the parameters of a packet change, the parameters are updated by sending additional information.

Figure 5.7 contains a graph illustrating the effectiveness of ROHC on wireless LAN. ROHC was implemented on IPv6, UDP, and RTP for 802.11b wireless LAN. The experiment was run by reducing the bandwidth on the wireless LAN access point down to 1 Mbps in

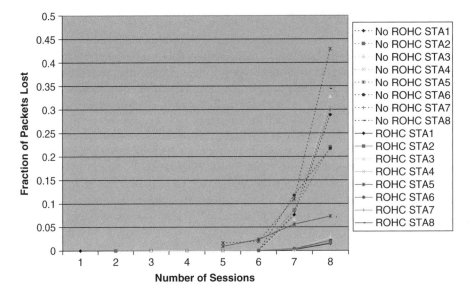

Figure 5.7 Robust header compression (ROHC) example

order to make it easier to saturate the link. Individual flows were introduced into the wireless LAN cell by sequentially introducing a new station into the cell and running a 6-min unencoded audio sample between a separate correspondent host per station and the station. As the figure shows, congestion on the link, as measured by dropped packets, increases above 10% (generally considered the maximum acceptable value) around 6 flows for uncompressed headers, whereas for ROHC compressed headers, congestion remains below 10% even for eight flows, which was the maximum number of stations available. The basic ROHC algorithm does not contain any explicit provision for compression of tunnel headers in Mobile IPv4 or IP header options in Mobile IPv6, but the same technique can be applied.

5.4.2 Context Transfer

Header compression maintains context on both the access router and mobile host. This context requires the exchange of a certain number of uncompressed or partially compressed packets before the header can be fully compressed. When a mobile host hands over from one link to another, in the absence of any other measures, the compression context on the new router must be reestablished from scratch. The headers may change only slightly, so it should be possible to reuse some of the header compression context on the new router. For example, in Mobile IPv6, the source address on outgoing packets may change if the mobile host is using its care-of address as the source address. The compression can be backed off slightly to accommodate the changes, resulting in partially compressed packets. But some method is required to transfer the context between routers.

A context transfer protocol, CTP is used to transfer the context associated with some type of feature or routing treatment from one access point or access router to another. Header compression context is an example of a feature that can be transferred using a CTP.

The mobile host could reestablish the header compression state from scratch on the new access router, but the signaling required would be time and bandwidth consuming. Context transfer is typically done to optimize handover, so the mobile host is provided with the same level of service on the new access router as on the old without having to go through the preliminary signaling in order to achieve that level of service.

Other examples of context that can be transferred are multicast routing, AAA, IPsec security association, QoS, access control, and so on. Not all of these are appropriate for simple transfer; for example, there are some potentially serious security issues surrounding IPsec security association context that have yet to receive intensive study. However, there are enough credible proposals for contexts so that a generalized protocol for context transfer seems useful for making handover more seamless.

Various kinds of feature contexts have different requirements for synchronization with external events. Header compression is an example of a feature context that is tightly synchronized with outside events, namely, with handover. If the header context is not transferred as soon as routing starts through the new router, the context becomes unusable and the mobile host must reestablish the context from scratch. Authorization credentials are an example of context that is not tightly synchronized. The mobile host establishes its authorization credentials only when it enters the network, and these credentials are typically not changed while the mobile host is using the network, though they may be revoked or updated if some change occurs in the mobile host's service profile.

Nonsynchronized context can be transferred proactively at any time prior to handover. The mobile host's current access router can proactively flood the context to nearby routers in order to avoid having to tightly couple sending the context together with the handover signaling. However, changes may occur in the feature on the current access router prior to handover. Any changes must be propagated to those access routers that received the context. A CTP for nonsynchronized context must provide update provisions. In addition, management of the state on multiple routers can be difficult, suggesting that context flooding might be limited only to very specific applications.

The IEEE InterAccess Point Protocol (IAPP) (IEEE 1999d) defines context transfer for AAA information between access points that support the IEEE 802.11 wireless LAN protocol (IEEE 1999e). The AAA protocol used by 802.11 is RADIUS (Rigney et al. 2000). IAPP AAA context transfer is based on equivalency. To be effective, context transfer must result in the exact same context on the new access point as would occur if the host had performed signaling with the new access point. The state cannot depend on the number of hosts on the access point nor on the state that may exist on the old access point but not on the new. This means that the AAA environment must be relatively homogeneous, so that features supported by the RADIUS server on both the old and new access points are the same. However, IAPP contains no procedure for defining context formats to transfer.

IETF has also developed a CTP that is not specific to a single wireless link type (Loughney 2004). If the host knows beforehand the last hop router to which it will hand over, the host indicates that it wants to activate context transfer to a new router by sending a Context Transfer Activate Request (CTAR) message to the current last hop router. The CTAR message contains an indication of which contexts the host wants transferred, a token authorizing the transfer, and the IP address of the host on the new router, if known. The current last hop router then transfers the contexts indicated in the message to the new last hop router by sending a Context Transfer Data (CTD) message. If the host finds out about its new last hop router only after moving, it sends the CTAR message to the new last hop router, and the new last hop router signals the old with a CT Requests (CT-Req), causing the transfer. CTP contains a procedure for extension by defining new context formats for transfer.

5.4.3   Intertechnology Handover

With the growing number of wireless media choices, many mobile hosts may be expected to support more than one wireless interface. Wireless media differ in a variety of characteristics, and users will typically be interested in using only a single medium at a time. For example, a user might start a Voice over IP session on GPRS, then switch to wireless LAN when she enters her office. Later, perhaps she decides to use a wireless headset supporting Bluetooth and would like the session switched there. These changes require that the host maintain session continuity while switching from one interface to another.

Intertechnology handover is fairly easy to perform using Mobile IP. Presuming that the second interface is active and a wireless link beacon has been discovered for an access point, the following steps can be used to transfer the Mobile IP binding:

- Configure the second interface so that it has an IP stack.

- Perform Router Solicitation to obtain a Router Advertisement on the second interface.

- Obtain a care-of address on the second interface. If this is Mobile IPv4, the Agent Advertisement should have the care-of address. If this is Mobile IPv6, either stateless or stateful address autoconfiguration is used.

- Send a binding update to the Home Agent on the new interface containing the new care-of address.

- If Mobile IPv6 is in use, send a binding update to correspondent hosts to perform route optimization.

Note that packets in flight to the old interface when the binding is changed at the Home Agent may show up at the old interface after the change if the old link is still usable. These packets can be tunneled to the new interface on the mobile host without performing signaling other than standard Mobile IP care-of address change. For this reason, the host should keep the old interface active until in-flight packets drain.

5.4.4   Candidate Access Router Discovery (CARD)

As described above, intertechnology handover requires the mobile host to maintain the second interface card in an active enough state that it is periodically scanning for beacons. Scanning for beacons uses power, so keeping the second interface active may negatively impact power consumption if the coverage areas for a particular wireless medium are discontinuous. A more efficient way of arranging to discover the currently available wireless link possibilities is Candidate Access Router Discovery (CARD). Information on available handover candidate routers is provided through the active interface by the CARD protocol (Leibsch and Singh 2004).

For CARD, available access routers advertise their link layer identifiers and important characteristics through access routers in neighboring subnets. One required characteristic is the type of wireless link layer protocol supported by the router and the access port link layer identifiers for the protocol. Other characteristics are the service levels available on the router and the cost of using the link. An example of the former is when the router provides expedited forwarding in addition to best-effort service.

For Mobile IPv6, the primary function of the CARD protocol is to provide the mobile host with enough information to begin configuration of a care-of address prior to moving to the new subnet. A secondary function is to allow a mobile host to determine whether an access router is a good handover candidate. The mobile host obtains the advertised characteristics by exchanging the protocol with its current access router. The mobile host can decide, on the basis of the characteristics, whether a router is a good choice for handover. If so, the mobile host can activate the second interface if intertechnology handover is being performed, and begin transferring the care-of address. Routers can also use the CARD protocol to choose an access router during intratechnology handover if the link layer technology on the interface allows the IP stack to trigger a handover. The information on the neighboring subnet routers returned by CARD includes the subnet prefix, router IP address, and router link layer identifier. The mobile host can use this information for fast handover.

A prerequisite for CARD is that access routers maintain a database of characteristics for access routers in neighboring subnets and a mapping between access router records and the access point link layer identifiers of access points in the subnet. This database can be either statically configured or maintained by a protocol that allows the access routers to exchange the information or obtain it from a centralized source. The mobile host uses the link layer identifier of access points that it hears in beacons to request information on the subnets served by the access points. The mapping between the access point link identifiers and the router's IP addresses is a kind of reverse address translation, similar to the RARP

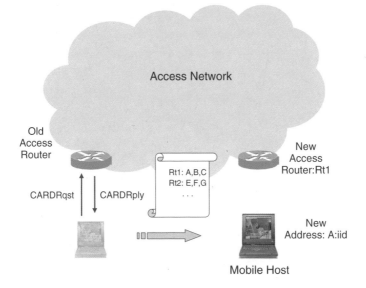

Figure 5.8 Candidate access router discovery (CARD)

(Finlayson et al. 1984) protocol on IPv4, but across subnet links instead of within a subnet link, and including more information than just the IP address.

Figure 5.8 illustrates how CARD works. The mobile host issues a CARD Request to the old access router and obtains information on routers to which it can hand over. In this case, the information consists of a list of subnet prefixes supported by the access routers. When the host hands over to Rt1, it utilizes the subnet prefix A to form a new address, removing the need for any signaling to the new access router. This allows the mobile host to come up on the new link more quickly.

5.5 Summary

The protocols for IP mobility in XG all-IP wireless networks are in various stages of completion. Standardization of header compression and the base Mobile IP protocol for IPv4 and IPv6 is complete, but development continues on issues that arise during implementation and deployment. However, the details of how to carry out AAA for Mobile IPv6 are yet to be worked out. The design of the SEND protocol is, as of this writing, about complete, though the standardization process has yet to be completed. The design of FMIP, CTP, and CARD is complete, but the protocols are not being standardized because there are many open issues about how the protocols interoperate with each other and with various aspects of wireless link media that need more research. The IEEE IAPP protocol has been designated a recommended practice by IEEE, because it is not a MAC or PHY layer protocol, but the design has been complete and published. Thus, researchers have a large toolkit of protocols from which to continue investigating the best way to provide IP mobility in XG all-IP networks.

6

APIs and Application Platforms for Next-generation Mobile Networks

Ravi Jain, Xia Gao

6.1 Introduction

The future growth of mobile wireless networks rests fundamentally on the ability to offer users innovative and commercially profitable applications rapidly and efficiently. Voice is already becoming a commodity application, with declining revenues on a monthly or per-minute basis, and it is likely that simple wireless Internet access will follow suit in a few years. Unfortunately, introducing new applications into the mobile wireless network is currently a slow, costly, and complex process. In addition, it is not clear that service providers can continuously create and develop killer applications that provide differentiation and seize the market by themselves.

Open application programming interfaces (APIs) offer a possible solution to this problem. They can allow service providers to tap into the energy, creativity, and diversity of the vast third-party software vendor community. APIs can provide a means to overcome these limitations, in a manner similar to the way applications are developed for PCs.

For future value-added services, high-performance, high-reliability application platforms are required to form the interface or hosting environment for third-party applications. The design and capabilities of these application platforms will also become an important differentiator for service providers. The functional aspects of these platforms, however, will need

Next Generation Mobile Systems. Edited by Dr. M. Etoh
© 2005 John Wiley & Sons, Ltd

to be represented, perhaps in a composite form, in terms of the APIs used to create services. Thus, the API specifications will form a high-level functional specification of critical application platforms (Jain 2003).

One of the key goals for developing open APIs for the next generation of telecommunications systems is to hide the heterogeneity of the underlying networks. The desire to be connected "any time, anywhere, and any way" has led to an increasing array of heterogeneous communication systems. Among these, the legacy fixed PSTN, different generations of cellular networks, and the Internet are the ones with the widest coverage area and largest use. This heterogeneity is unlikely to disappear in the foreseeable future, if ever. To obtain the most revenue potential, next-generation services must be able to operate over these heterogeneous networks in a seamless manner, not only to reach the broadest market and to serve users better but also to attract the largest number of service developers. As this chapter describes, many current API development and standardization efforts attempt to hide this heterogeneity and to offer a uniform interface to service developers. Services developed using a uniform API are also likely to be easier to adapt to different deployment scenarios, easier to maintain, and easier to integrate with other services.

This chapter provides background on telecommunications service creation, types of APIs, and the difference between APIs and protocols. It also discusses existing API standards efforts (for example, Parlay/OSA, JAIN, and OMA), followed by other more recent API approaches that have been suggested for advanced network architectures. Finally, this chapter presents a short description of our approach to developing a layered API model for next generation (XG) mobile networks, illustrating it with an API design for Content Distribution Networks (CDN), and ends with a brief discussion and conclusions.

6.2 Background

6.2.1 Service Creation in the PSTN

The Intelligent Network (IN) is a framework designed to make the implementation and deployment of value-added services in telecommunication networks faster, easier, and more efficient. Originally designed for the PSTN, the IN approach has also been applied to cellular networks. IN represents a significant advance over the previous integrated PSTN architecture. IN separates the service intelligence from the circuit switching, placing the former in the Service Control Point (SCP) and the latter in the Service Switching Point (SSP), and defining a set of protocols (SS7) for exchanging messages between them.

Although IN distributes functionality between the SCP and SSP, the interface between them is typically not open, so developing new services requires programming the SCP. Programming environments do exist for the SCP. However, these are often tied to particular vendors' implementations of the SCP itself, and are not generally available to third parties.

The programming interface provided by IN defines standardized Service Independent Building blocks (SIBs) that can be reused and composed to form new services. Applications reuse SIBs by composing them into service scripts. Typically, SIB functions are mainly limited to low-level, call-related intelligence. Furthermore, the assumption of a "dumb" terminal implies a very simple user interface, such as a telephone keypad, restricting the services that can be developed. Thus, even if the SCP were open, the types of services that can be developed are limited.

6.2.2 Service Creation in the Internet

The Internet is built on principles completely different from those of IN. The Internet Protocol (IP), which is the universal network layer for interconnecting heterogeneous networks, provides unique addressing and packet routing/forwarding services to upper-layer applications. Unlike IN, which has centralized logic and service intelligence at the SCP, the Internet's routing intelligence is distributed among routers all around the world. These routers have no common administrator and use the standardized signaling protocols to communicate and cooperate with each other in a loosely coupled fashion.

The Internet allows easy development of services by third parties because in most cases they can be deployed on servers located at the network periphery.

6.2.3 Service Creation in Converged Networks

In the late 1990s, a new telecommunications architecture was developed to integrate the PSTN and the Internet. This was often called a *Next-generation Network (NGN) architecture*, an unfortunate name because, of course, any new architecture is a next-generation architecture. Generally, the architecture attempts to bring about a convergence of circuit-switched and packet-switched (IP or ATM) networks, as well as wired and wireless networks; for this reason it is called a *converged network architecture*. A converged network combines the PSTN and an IP network by means of signaling and media (or trunking) gateways between the two. Its principal new component is the *soft switch,* or *call agent,* A soft switch is the "brains" behind the convergence in the converged network. It maintains the state of each call or session and employs special protocols to issue appropriate commands to the gateways. In some sense, the routers in the IP network perform the low-level switching functionality provided by the switching fabric of a traditional telephone switch, while the soft switch provides some of the higher-level functions, such as maintaining call state and related information.

A traditional telephony switch typically combines the hardware (a switching matrix), control software, and signaling termination in one box. In the case of IN, the switching hardware (SSP) requires a large investment in hardware and needs to support complex SS7 signaling protocols to communicate with the SCP. In contrast, by relying on a commodity router fabric and using modern software and hardware technology in the soft switch itself, a soft switch solution can, in principle, require a lower investment and yet lead to a more flexible solution.

Figure 6.1 shows a simple example of how a soft switch interconnects an IN network with the Internet through a media gateway. A media gateway usually exists between these two heterogeneous networks to take care of translating data in the format required in each network and to terminate two different types of connections (switching or packet based). Thus, the gateway should understand two different types of signaling protocols (e.g., INAP and SIP).

Instead of building network-specific control logic within the code, applications use the generic API interface provided by the soft switch, and the soft switch is in charge of translating the generic signaling message into the network-specific signaling message. In this way, the application is easier to develop, to migrate, and to maintain.

Parlay and JAIN can be seen as APIs for programming such a soft switch. By providing a generic interface to upper-layer applications and hiding lower hardware and signaling

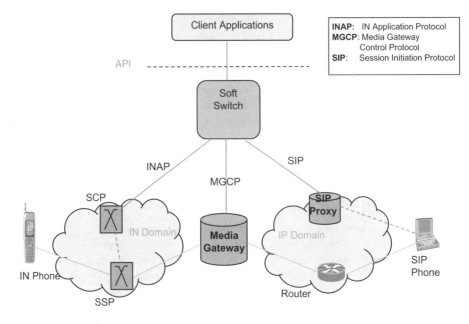

Figure 6.1 Internetworking of heterogeneous networks through soft switch

complexity from them, Parlay and JAIN allow applications built on this API to control network elements of different types of networks.

6.2.4   Types of APIs

Several different types of telecommunication APIs have been developed and standardized. This chapter focuses primarily on service APIs, namely APIs for developing end-user services.

APIs have also been developed for individual protocols, like SIP (Jepsen et al. 2001), Mobile IP (Yokote et al. 2002), and others (Jepsen 2001a). Protocol APIs are at a lower level of abstraction than higher-layer service APIs like the call-control APIs of JAIN or Parlay. As a result, they offer programmers finer-grained control (e.g., over the content and timing of individual messages) and, probably, better performance than the call-control APIs. Programming with the call-control APIs rather than the SIP API is roughly analogous to programming in high-level languages rather than assembly in terms of the tradeoffs involved. This chapter does not discuss protocol APIs.

Finally, APIs have been developed for individual network elements, such as ATM switches (Lazar et al. 1996) and IP routers. APIs for routers present a very different paradigm for making networks more flexible and, hence, enabling creation of novel value-added services, because, in principle, arbitrary pieces of software could be downloaded to the router dynamically or even on demand. Currently, this is largely a research topic, which is briefly reviewed in Section 6.4.

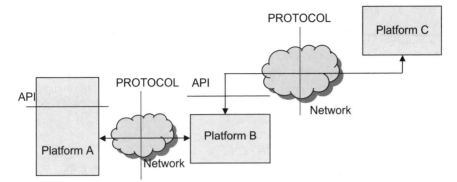

Figure 6.2   APIs and protocols (R. Jain et al. 2005). Reproduced by permission of John Wiley & Sons, Inc.

6.2.5   APIs Versus Protocols

APIs and protocols are often confused in the industry. This section reviews these concepts and attempts to clarify their differences (see Figure 6.2).

APIs provide applications with a well-defined mechanism for accessing the underlying resources of a system. The hardware and software implementation of the API is called *the platform*. An API represents a horizontal separation between different layers of a software stack (see the left side of Figure 6.2). An API provides an abstract representation of the commands that the application can issue to the platform. An API is inherently asymmetric, because it (logically) involves a situation where the application issues commands to the platform and makes decisions on the basis of events reported by the platform.

In contrast, a protocol can be regarded as a vertical separation between communicating entities, typically entities on different machines communicating over a network. Unlike APIs, the goal of standard protocols is interoperability. If an entity uses a standard protocol, it can interoperate with any other entity that uses the same standard. Also, unlike APIs, protocols may be symmetric or asymmetric. For example, an Internet routing protocol like Open Shortest Path First (OSPF) is symmetric in that every router runs the same protocol software and has the same role as any other. On the other hand, a protocol like the Session Initiation Protocol (SIP) for setting up Internet communications sessions is asymmetric in that the entity that initiates the session is always logically distinguished from the other side.

As mentioned above, protocols themselves can have APIs. In that case, the protocol API, as distinct from the protocol, can be regarded as a programming abstraction that creates a horizontal separation between software layers. The protocol itself remains a vertical separation providing interoperability.

6.2.6   Programming Languages

IN telecommunication services are mainly programmed in procedural languages based on proprietary hardware and software. The IN service model introduces the concept of service scripts that specify the logic to form a new service from elementary building blocks. But some important problems still exist. One problem is that the model does not take advantage of modern object-oriented (OO) techniques, as discussed below. Another problem is that the granularity of basic IN building blocks is inconsistent. Some blocks, such as Service Data Management, represent very complex operations that can treat any kind of service data. Other blocks, such as Comparison, represent simple calculation capabilities. The difficulty of reusing these building blocks makes many telephony vendors develop their own proprietary building blocks (Zuidweg 2002).

Modern OO programming techniques have proven to be more modular than procedural programming techniques. Objects are reusable software building blocks that encapsulate both data and related processing codes. Objects can be envisioned as service providers in the sense that they provide services to other objects, and to achieve this, they utilize services from other objects. Well-defined objects have atomic functionalities and maintain a clean interface for controlled access and easy service provisioning. Furthermore, because real implementation is hidden from external access, by keeping the interface unchanged, the implementation of objects can be changed without affecting the overall program. Such characteristics greatly benefit application modification, adaptation, and maintenance.

The service APIs discussed in this chapter mainly use OO techniques. JAIN is specified in Java, so it is OO by design. Parlay is specified in the Unified Modeling Language (UML), so it has the advantage of being language independent. Automatic tools exist to convert the UML to specific Interface Definition Languages (IDLs), like CORBA IDL and Microsoft IDL. Translations from UML to specific programming languages like C++ and Java are also possible. While in principle these translations could be done automatically, the resulting specification is likely to be very hard to read and understand. Nonetheless, translation rules have been devised to make the translation process fairly straightforward.

The various standardization bodies and industrial interest groups have produced a considerable amount of work in this area. This chapter first discusses the strengths and limitations of existing API efforts (for example, Parlay/OSA, JAIN, and OMA) as well as other API approaches that have been suggested for advanced network architectures. Following that is a short description of our approach to developing a layered API model for XG mobile networks. Finally, the chapter concludes with some observations on future work.

6.3   Standard Telecommunications APIs

In traditional telecommunications networks, network operators play two distinct roles: the owner and operator of the communication networks and the provider of value-added applications and services. By securing a trusted environment and judiciously managing underlying switching and control components, network operators can implement, deploy, and manage carrier-grade services for mass markets with a scale of millions of users. However, because of the long development cycle and large R&D overhead, this business model is not suitable for niche market applications. At the same time, facing the increasing competition from the deregulation of the telecommunications market and shrinking profit margin of traditional voice services, network operators have to rely on new services to differentiate themselves and produce more revenues.

This context forms the background for the development of the standard telecommunications APIs discussed in this section. (The term "standard" is used loosely here. In some cases these APIs have not been developed by nationally or internationally authorized standards bodies, but by industry forums.) Note that this discussion focuses on the service APIs, although in some cases, like JAIN, protocol APIs have also been defined.

6.3.1 Parlay

The Parlay[1] Group was established to develop an API to allow third-party service providers access to telecommunications network functions. The creation of the Parlay group was largely driven by pressure from British regulators on British Telecom (BT) to open up the network, enabling more competition in the telecommunications industry. The Parlay API is designed for the easy creation and rapid delivery of innovative services as well as the maintenance of the integrity and security of the network, so this API provides both an extensive framework to authenticate and authorize third-party services and a set of standardized interfaces to access network functionalities.

The first part of this section provides an overview of Parlay architecture, followed by a detailed discussion of two key components of the architecture, namely, the framework and the service capability features (SCFs).

Architectural Vision of Parlay

There are three main entities considered in the Parlay architecture model: the third-party client applications, the Parlay Framework, and the Parlay SCFs; these are illustrated in Figure 6.3. Multiple interfaces are defined to connect these entities. Third-party service providers deploy and manage client applications running outside of the trusted telecommunications network domain. These client applications have controlled access to the telecommunication SCFs through well-defined interfaces.

The Parlay Framework and SCFs run within the trusted telecommunications network domain. By defining the interface between the framework and the SCFs, the Parlay architecture has a somewhat flexible configuration that allows the provider of the framework to be different from the provider of the SCFs. Of course, this does not exclude network operators to be the provider of both the framework and the SCFs. A Parlay Gateway typically has all the functionalities of the Parlay Framework and perhaps part of Parlay's SCFs.

The current Parlay API specification (ETSI 2002a) has covered interfaces from 1 to 4 and will extend to interfaces 5 and 6. Interfaces 7 to 11 are considered out of the scope of Parlay, and can be either vendor-proprietary techniques or standardized interfaces of other interest groups, such as JAIN[2].

Interfaces 1 and 4 between the third-party applications and the framework provide the application with basic mechanisms (such as authentication, authorization, service discovery, service subscription, service negotiation, and integrity management) that enable the applications to make use of the service capabilities in the network. Interface 2 between the applications and the SCFs allows the applications to have real-time access to network features (such as call control, user interaction, messaging, and mobility management) and integrate these telecommunication features into their own IT business logic. Interface 3 between the SCFs and the framework allows new service features from different vendors

---

[1] See http://www.parlay.org for the Parlay Group's web page.
[2] See http://java.sun.com/products/jain for information on the JAIN APIs.

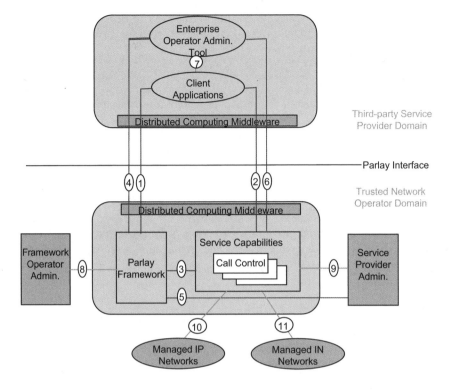

Figure 6.3 Architecture of Parlay API

to be registered by the framework and later be discovered and accessed by third-party applications.

The Parlay API defines OO interfaces on both the client application side and the network side, but does not specify how the communications are carried on between these two parties. These communications could be over Internet, data VPNs, or private lines and can use any protocols.

Distributed OO computing middleware, such as CORBA, allows objects on different computers to communicate as if they were located on the same machine. In this way, a programmer of client applications can concentrate on developing business applications with familiar OO tools and rely on CORBA to handle remote method calls and message exchanges among distributed objects. CORBA could also be used between Parlay Framework and Parlay SCFs if these two components do not reside on the same gateway.

Although CORBA is a popular paradigm to support the Parlay API, it has its own limitations, such as a large communication overhead and difficulty supporting mobile applications. With different application scenarios, other technologies, such as Java Remote Method Invocation (RMI) and SOAP, could also be used.

Functionality of Framework

The Parlay Framework API provides applications with controlled access to services offered by the network. This API consists of three different interfaces between applications and the framework, between SCFs and the framework, and between the enterprise operator and the framework. The basic framework capabilities are as follows (ETSI 2002b):

Authentication. The authentication model of Parlay is a peer-to-peer model. The application must be authenticated before it is allowed to use any other Parlay services. Applications can choose to authenticate the framework before sending any sensitive information to it. The authentication procedure is similar for both third-party applications and Parlay SCFs.

Authorization. After authentication succeeds, the authorization procedure ensures that an application is able to access services that it is entitled to.

Service Discovery. After successful authorization, applications can obtain the service discovery interface from the framework to retrieve information on authorized available network SCFs. This information includes both service types and more detailed characteristics of instances of each service type, which typically indicate the constraints of the underlying network's capabilities.

Establishment of Service Agreement. A service agreement must be established before an application can interact with any network capability features. A service agreement may consist of an off-line (e.g., by physically exchanging documents) and an on-line part (e.g., by digital signature).

Access Control. After an application and the framework agree on the usage term of a service and a service agreement is signed by both sides, the framework provides the application with a reference to the requested service with the specified security level, context, property, and so on.

Service Registration. This interface is used by Service Capability Servers (SCS) to register SCFs that they offer so that the framework can inform the applications upon request in the service discovery phase. SCSs operate within the trusted telecommunication network domain and are operated by either telecommunication operators or third-party framework service providers. The framework supports a selected set of service types, so only services of these types can be registered. These service types are not specified by Parlay standards and, as such, are currently defined by the framework operators. This allows for multiple vendors and even the inclusion of nonstandardized services, which is crucial for innovation and service differentiation.

Service Subscription. This interface represents a contractual agreement between the framework and an enterprise operator. In this subscription business model, the enterprise operator takes the role of the subscriber/customer of services, and client applications take the role of users or customers of services. The framework itself takes the role of the retailer of services.

Figure 6.4 shows the basic interaction sequence among these functionalities in an example scenario where a SCF is first registered and then accessed by an application. In step 1, the service capability operator registers new SCF implementations to the framework to enable the service to be discovered and used by later applications. In step 2, the application must contact the initialization interface of the framework using a well-known

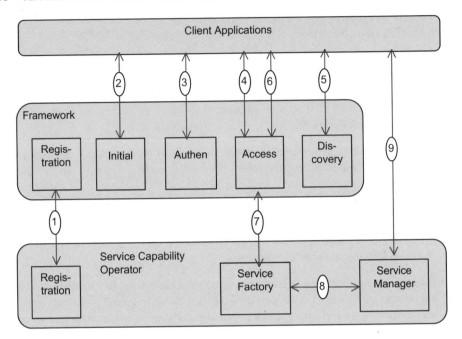

Figure 6.4 Interaction sequence of an example scenario

address to get the reference to the authentication interface. In step 3, the application and the framework use a negotiated algorithm such as MD5 to carry on the mutual authentication. Security keys are exchanged offline. Authentication occurs not only at the initialization phase of a service but also during the service period at random intervals. In step 4, the application passes the authentication phase and asks the service discovery interface to seek the appropriate services. After the desired services are chosen in step 5, the applications and the framework sign the service agreement using mutually agreed upon digital signature schemes in step 6.

Upon successfully receiving the signed service agreement, the framework instructs the service capability operator to produce an instance of the service manager to serve the application in step 7. In step 8, a factory pattern is used to produce such an instance. The reference to this instance is returned to the framework, which in turn passes the reference to the client applications. Finally in step 9, the client applications can utilize the reference to get services from the required SCFs.

Functionality of Service Capability Features (SCFs)

A number of SCFs have been standardized. These are the logical entities to implement the API and to potentially interact with core network elements. These SCFs are briefly summarized in Table 6.1 (Moerdijk and Klostermann 2003). With these SCFs, Parlay provides a rich set of service features in the collection of APIs that it defines. For more information on these APIs, refer to the Parlay specifications or (Hohler et al. 2000; Zuidweg 2002).

Table 6.1   Summary of SCFs

API	Descriptions
Call control	Four call-control APIs are defined: Generic call control, Multiparty call control, Multimedia call control, and Conference call control. The functionalities supported range from setting up basic two-party calls to manipulating multiparty, multistream calls.
User interaction (UI)	It is used by applications to interact with end users. It provides functions to send information to or gather information from users to whom a call leg is connected; or it can initiate a user interaction session with an end user who is not participating in a call.
Mobility	It has two separate APIs, mobile terminal status and mobile terminal location. The mobile terminal status API allows applications to obtain the status of fixed, mobile, and IP-based telephony users. The mobile terminal location API allows applications to request the location of end users, to register location events, and to ask for location notifications.
Terminal capability	It allows an application to retrieve the known capabilities of an end-user terminal.
Data session control	It controls nonvoice value-added data services that allow information to be sent and received across a telephone network.
Generic messaging	It supports the management of mailboxes and allows applications to send, store, and receive messages.
Connectivity management	It allows applications to specify the QoS parameters for packets traveling through the network provider's network. It can be used to set up and tear down QoS virtual private pipe.
Account management	It allows applications to query account balances, retrieve transaction histories, and enable/disable charging-related event notification.
Charging	It allows applications to charge parties for the services offered.
Policy management	It is used to create, update, and view polices. It allows applications to subscribe to policy-related events and obtain the statistics of the events.
Presence and availability management	It allows applications to obtain and set information about a user's presence and availability.

3GPP Open Service Access

Around the same time that work on Parlay was initiated, similar work began within the 3GPP to deliver value-added services in UMTS networks. UMTS aims to unify the features of mobile telephony and Internet, and for this purpose, it needs a model to accommodate both centralized and distributed network intelligence. This model is Open Service Access (OSA).

Instead of defining a new architecture and requiring a complete overhaul of the existing network infrastructure, OSA is defined in a bottom-up fashion, built on the functions already provided by underlying networks. By defining a set of standardized object-oriented APIs on top of existing network features, OSA allows platform-independent applications to be developed by integrating these features through such interfaces. Because the goal of the OSA programming interface is in line with that of Parlay, instead of defining two parallel but possibly incompatible interfaces, the Parlay Group and 3GPP decided to combine their work and produce a unique Parlay/OSA interface. Previous sections have already described the Parlay interface, so this part of OSA is not discussed here.

Note that Parlay only specifies a set of interface APIs and refrains from specifying any requirements on the implementation of interfaces. OSA, on the other hand, is designed for use within UMTS, so it also makes recommendations for mapping the OSA interface to specific network protocols. In this sense, OSA is considered an integral part of the service architecture of UMTS.

OSA can be implemented in a distributed or centralized manner. In the centralized approach, the network operator implements all OSA interfaces in one central service that acts as the gateway between the application and the underlying network components. In the distributed approach, each interface is directly mapped to its corresponding network component. As illustrated in Figure 6.5, the call-control interface is implemented by a soft switch; the mobility interface is mapped to HLR; and the data session control interface is mapped to a 3G SGSN. In the OSA architecture, existing network elements are called *Service Capability Servers* (SCSs); the element service building blocks, called *Service Capability Features (SCFs)*, and their interface (Parlay/OSA interface), is defined on top of the SCS.

6.3.2 JAIN

The aim of the Java API for Integrated Networks (JAIN) is to bring service portability, network convergence, and secure network access to telephony and Internet networks. JAIN standardizes the information flow between heterogeneous network elements in the form of an API and defines a communications framework for services to be created, tested, and deployed over this API. JAIN is defined within the context of Sun Microsystems' Java Community Process (JCP). JAIN thus forms an industry forum that issues Java APIs. All the JAIN APIs, including its service APIs, are defined in Java.

Java, as a highly structured and strongly typed OO language, has unique advantages.

1. Java is built on the virtual machine concept so that it is platform independent. Java programs are not compiled to a native machine code, but rather to an intermediary byte code. At the time of execution, the byte code can run on any physical machine that has the byte code interpreter (Java Virtual Machine) installed. This ensures high program portability, which is also why Java is sometimes referred to as "Write Once, Run Anywhere" language.

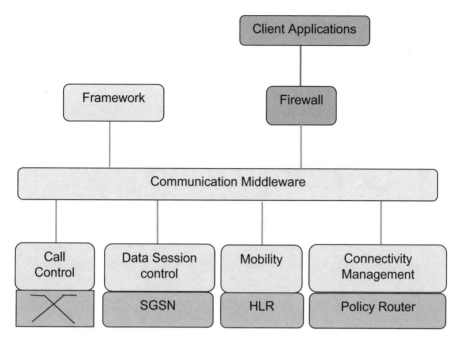

Figure 6.5 Distributed implementation of OSA architecture (source of (Zuidweg 2002))

2. Java's component models, JavaBeans and Enterprise JavaBeans (EJBs), allow components to be added, taken away, enhanced, assembled, shared, or redistributed in a
dynamic running system. This allows services and features to be added, updated, or
removed in a live environment. Some other merits include fine granularity of security
management, inherent support of design patterns, automated resource collection, and
compatibility with Internet technologies.

On the other hand, Java typically has lower performance than other languages because
it is interpreted and not compiled, although the use of optimizers and just-in-time (JIT)
compilers can mitigate this performance penalty. A more difficult problem is that Java's
execution model relies on a garbage collector, which makes it difficult to provide real-time
performance guarantees, because the scheduling and execution of the garbage collector cannot be precisely controlled. Real-time Java (Bollella et al. 2002) addresses these difficulties
to some extent, although there is more work to be done.

Architectural View of JAIN

Figure 6.6 shows the overall layering architecture of JAIN. Within this framework, there
are multiple abstraction layers to handle different functionalities. At the lowest layer is the
Protocol API layer, which standardizes the interface to signaling protocols used in different
networks. This allows applications and the protocol stack to be interchanged dynamically
and applications can be easily ported to products from different vendors. The next layer
up is the JCAT Call Control JCC layer, which provides a call-control API to hide the

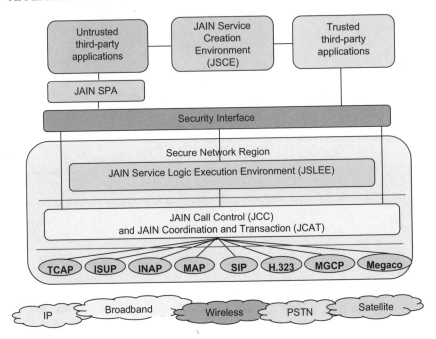

Figure 6.6 Overview of JAIN architecture. Reproduced by permission of John Wiley & Sons, Inc.

heterogeneity of underlying networks from the upper layer and to define the uniformed interface for applications to set up, manipulate, and release calls.

The next layer up is the Java Service Logic Execution Environment (JSLEE). Although the generic call-control API provides applications with a powerful and convenient abstraction to manipulate calls, applications generally require far more network support than just a call model, such as database access, transaction support, and name resolution. To facilitate the production of applications and allow easy adaptation, such network-specific services are abstracted and encapsulated by Java objects (such as EJBs) in a platform-independent manner. These objects, together with other common utility objects, are integrated into the JAIN Service Creation Environment (JSCE), where these predefined and well-tested objects are reused and composed to produce JAIN- and JSLEE-compliant applications. JSLEE then behaves as an integrated network service container that provides application context for these objects to be interpreted and executed.

The Protocol API, Call-control API, and JSLEE API layers reside inside the trusted network domain and have direct access to underlying network resources. To access these APIs, both trusted and untrusted third-party applications must undergo an appropriate security interface. After passing the security check, trusted third-party applications typically execute inside the JSLEE. In Figure 6.6, note that the untrusted third-party applications must go through a more rigorous authentication and authorization process than the trusted third-party applications before utilizing functionalities within the network. Rather than designing the interface from scratch, JAIN adopted the Parlay security interface, which is then named *JAIN Service Provider Access* (JSPA) to avoid incompatibility and duplicate work. In this case, untrusted third-party applications usually execute inside the third-party SLEE environment and have controlled and limited access to network resources through SPA interface.

The JAIN architecture is modular in the sense that applications are not restricted to use APIs of just one layer. On the basis of an application developer's judgment, an application running in the JSLEE can directly access the APIs of the lower layers (for example, the SIP or ISUP API), can use the Java call-control API, or both. The fundamental tradeoffs are programmability, adoptability, and performance. Applications using upper-layer APIs are easier and quicker to develop and have better adoptability across different platforms by leveraging higher levels of abstraction. The disadvantage is that their performance may not be as good as that of applications with finer-grained control over lower-layer protocols. On the other hand, applications using lower-layer APIs can achieve better performance by avoiding software manipulation overhead and having more precise control over the sequence of actions of lower layers. The disadvantage is that these applications must deal with certain lower-layer complexities. This not only leads to a longer development cycle but also weakens the adoptability of applications on different platforms. With this in mind, the rest of this section will discuss the Protocol API and Call-control API of the JAIN architecture in more detail.

JAIN Protocol API

The JAIN Protocol API specifies the interface for the most-common signaling protocols used in telephony, IN, wireless networks, and the Internet. As shown in Figure 6.1, the goal is to enable an application, such as those in the soft switch, executing in the Java space to access the functions provided by underlying signaling protocols written in a native language. Instead of directly accessing the protocol stack, applications can interact with Java objects that represent these protocols.

Protocol APIs can be divided into two groups, SS7 APIs and Internet APIs. SS7 APIs include Transaction Capability Application Part (TCAP), Integrated Services Digital Network User Part (ISUP), Intelligent Network Application Part (INAP), and Mobile Application Part (MAP). Internet APIs include SIP, Media Gateway Control (Megaco), MGCP, and Session Description Protocol (SDP). This list can also be expanded to include future protocols. For a detailed description of each protocol interface, see http://java.sun.com/products/jain.

Protocol APIs are based on the Java Listener pattern. As shown in Figure 6.7, each protocol API should define three components, namely Stack, Provider, and Listener. The Stack component abstracts the underlying native protocol stack and provides a factory to create a Provider instance when requested by an application. The Provider component is in charge of translating a local method call to the native message of the underlying signaling protocol. The application creates an instance of the Listener component to communicate with the Provider. The communication between the Provider and Listener uses the Event pattern, the standard communication mechanism used between JavaBeans. Before receiving any Event component from the Provider, the Listener has to register with the Provider. Figure 6.7 shows the communication between an application and the JAIN protocol API. When an incoming message arrives from a lower layer, a Java Event is produced by the Protocol API software and then uploaded from the Provider to the Listener. On the other hand, when the application has outbound messages to send, it calls the corresponding method of the Provider directly.

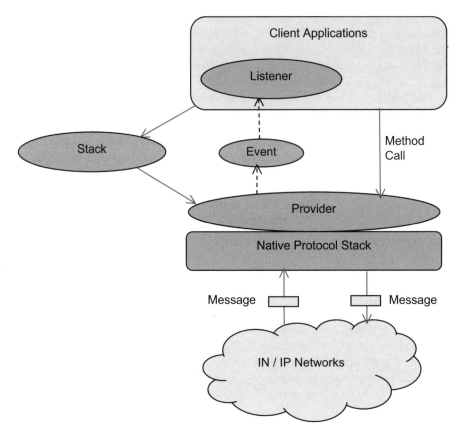

Figure 6.7 JAIN protocol API architecture

The abstraction at the Protocol API level does not provide a single uniform signaling interface to hide the underlying heterogeneous signaling protocols. So applications at this level should understand each specific signaling protocol of the networks they want to support. However, the standardized API for each signaling protocol does hide the heterogeneous implementation of different vendors. A vendor must run and pass the Test Compatibility Kit (TCK) compatibility test suite before it can claim JAIN compliance.

JAIN Call-control API

With a higher level of abstraction than the Protocol API, the JAIN Call-control API aims to hide the heterogeneous underlying network architecture and signaling protocols from applications, and provides application programmers with a unified platform to set up, manipulate, and tear down communications between multiple parties.

The core concept of this set of APIs is *Call*. The concept of a Call, which is originally used in the PSTN network to stand for two-party point-to-point voice call, is largely extended here to stand for "a multiparty, multiprotocol, and multimedia communication session over the underlying integrated networks" (Jepsen 2001b). Multiparty means that each call can have more than two parties, each of which is called *a leg of a call*. Multiprotocol means that each leg of a call could reside in a different network and, therefore, different underlying signaling protocols could be used within one call. By using a unified call model, such complexity is shielded from applications. Multimedia means that multiple substreams (voice and video) of a multimedia stream can be served on one leg.

Because the call model is such an important abstraction, several call models have previously been defined, including IN, the Java telephony API (JTAPI), and the Parlay Call-control API. Because of the architecture or application for which they were intended, these call models have great differences, but their overall goal is similar: to initiate, control, and manipulate calls, and to facilitate the development of applications that execute before, during, or after a call. A good review and comparison among these models has been given elsewhere (Jepsen 2001b) and will not be repeated here. This section briefly introduces the JCC and JCAT call models adopted in JAIN, which are more generic interfaces and capture the essential aspects of three call models mentioned above.

Both the JCC and JCAT call models define four basic objects, namely Provider, Call, Connection, and Address. In addition to these, the JCAT call model has two extra objects, Terminal Connection and Terminal, to capture the physical properties of a terminal. The association relationship among these objects is shown in Figure 6.8. A Provider object is the entity that a call-control application has to access in order to initiate the placement of a call. A Call object represents the temporal association between two or more parties. A Connection object represents the specific status of the relationship between a Call object and

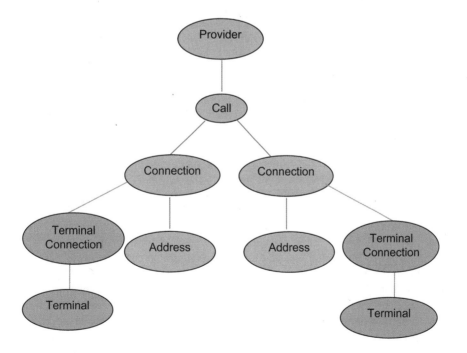

Figure 6.8 Object model of JCC and JCAT call model (two-party). Reproduced by permission of John Wiley & Sons, Inc.

an Address object. Each Connection object has variables to maintain the current call state of one call leg. The states of a Connection and their transition conditions are expressed by a Finite State Machine (FSM). Conceptually, JCC and JCAT's Connection object resembles the call states of IN to provide applications with detailed control points to interact with a call. An Address object represents a logical end point of a call leg, such as a telephone number or an IP address. A Terminal object represents the features of a physical end point of a Connection, for example a telephone set or wireless PDA. It also models the relationship between Terminals and Addresses, for example, allowing multiple addresses per terminal, and vice versa. A Terminal Connection object represents the relationship between a Terminal object and a Connection object.

Applications and call-control APIs communicate with each other in the same manner as the Protocol API. In one direction, applications make synchronous calls to API methods to evoke services. In the reverse direction, call-control APIs use the Event-listener pattern to inform applications of lower-layer events. FSMs of the call-control objects have defined the types of events accompanying each state transition. Applications can choose to register the events they have interest in and filter out irrelevant events. Because a hierarchical object model is used in call control and each object is able to produce events, the Listener objects of applications have the same hierarchical architecture in order to receive events from their counterparts in the call-control model.

In this way, JCC provides the basic facilities for applications to be invoked and return results before, during, or after the call, and the basic facilities to process call parameters or subscriber-supplied information. JCAT uses these facilities and extends them further.

6.3.3 Open Mobile Alliance (OMA)

The Open Mobile Alliance (OMA) was formed in June 2002 with a nucleus of supporters and members of the Wireless Application Protocol (WAP) Forum. It later integrated other industry fora such as the Location Interoperability Forum (LIF), SyncML Initiative, Multimedia Messaging Service Interoperability Process (MMS-IOP), Wireless Village, Mobile Gaming Interoperability Forum (MGIF), and Mobile Wireless Internet Forum (MWIF). OMA currently has more than 320 members that cover the entire mobile value chain, including application and content providers (CPs), mobile operators, device and network vendors, and IT companies. The growth rate of OMA has exceeded that of earlier industry fora, such as Parlay and JAIN. Since its birth, OMA has focused on the consolidation of standard forums and works closely with other existing standardization organizations such as IETF, 3GPP, 3GPP2, and W3C.

The goals of OMA are to become the central method for specifying mobile services, to promote service interoperability, and to ensure seamless user experience. The benefit for users is that no matter what device or operating system you have, no matter what service you have, no matter what carrier you use, you can communicate and exchange information. Thus, unlike the Parlay/OSA interface that is focused on the interaction between telecom operators and third-party service providers, the user-centric OMA is focused on the seamless end-to-end user experience during user mobility, user terminal mobility, and user service mobility between multivendor and multidomain environments regardless of the underlying network infrastructure.

To this end, OMA promotes open interfaces, open standards, and common platforms to guarantee the interoperability between products from different vendors on the mobile value chain. More specifically, OMA requires that

- products and services be based on open, global standards, protocols, and interfaces and not locked to proprietary technologies;

- the application layer is bearer-agnostic (for example, GSM, GPRS, EDGE, CDMA, UMTS);

- the architecture framework and service enablers are independent of a particular operating system;

- applications and platforms are interoperable, providing seamless geographic and intergenerational roaming.

The work of OMA is divided into two parts, a common OMA system architecture and a set of OMA enabler services. The main objective of the common OMA architecture is to integrate OMA enabler services into a single framework to avoid monolithic enablers, to identify common functions, to handle heterogeneity of underlying networks, and to guarantee that all entities conform to the OMA principles. The OMA enabler services standardize a set of basic services on which applications can be built. Enabler services can be accessed by a standardized interface in a way similar to the service components of the Parlay/OSA architecture.

Some enabler services are Billing Framework, Browsing, Client Provisioning, Digital Rights Management (DRM), DNS, E-mail notification, Instant Messaging and Presence Services, Multimedia Messaging (MMS), User Agent Profile, Device Management, Games Services, Data Synchronization, and SyncML Common Enablers. More services such as location, web services, and m-commerce will be defined in the near future.

The OMA architecture is currently being defined in the OMA Architecture WG in both top-down and bottom-up fashion. The top-down direction is designed to define system requirements (OMA 2003a) based on user scenarios, and the bottom-up direction is designed to summarize existing architectures and map OMA requirements to them. In this way, it is ensured that market requirements are satisfied and existing solutions are not reinvented.

The vision of OMA and Parlay/OSA overlaps quite a bit, so it is natural that many system requirements of OMA have been embodied in the framework of Parlay/OSA (OMA 2003a). The architecture of OMA has not been finalized yet, so it is not clear how the work of Parlay/OSA will be reused. On the other hand, it is also pointed out that many system requirements of OMA cannot be answered by the Parlay/OSA framework (OMA 2003a). The main reason is that OMA requires strong support for user interaction and different business models, which are not the focus of Parlay/OSA. For example, OMA supports single sign-on/log-out, trust delegation, and identity federation and de-federation authorized by the user. OMA allows different trust relationships as envisioned in the Liberty Alliance Project (LAP 2003) to be built among service providers. OMA provides end users with service registration and discovery, QoS management, policy management, device management, and so on. These functions are either out of the scope of Parlay/OSA or only include limited features.

From the bottom up, the current picture of the OMA architecture is a conglomerate of the existing architecture used in OMA enabler services. The detailed graph is shown in the OMA document (OMA 2003b), and the corresponding discussion is out of scope of this chapter. From the conglomerate of existing architectures, more work should be done to modify it according to the principles of OMA and to reflect the architecture requirements defined beforehand.

Up to now, OMA work has been underway for about a year, and most of the work is still in the stage of requirement definition or standardization proposal. Although the mission of OMA is on target, because of the large question domain and the number of interest groups involved, it is still unclear how influential the OMA effort will be.

6.4 Advanced API Efforts

6.4.1 Parlay Web Services and Parlay X Web Services

The Parlay API offers a secure, controlled, and accountable common interface across networks, allowing applications to be created that are both network transparent and implementation transparent. It is envisioned that the introduction of the Parlay API into the monolithic telecommunication networks will greatly stimulate third-party service providers to create a wide range of innovative value-added services rapidly and inexpensively. Further down the road, the next step of the integration of telecom networks and IT applications is to provide access to these direct programmatic interfaces using web services, which will bring in a much wider application developers base and additional business potential.

It has been claimed that the integration of Parlay API and web services is both commercially profitable and technically feasible (Parlay 2002a,b,c). We discuss this briefly here.

In terms of business motivations, by tapping into web services, the large web developer base and richer development tools and skill sets can be leveraged to greatly reduce the development time and cost of telecommunications or converged network services. Furthermore, niche third-party service providers are often focused on delivering high value-added services to a particular set of customers. In many cases, such niche services (for example, a stamp exchanging site that allows collectors to exchange postage stamps) are built on top of lower-level telecommunication services (such as billing and call forwarding). Compared with basic telecommunication services, these niche services have a much smaller and more focused customer base and may have a shorter lifetime. The key factors of the success of these services are fast deployment and flexible service profiles in order to seize market share quickly. To this end, a high-level web-based interface is more suitable than a lower-level interface, particularly one that is based on the details of individual protocols. At the same time, web services are becoming a core technology used in a large number of internal or external enterprise business integration projects. As telecom network capability becomes a larger part of enterprise applications, the ability to tie these applications into business process integration tools and systems will become increasingly important. By using web services, Parlay may be represented within the tools and applications as another resource available to the enterprise developer.

In terms of technology, the Parlay model is consistent with the goal of web services. Both models provide an application with an implementation-neutral interface to access network services without being exposed to unnecessary complexity.

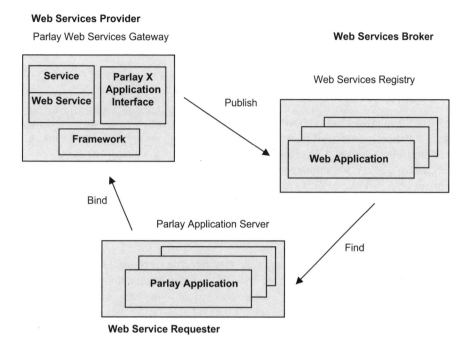

Figure 6.9 Parlay web services architecture (Parlay 2002b)

The Parlay web services effort aims to provide the interface definitions and infrastructure definitions for the use of web services within a telecommunications network. The Parlay web services interface is defined using a WSDL version of the Parlay API, which is automatically translated from the UML version of the Parlay API. The typical Parlay web services infrastructure is shown in Figure 6.9.

Typically Parlay web services are implemented within the Parlay Web Services Gateway, which can be integrated with, or be separate from, the Parlay Gateway (which has all the functionalities of the Parlay Framework and perhaps part of Parlay's SCFs). The Parlay Web Services Gateway provides web-based access to underlying network services. To make itself available for discovery by a Parlay application, the Parlay Web Services Gateway publishes its WSDL interface in the Web Services Registry, which is typically implemented in a UDDI format. The content of this publication includes service types and parameters, as well as the detailed location where the service can be invoked. To use the service, an application first uses a Find operation to retrieve related services information from the Web Services Registry using the SOAP/HTTP (W3C 2003a) protocol. After selecting the desired services and the corresponding Parlay Web Services Gateway, the application uses a Bind operation to connect to the gateway and begin the services.

Also shown in Figure 6.9 is Parlay X. Instead of a full-fledged interface like Parlay Web Services, Parlay X Web Services represents a highly abstracted and simplified version of the Parlay API. The purpose of defining such an interface is to stimulate the development of next-generation applications by web developers who are not necessarily experts in telecommunications. Thus, Parlay X is suitable for those value-added services that are focused on high-level business logic, mainly interpreting the telecommunications network as an internetworking pipe, and as a result, it does not require fine-grained control on connection states. The Parlay X Web Services version 1.0 specification has been published, but further development is in process.

6.4.2   Router APIs

The core of the Internet lies in robust and scalable router design. Routers are mainly designed to optimize the performance of packet forwarding, focusing on the traditional naming, routing, and best-effort packet delivery functions of the Internet. Thus, many functions on the critical path of a router are implemented with hardware. As the Internet gradually becomes an indispensable part of people's daily lives, more and more applications, such as multimedia streaming and interactive gaming, begin to appear. These applications, combined with other Internet-based computation models, including peer-to-peer networking and overlay networks, require new designs of router architectures so that routers can be easily extended to include new types of services.

Compared with the hardware-based router architecture, extensible router architectures are designed as software and hardware hybrids in order to maintain a balance between performance and flexibility. At the service level, a virtual router abstraction is used to model the main functions of a router. Such an abstraction allows the service to be developed in an implementation-independent way so that the only interaction between services and underlying hardware is through the API.

There have been a substantial number of research projects on extensible router architecture design and implementation. Among them, the Scout (Mosberger and Peterson 1996), Click (Hohler et al. 2000), and Router Plugins (Decasper et al. 2000) projects are among the most influential. These architectures share the same motivation but differ in their definitions of key abstractions and, as a result, the extensibility and complexity of real implementations. A comparison of these architectures has been done by Gottlieb (Gottlieb and Peterson 2002) and will not be discussed here. The rest of this section uses Scout's abstraction as a working example to show how a virtual router is modeled and used to support extensible services. This section also introduces an extension (Zerfos et al. 2003) of this abstraction model, which works as an edge router in the wireless domain and facilitates the deployment of adaptive applications and protocols requiring close interaction between link layer and IP layer.

The extensible router abstraction shown in Figure 6.10 has four main components: Queue, Classifier, Forwarder, and Scheduler.

Queue has the same meaning as a common queue object, having storage to store packets in a FIFO manner and to enqueue and dequeue methods.

Classifier has functions to dequeue a packet from the head of the single input queue, examine the packet headers, classify the packet according to predefined policies (such as routing table, access control lists, etc.), and send the packets to zero, one, or multiple queues. The packet is put in zero queues (no forwarding) to implement certain filtering actions. The packet is copied to multiple queues if multicast is supported.

Forwarder removes a packet from its own input queue, performs some specifiable computation, and puts the packet into its output queue. The Forwarder component encapsulates

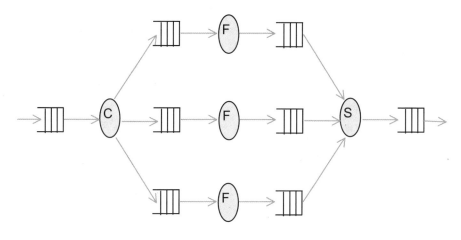

Figure 6.10   Abstraction of an extensible router (Gottlieb and Peterson 2002)

most of the add-on services. However, it also supports traditional IP services such as manipulation of the Time-to-live (TTL) field in IP packet headers, checksum computation, and link-layer header modification. Some more advanced services include tag insertion (such as tag used in MPLS), overlay support (by manipulating the packet destination address to reroute the packet), and denial-of-service detection (by monitoring the traffic statistics from each individual source address).

Scheduler chooses one packet from the header of its multiple input queues and places it into its single output queue. Instead of serving packets in a FIFO way, a router can differentiate the QoS of each packet (or flow) by altering their serving order. Thus, a more complicated queuing algorithm could be used in the scheduler. The architecture itself is extensible in the sense that each component is an independent module so that the behavior of the router could be dynamically changed by inserting a new component, modifying an existing component, or replacing an old component with a newer version.

The extensible router architecture provides applications with a mechanism to add new services to the router. This architecture defines a new component, *path*, which is the instantiation of the Forwarder, along with its associated input and output queue. Three method calls, createpath(), removepath(), and updateclassifier(), allow applications to add, remove, and modify services, respectively. As implied by the name, createpath() allows a new processing path to be generated and linked to input and output ports through new queues. It also contains the parameters to set up the classifiers and schedulers. The methods removepath() and updateclassifier() remove or modify the path in the router, respectively. The service is removed when the path object is deleted and the service is modified by changing the behavior of the Forwarder object on the path.

DIstributed Router ArChitecture (DIRAC) for wireless network (Zerfos et al. 2003) is a distributed software-based router architecture that facilitates the quick deployment of wireless link-aware protocols and services. In the last few years, several protocols have been developed to combat two big challenges of mobile computing, the error-prone wireless transmission and the handovers. The core of many of these protocols is the idea of cross-layer adaptation, which adjusts the IP layer protocol behavior by utilizing link-layer feedback or statistics. Some examples include wireless TCP, adaptive source coding or channel coding, or link-layer-assisted fast handover. But such protocols are not easy to implement on any current core router architecture that has not taken link-layer information into account. DIRAC is proposed to solve this problem.

DIRAC uses a distributed architecture that consists of two parts: a Router Core (RC) that resides on the access router and is shared by many access points and a Router Agent (RA) at each access point. RC interacts with RA through a lightweight signaling protocol. Each RA collects and reports link-layer information back to RC and accepts and enforces router commands sent by RC. RC carries out most of the traditional functions of a router and also makes adaptive decisions based on link-layer feedback. The data forwarding abstraction of RC has the same model as Scout (Figure 6.10). The difference is that the Classifier, Forwarder, and Scheduler are now managed by the controlling components within the RC on the basis of link-layer information.

DIRAC defines three communication mechanisms between RC and RA: Event, Statistics, and Action. Event denotes occurrences of asynchronous link-layer activities in the cell, detected by the access point and reported to RC. Statistics reports the latest information on channel quality or user information such as location-dependent information. This is not the general location information, but more in the context of wireless scheduling when receivers at different location experience different error patterns at the same time. Such a difference could be observed by AP and reported to AR. Action is sent from RC to RA in order to express RC's command. The deployment of new services within DIRAC involves the instantiation of a path object with possibly service-specific Classifier, Forwarder, and Scheduler components. It also includes the step of setting up a link-layer interaction channel through APIs provided by DIRAC. A set of methods is designed for each communication mechanisms between RC and RA. For example, the register() method is used to register an event listener to listen to a specific link-layer event; the get_stat() method is used to collect statistics of some variables from link layer; and the execute() method is used to send out commands. By leveraging these method calls, any link-layer-aware modules along the path are able to make a knowledgeable adaptation based on collected feedback.

From this discussion of the APIs of the Scout and DIRAC systems, it is clear that these APIs are still primitive and do not catch all the necessary interaction between services and routers. So a lot of work still needs to be put into the design of every single component and their interaction in order to implement a service.

6.5 Our Approach

6.5.1 Layered API Design

Our view of programmability for XG networks is structured as a set of layered APIs (Jain et al. 2004b). In fact, this set of APIs can be viewed as an application-oriented abstract specification of the architecture. For this reason, we have introduced this architecture in Chapter 2 of the book.

The highest layer API is what we term the *middleware API,* which offers applications with the required security and billing credentials access to powerful, coarse-grained functions, such as access to overlay network elements for content distribution location servers and security certificate authorities. Below this layer, the core network API offers interfaces to Internet signaling and coordination protocols such as SIP, Mobile IP, and so on. The lowest layer of APIs provide security, QoS, and billing functions, especially those involving real-time constraints; third-party access networks should provide the service provider with appropriate levels of programmability.

To briefly illustrate usage of these APIs, consider a buddy-alert application. A user is provided a web interface to upload her buddy list and related rules; she gets an alert if a buddy is in the geographic vicinity and also available to chat online. One design for the application deployment is shown in Figure 6.11.

The buddy-alert application is developed and hosted by a third-party service provider. It interacts with the operator (and also service provider) through a predefined open API. Depending on the technologies used by the third-party application, different levels of the operator's open API might be used.

For example, if the third-party service provider developed the application by composing primitive services provided by other parties (such as the network operator, local ASPs, or even other service providers), the third-party service provider can use the high-level middleware API to improve service portability and to avoid the exposition of unnecessary low-level complexity. Such a middleware API could be implemented as a standard web service interface, which allows fast service deployment and easy integration with other business logic. On the other hand, because of the coarse grain control of the network functionalities, the performance of the middleware API might not meet the requirements of

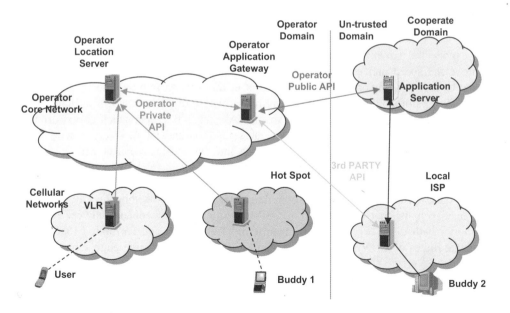

Figure 6.11 Design of buddy-alert application

the third-party service provider in some scenarios. For example, if the buddy-alert application is part of an on-line chat application that supports seamless chatting as buddies move, the third-party service provider can use the open API to access location information available at the IP level, such as from mobile IP or SIP messages. By bypassing the high-level middleware API, the third-party service provider can improve the accuracy of location information but at the same time will need to process more protocol details. Furthermore, even though the API at the core network level might provide a certain level of abstraction for underlying access networks, because of the heterogeneity of the next-generation access systems, APIs at the access network level might be useful for developing location-aware adaptation schemes, such as, adaptive codecs for different access network bandwidths.

No matter what level of API is used, the third-party service provider must pass security checks of the network operator before it can access the functionalities supported by the network. In the deployment here, the security check is handled by an operator application gateway that separates the trusted network operator's domain from the untrusted outside network domain. In any case, part of the user data or business logic might be deployed within the operator's domain in a secure and controllable manner. In the deployment here, location services are provided by a Location Server offered by the operator, accessed via the operator's application gateway through a closed API. The collected location information is then parsed and provided to third-party service provider via a chosen open API. To collect buddy location information, the Location Service uses an internal closed API to contact servers at different access networks. If some buddy is outside of the operator's network, the Location Service can use a third-party open API to interact with servers of other providers, for example, the local ISP if required by the application.

The following section briefly describes an example of a high-level API (conceptually, at the middleware level) that we have developed in accordance with this overall design.

6.5.2 Content Delivery Network API

The exponential growth of Internet traffic has created a huge demand for techniques that improve the speed and efficiency of data delivery. Content Delivery Networks (CDN) is one of the most important emerging efforts in this area to improve the scalability and responsiveness of applications. The general infrastructure of CDN is shown in Figure 6.12. CDN service providers (CSPs), such as Akamai (Akamai 2002), operate a highly distributed overlay network consisting of a few data centers and thousands of edge servers worldwide. These overlay network nodes run on top of the Internet and interconnect with each other through dedicated links obtained from local ISPs. The content from CPs is first distributed among edge servers, and subsequent content requests from clients are then fulfilled locally by the closest edge servers instead of remote original servers within the CP's network. The main benefits that CPs gets from CSP are better QoS of content delivery (because of shorter path distance), robustness against link congestion and corruption (because of duplicate routes of CDN), and better scalability (because of the large edge server pool of CDN). By leveraging its large footprint and better knowledge of user location, the CSP can provide CPs with additional value-added services such as traffic management, denial-of-service protection, geo-targeting, and location-aware computing. For streaming services, which are becoming popular over the Internet and presenting more stringent bandwidth and delay-bound requirements, CDN can provide more value-added services to alleviate

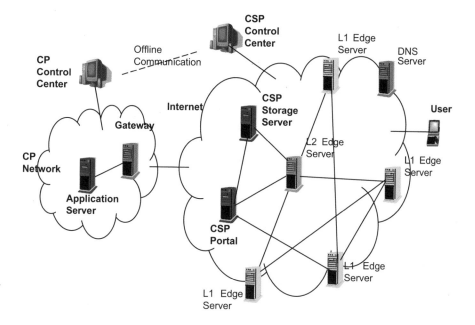

Figure 6.12 Overlay CDN architecture

these problems. Some of these services include content segmentation, prefetching, session handoff, and transcoding (Jain et al. 2004b; Yoshimura et al. 2002).

We observe that current CDN implementations lack or do not focus on programmability. Many configurations are made offline through human intervention and interaction and static configuration information is kept in a configuration file in each surrogate. This makes both the maintenance of the CDN and the introduction of new functionalities very difficult. A powerful API will enable CPs to seamlessly and dynamically control how their content is served by CDN networks and will integrate CDN services within their business strategies better.

For example, consider an on-line merchant, such as Amazon.com, which maintains a complex website to sell products to millions of users. The web pages of the website have different characteristics. The product introduction page has a long lifetime and at any given time multiple versions of the page could coexist in the CDN without much harm. Product-pricing pages can be updated at any time and multiple versions of the page are undesirable to avoid discrepancy in prices. Finally, some sensitive pages, such as those confirming user account information or credit card information, are not cacheable at CDN proxies at all. Such differences among various web pages require that the CP have the tools to inform the CSP whether the content should be cached, how long the content can be cached, or when a content update should occur to guarantee synchronization between the cached copy and the original copy. Other functions might include purging the outdated or invalided pages from the CSP's local caches or controlling who has the right to access certain content.

Such functionalities are mainly supported in two ways. The first is by organizing pages into different directories according to their access characteristics and circulating the directory configuration information among all CSP surrogates in the form of a configuration file. This static configuration requires manual intervention and is very labor intensive to update when the directory configuration changes. The second way is by appropriately setting up header fields in the HTTP response packet so that the CSP can make the right decision by parsing these headers. Because HTTP is the standard web access protocol and has well-defined header fields for these purposes, the second method proves to be more flexible and powerful. But this requires detailed knowledge of and dependency upon the HTTP protocol and the ability to manipulate low-level HTTP packets. Usually web page producers are not – and should not need to be – experts on networking, so web page publishing must be handled by specialized personnel. So, the whole producing, publishing, and, later, maintenance process is both time consuming and error prone.

This motivates the abstraction of CDN functionalities. By encapsulating the lower-layer protocol details into high-level interfaces that are easier to understand and manipulate, our CDN API makes it possible to automate the publishing process and have finer control over the content. The CDN API can be utilized by different applications and integrated into different business logic. The GUI can be built on top of this set of APIs so that even users with no technical background can publish and maintain their content through the GUI. CSPs also have more flexibility to modify their network functions and guarantee the backward compatibility of previous applications by keeping the CDN API unchanged. Therefore, our CDN API is positioned between the CP and CSP and provides an application-level abstraction of CDN functionalities. Through this API, the CP can have more programmability and flexibility to fully exploit the power of the CDN network.

Figure 6.13 shows the structure of our CDN API design, which is summarized below. This API's functionalities are divided into two parts, in-band interface and out-of-band interface.

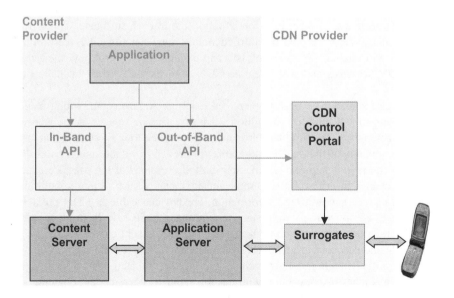

Figure 6.13   Internal and external interfaces of the CDN API

The in-band interface has defined a set of metadata that describes the intrinsic characteristics and service requirements of each object, such as cacheability, consistency, and security. It also defines simple function calls for applications to set and retrieve metadata values. When fulfilling a content request, an application server retrieves the content as well as the metadata of the content. On the basis of the metadata, protocol-specific control headers (such as an HTTP header or RTSP header) are produced and embedded into the application layer data packet. After the packet is injected into the CDN, the intermediate CSP surrogates and proxies begin to serve the object according to these controlling headers. Thus, by changing the value of the metadata through interface function calls, an application can indirectly control the service received by each object in the CDN. Because the control message and data message share the same data channel in this case, the interface is called *in band*.

The control functionality supported by the in-band interface is considered to be asynchronous. This is because when the metadata is changed because of updated requirements, it does not immediately alter the behavior of the CDN. When new requests for the content are received by the application sever, the changes of the metadata are taken into account to prepare the corresponding response headers. Thus, the time delay between the setup and usage of the metadata is unpredictable and can be quite large.

Another characteristic of the in-band interface is that its control has the granularity of a single content object. Because current application layer transport protocols are for single content object retrieval and the response header is also designed for the same purpose, the control information that can be contained in response headers is also content specific and has very small granularity.

The out-of-band interface, on the other hand, provides applications with a dedicated control channel to control the way CDN surrogates serve the CP's content. Instead of using embedded response headers, the out-of-band interface may use proprietary protocols and message formats for communication. This set of function calls provides applications with the ability to directly communicate with the CSP's portal or different servers to negotiate CDN services.

From the discussion on limitations of the in-band interface, it is clear that there are two kinds of control information that should be exchanged through the out-of-band interface. The first type is control information that requires a real-time response. An example of this is immediate content invalidation when a content error is detected. Quick content invalidation can prevent incorrect copies of content from being accessed in the CDN. The other type is control information that has coarser granularity. An example of this type is information about the whole site or a subset of the whole site. Such control information normally cannot be represented by the underlying transport protocol for a single content.

6.6 Discussion and Conclusions

The benefits of APIs for rapid service creation are becoming widely recognized in telecommunications circles, as shown by the large number of standards and efforts by industry forums to develop them. Figure 6.14 shows a rough view of trends in programmability of networks in terms of the APIs discussed in this chapter. Obviously, one can argue about the placement of each API along the vertical axis, but that is not the point we wish to make here. The general trend is clear: there is a movement toward greater flexibility in

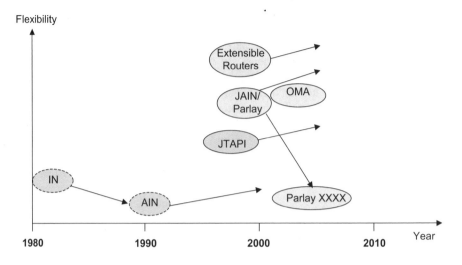

Figure 6.14   API technology trends

terms of service creation using programmable interfaces. We have included the original IN concept defined by Bellcore in the early 1980s (not to be confused by the standard, of the same name, developed by ITU-T later), as well as the AIN architecture. Although IN and AIN do not have APIs as such, we have included them for completeness and comparison purposes. Also included is the JTAPI, which is intended for CTI environments; although it is not discussed in this chapter, it has greatly influenced the design of later APIs like JAIN and Parlay.

There is also observed another trend, namely simplification of APIs, to make the APIs more easily implementable or usable. This can be seen in the case of the simplification of IN to AIN and also the simplification of Parlay to Parlay X. This tradeoff between power and simplicity will continue to be negotiated within the industry.

In some sense, JAIN, Parlay, and OSA can all be regarded as first-generation APIs. They represent a significant advance over the service creation methodology available via IN in the PTSN, which lacked a well-defined API. They also represent a significant advance over the traditional service creation methodology in the Internet, which relies on either low-level abstractions at end hosts (such as sockets) or application-specific abstractions (such as MAPI or TAPI) that do not form a coherent, well-defined, extensible, and secure programming system.

Nonetheless, while JAIN, Parlay, and OSA differ in scope, approach, and details, they share some fundamental similarities:

- They implicitly assume a network-centric, or *smart network*, architecture. In principle, PC applications can be written using, for example, the Parlay API if a suitable underlying PC implementation was carried out and made available. However, the Parlay API is meant for implementation at tandem switches (Class 4 switches, in North American terminology) or gateways and, in practice, it would be difficult to use in an end-to-end scenario with smart terminals and a dumb network. The JCC/JCAT API is more flexible in this regard, because it can be implemented at end offices (Class 5 switches) or tandem switches, but its use on a PC is likely to be problematic. (The design of an API that can be used at terminals and in the network has been proposed in (Anjum et al. 2001)).

- They fundamentally assume that the central abstraction to be manipulated is a *call*. For all these APIs, a call can, in principle, be a multimedia, multiparty communication session of any type. However, the nature of these APIs makes it difficult to manipulate call objects that represent extremely short-lived interactions because of the overhead involved. Parlay is more flexible than JCC/JCAT in this regard, because it includes a messaging API. However, a Parlay application must still abide by the rules of the framework and the Parlay programming methodology, which can be inefficient for short-lived applications.

- They are complex and their implementation is likely to be heavyweight. The JAIN APIs tend to be simpler than the Parlay APIs, partly because they are targeted for a single application language (Java) with rich supporting features and partly because they do not have the equivalent of the Parlay Framework APIs. However, the latter feature also means that support for security is limited.

- They are oriented towards programming in procedural languages like Java, C, and C++.

The Parlay X API is intended to overcome some of the complexity of programming using the Parlay APIs by offering a simpler, higher-level interface. It also allows programming using XML, which may be simpler than using procedural languages in some cases. Thus, it represents an interesting approach, but it remains to be seen how effective it is for building practical applications. It is not clear that the right balance between simplicity and expressiveness has been found; only actual experience will tell.

The router APIs discussed in Section 6.4.2 represent another approach to service development, and one that is, in some sense, orthogonal to that of Parlay X. Compared with network-centric frameworks such as Parlay, OSA, and JAIN, router API provide service providers more power programmable interface to control the forwarding behaviors of underlying network components. On the basis of the end-to-end service model, the service logic could be deployed on any intermediate nodes along the path between a server and a user. However, these approaches are still in the research stage. And it is not clear how services could be composed in this framework in an efficient and robust way.

One limitation of the Parlay, OSA, and JAIN APIs is that they are inflexible. Offering a radically new service might require modifying the API itself. Because the APIs are extensible, they can, in principle, be modified to provide new functionality, but it is likely to be a slow and complex effort, even if the standards (or industry forum) process is bypassed or carried out in parallel.

It is fashionable in the industry to talk about *programmable APIs* or *self-modifying APIs*. Presumably, these are APIs that can be modified quickly (or modify themselves quickly) as application developers or marketing departments dream up new services. We tend to be skeptical of these notions; the fundamental conceptual tools for reasoning about self-modifying abstractions, let alone implementing them, seem to be far in the future. (Of course it is easy to modify the actual FSM at the heart of an API, but what happens to the semantics of methods defined using that machine?) Nonetheless, it is clear that there is a need for more adaptive and flexible APIs, and developing methodologies for specifying such APIs is an interesting research topic.

We briefly comment on a technology trend that indicates required changes in next-generation APIs: the rapid proliferation of PDA-type mobile terminals equipped with fast CPUs, large memories, and flexible input/output devices. Compared with legacy mobile terminals equipped with only a 10-digit keypad, this new generation of terminals is far more powerful. This has two implications. The first is that more and more functions will be integrated into the mobile terminals, such that they become the user's personal data center and remote control to virtually any service. APIs that conceptually seamlessly span the client and the server side of application development are thus required. The second is that user interaction APIs must become more sophisticated and flexible. For example, Parlay/OSA's user interaction API is fairly limited, partly because of its network-centric methodology. How to handle a flexible division of functionality among different system segments (user devices, core networks, application servers, etc.) and how to handle the heterogeneity of user devices with a wide range of UI capabilities are two important issues that need further attention.

Finally, there is a recent trend in the design of mobile wireless networks, particularly ad hoc networks, to take advantage of information that is passed across multiple layers of the protocol stack. For example, information from the wireless link layer can be used to make the behavior of TCP more efficient. Even though separation and transparency between different layers are the general design principles to guarantee the interoperability and independence among different system components, at the moment, these cross-layer optimizations are being explored in response to specific problems that are observed in network performance or QoS. A more systematic and flexible approach would be to define APIs that allow information to be exchanged across multiple layers in a disciplined, secure, and efficient manner. Thus, developing cross-layer APIs for XG networks is likely to be an important area for future research.

Part III

Middleware and Applications

As the next-generation core network technologies evolve so will the middleware infrastructures that manage them and the application that run on them. In this section, we examine the evolution of middleware technologies and applications on next-generation networks. Middleware technologies typically straddle the terminal and the network and provide services for other applications. In Chapter 6, we discuss middleware technologies for the network. In Chapters 7 and 9, middleware technologies for the mobile terminals are examined. In Chapter 8, we discuss media processing and transport technologies for multimedia applications. Several chapters in this part discuss standards that are often developed and agreed upon by a standards committee prior to their deployment in real networks.

The growth of XG networks will be driven by two several trends:

1. Opening of the carrier network through Application Programming Interfaces (APIs) that enable third parties to provide new services, such as billing, and sophisticated messaging and thus dramatically reduce the time for carriers to introduce new products to their customers.

2. As devices become more capable with sophisticated processors and more memory, sophisticated multimedia-based and web services-based applications become possible. In addition, the device can be used to mask the vagaries of the network and services by caching services on the device itself.

The opening of the network is enabled through a set of emerging standards, such as Parlay/OSA, JAIN, and OMA. Our view of the network is that open API will be enabled at the core network as well as at the application middleware layers. A key element of the application middleware layer is a set of APIs for content distribution networks. This approach can be viewed as an elaboration of the successful model deployed in the I-mode services of DoCoMo, where third-party content providers sell content by plugging into our network APIs. In the longer term there may be more radical approaches to opening up the network including programmable APIs that would enable even further service differentiation for carriers.

The sophistication of the mobile terminals enables better end-user experience by caching services on the handset. For example, when a network connection is lost, sufficient data may be stored on the mobile terminal for the user to continue to access the application. To this end, new application-structuring techniques, such as Mervlets and replets, are proposed.

Next Generation Mobile Systems. Edited by Dr. M. Etoh
© 2005 John Wiley & Sons, Ltd

As the networks are opening up, so are the terminals, but in a slightly different manner. In particular, XML-based web services enable application developers to use many existing services over the Internet to quickly create new applications. Since mobile terminals are constrained with respect to memory and processor power, new versions of XML processing had to be developed. The most promising approach in this area is kXML, kSOAP, and kUDDI. These approaches to XML processing are specifically designed for mobile terminals. As XML web services evolve in the enterprise computing arena with work flow management techniques, so will their mobile terminal counterparts, thus increasing the classes of applications that can quickly be enabled on XG mobile networks.

With 3G networks we are already seeing new applications such as video conferencing. With next-generation networks, the trend toward the use and acceptance of multimedia applications will accelerate with applications such as mobile TV (television channels from the mobile phone) becoming a reality. The focus of a lot of the work in media technologies is codec development for speech, audio, and video. The future of this work will progress in a variety of directions including the unification of codecs for speech and audio. In video, new coding techniques, besides Huffman coding, will need to be developed to make progress beyond the current state of the art. For applications such as mobile TV to become reality, we will need to stream data in real time over the air to mobile terminals.

7

Terminal Software Platform Technologies

Manuel Roman, Dong Zhou, Nayeem Islam

7.1   Introduction

The functionality of mobile terminals has evolved tremendously over the last 10 years. Initially, there was just voice transmission. Then, short message service (SMS) and web browsing (WAP and i-mode) were added. Later, interactions with vending machines (c-mode) and multimedia messaging services (MMS) became available. Most recently, video conferencing and interaction with the surrounding physical environment (i-area) became possible. This trend indicates a clear evolution toward Machine-to-machine communication, which requires a sophisticated software infrastructure running in the mobile terminals.

The evolution of mobile terminals and wireless-enabled handheld devices as well as the increasing proliferation of wireless networks are changing our traditional understanding of computers. The notion of desktop computing is slowly evolving into a more dynamic model. Handheld computers do not sit on a desktop, are not disconnected from the surrounding environment, and are not immobile anymore. These devices are capable of connecting to wireless networks, they have enough processing power to perform tasks previously reserved for servers and workstations, and they are carried by users on a regular basis. They are digital companions that operate in our own context and assist us with everyday tasks.

DoCoMo envisions a future in which handheld devices play a central role in users' activities. These devices will provide functionality to access and configure users' digital resources (applications and data), to monitor and leverage resources present in the user surroundings, and to store and manage users' confidential information. On the basis of the

Next Generation Mobile Systems.   Edited by Dr. M. Etoh
© 2005 John Wiley & Sons, Ltd

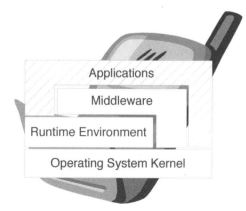

Figure 7.1 Terminal software platform building blocks

current utilization trend, phones are poised to replace our keys, identification cards, and money with digital counterparts. Some existing research projects and commercial applications have demonstrated that the real potential of cell phones and handheld devices is their ability to interact with remote services and applications. Considering handheld devices as standalone units disconnected from the network hinders their true potential.

7.1.1 Generic Terminal Software Platform

The evolution of cell phone functionality is the result of the sophistication of the supporting infrastructure running in the phones. Figure 7.1 illustrates the generic terminal software platform that includes the building blocks shared by most existing approaches. These building blocks are operating system kernel, runtime environment, middleware, and applications.

The *operating system kernel* is the software responsible for managing, exporting, and arbitrating the hardware resources provided by the terminals. The operating system kernel is a vital component that hides the underlying hardware complexity and heterogeneity and enables the construction of software. All remaining building blocks depend on the operating system kernel. Operating system kernels for terminals have matured significantly over the last years, evolving from simple hardware monitors to sophisticated designs, similar to existing desktops' operating systems. However, unlike existing desktop operating system kernels, mobile terminals impose restrictions on their operating system kernels, including low memory footprint, low dynamic memory usage, efficient power-management framework, real-time support for telephony and communication protocols, and reliability.

The *runtime environment* provides a safe and managed execution environment and contributes to the development of portable applications. Most existing runtime environments are based on a sandbox model that prevents applications from accessing unauthorized resources. Applications that execute in the runtime environment are coded in an intermediate language (for example, Java byte code or Microsoft Intermediate Language (MSIL)) and are interpreted or compiled on demand before execution. This execution model isolates applications from the rest of the software infrastructure, and therefore protects the terminal platform software from malicious or faulty applications. Furthermore, the use of an intermediate language allows execution of the same applications in heterogeneous environments. Only the runtime environment has to be customized for each device.

Middleware comprises a collection of services that provide functionality to simplify the construction of applications that execute across heterogeneous platforms. Middleware includes support for remote communication (for example, remote procedure calls and object request brokers), synchronization, caching, event notification, multimedia, security, and graphics. As illustrated in Figure 7.1, middleware can leverage the execution runtime or it can interact directly with the operating system for improved performance.

Finally, the *applications* layer includes user software that leverages the previously mentioned building blocks. The number of applications is growing exponentially and different applications have different requirements. For example, distributed applications require interaction and coordination with remote applications and, therefore, leverage the middleware services. Some applications require the runtime environment to execute, while other applications run natively and interact directly with the operating system kernel.

7.1.2 Terminal Software Platform Evolution

In this section, we present the evolution of the terminal software platform. We define four key features that correspond to the functional blocks presented in Section 7.1.1 and use a timeline based on phone generations to point out the key differences and additions. The features we address are operating system, runtime environment, middleware infrastructure, and security. The phone generations we cover are 1G, 2G, 2.5G, 3G, and the future XG. Although the boundaries between generations are sometimes blurred, the overall information provided here is accurate. Figure 7.2 illustrates the evolution of the terminal software platform.

1G and 2G phones correspond to the beginning of the era of cellular phones. The difference between 1G and 2G is the upgrade from analog to digital networks; however, phone functionality was identical and mostly voice oriented. Strictly speaking, phones did not have an OS as we now have, that is, a separate component running in privileged mode and supporting general-purpose applications. Applications and OS were packaged in a single binary image, customized to the hardware configuration of each phone. Furthermore, there was no runtime environment or middleware services. Users were not allowed to install software on the phones. Finally, regarding security, GSM phones were equipped (an still are) with a subscribers identification module (SIM) card that protects phone in case of theft.

2.5G/3G phones represent a major step forward in terms of data connectivity and functionality they provide. These phones are equipped with a real OS that exports functionality including thread, process, and memory management, scheduling, multimedia, and graphic acceleration. Furthermore, these phones allow users to download applications at runtime, and therefore have a runtime infrastructure that provides a safe execution environment for these applications. The runtime ensures that malicious applications cannot disrupt the basic phone functionality. These phones also provide middleware services to support personal information management applications, data synchronization, messaging, and secure transactions. Finally, most 2.5G/3G phones provide VPN, secure sockets, and encryption/decryption libraries to support the development of secure applications. The most noticeable feature with

Features

	1G / 2G	2.5G / 3G	XG
Security	For GSM phones: SIM (Subscriber Identification Module)	Safe application downloading, VPN, encryption/decryption, secure sockets	Hardware security attachments and standards (Trusted Computing Group) Secure data in device, transmitted data, and user identity
Middleware Infrastructure	Non existing	Data synchronization, Personal Information Management, Messaging	Distributed middleware and ubiquitous computing services: discovery, synchronization, mobility, publish/subscriber, etc. Runtime configurable and upgradeable.
Runtime Environment	Non existing	Support for safe user-level application execution. (Most popular: Java runtime)	Faster execution time, memory usage efficient, real time support.
Operating System	Non existing as such. Phone software was a single image including OS and application code.	OS becomes a basic component of the phone (kernel and user mode). Basic OS support: processes, threads, memory management, scheduling, exception handling. Multimedia support, real time communication services, graphic libraries, real time communication services	Runtime upgradeable (patches), configurable, and adaptable. Power efficient, Secure, and Fault Tolerant.
Phone Generation	1G / 2G	2.5G / 3G	XG

Figure 7.2 Terminal software platform evolution

2.5G/3G phones is the significant increase in data communication. In Japan, for example, the number of e-mails sent and received from phones is larger than the number sent and received from PCs.

XG phones will be clearly oriented toward data transmission. Their software infrastructure will evolve to provide advanced functionality in terms of OS, runtime infrastructure, middleware, and security. The operating system will be dynamically configurable and upgradeable. Furthermore, it will be highly optimized to reduce energy consumption, fault tolerant, and secure. The runtime environment will provide faster execution times and support for real-time applications. The middleware infrastructure will provide advanced services to enable seamless distributed computing and will be the basis for future ubiquitous computing. Finally, XG devices will provide hardware attachments to protect the data stored in the device in case of theft or unauthorized remote access. Also, these hardware attachments will protect the identity of the user, so it is not possible to use the device to impersonate the owner.

In this chapter, we describe the topics presented in Figure 7.2 regarding to 2.5G/3G and XG phones. We will discuss advanced operating systems (Section 7.3), runtime environments (Section 7.4), security (Section 7.5), and future research directions in terms of middleware (Sections 7.5 and 7.6).

7.2 Existing Terminal Software Platforms

In this section is a summary of the four most popular terminal software platforms: Symbian OS, Palm OS, Windows CE .NET OS, and Qualcomm BREW. Note that except BREW, the other three platforms follow the architecture presented in Figure 7.1. Note also that Symbian, Palm, and Windows use the term OS to denote the whole terminal software platform.

7.2.1 Symbian OS

Symbian OS[1] is licensed to a large number of handset manufacturers, which account for over 80% of annual worldwide mobile phone sales (Symbian 2003). The latest version of the OS, at the time of writing, is 8. This is the first version that provides a real-time OS kernel, and supports the following features:

- Rich suite of application services, including services for contacts, schedule, messaging, browsing, and system control

- Java support

- Real time

- Hardware support, including different CPUs, peripherals, and memory types

- Messaging with support for MMS, EMS, SMS, POP3, IMAP4, SMTP, and MHTML

- Multimedia, including image support, as well as video and audio streaming

- Graphics with a graphic accelerator API

- Mobile telephony, with support for most existing carriers and ready for 3G networks

- International support

- Data synchronization

- Device management/ Over the Air (OTA) provisioning

- Security

- Wireless connectivity, including Bluetooth and 802.11b.

Figure 7.3 depicts the architecture of Symbian OS, which includes six functional layers: kernel hardware and integration, base services, OS services, Java (runtime environment), application services, and UI framework. According to Figure 7.1, the first three layers (kernel hardware and integration, base services, and OS services) correspond to the OS kernel, Java corresponds to the runtime environment, application services corresponds to the middleware layer, and the UI Framework can be considered as part of the applications building block.

---

[1] See http://www.symbian.com for information about Symbion OS.

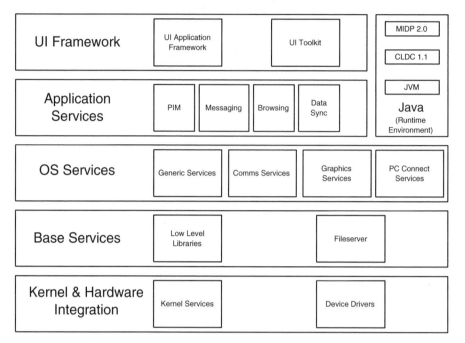

Figure 7.3 Symbian terminal software platform

The kernel and hardware integration layer includes the core kernel services (for example, memory management, process management, and power management) as well as the device drivers and kernel extensions. This layer simplifies portability across multiple platforms.

The base services layer consists of a collection of low-level libraries that export the OS kernel interfaces. These interfaces support the development of the remaining software components. Furthermore, this layer provides the file server, which exports functionality to manage the device's file system.

The OS services layer provides a collection of subsystems with functionality organized into four major categories: generic services (for example, database and security), communication services (for example, TCP/IP, PPP, and FTP), graphics services, and PC connection services (for example, synchronization).

The application services provides middleware functionality to manage user data. Symbian provides an API that allows programmers leverage the functionality and incorporate it into their applications. These services are personal information management, messaging, browsing, and data synchronization.

The runtime environment provides support for Java, more specifically for the CLDC 1.1 profile and MIDP2.0 configuration. See Section 7.3.1 for more information about the Java runtime.

The UI framework provides support to customize the user interfaces of applications. Customization is an important requirement for licensees and partners to differentiate their applications from those of the rest of competitors.

7.2.2 Palm OS

Palm OS is the popular software platform for PDAs, which has been extended with functionality for telephony. The latest version of the OS at the time of writing the book is 5, which provides the following features[2]:

- Multimedia, including high resolution display, video, and audio

- Wireless connectivity, including 802.11b, Bluetooth, GSM, CDMA, and 2.5G and 3G networks

- Security with different encryption algorithms and SSL

- Built-in support for ARM processors

- PIM programs built in

- Large collection of software and one of the largest software development communities

- PC synchronization.

Compared to Windows CE .NET and Symbian OS, Palm OS v5.0 is probably the less-sophisticated terminal software platform, both in terms of OS design and programming features. However, the simplicity of usability and programmability make it a popular terminal software platform (especially for PDAs where it has the largest market share). Palm is already working on a newer OS version (Palm OS v6.0) that addresses these issues.

7.2.3 Windows CE .NET

Finally, the Windows CE .NET platform is a new version of the popular Windows OS specifically customized to embedded devices. Windows CE .NET is part of the Windows Mobile initiative, which includes a built-in bundle providing PIM functionality, e-mail, and browsing capabilities. The operating system provides the following features[3]:

- Support for small-footprint optimization

- Hard real-time kernel

- Robust memory management

- Advanced power management

- Open communications platform (such as TCP/IP, IPv6, and OBEX)

- Remote and systems manageability (SNMP v2 Client, Device management client)

- Standards support (such as UPnP, Bluetooth, XML and SOAP, and USB)

- Extensive storage and file systems

---

[2] See http://www.palmos.com/ for more information about the Palm OS.

[3] See http://msdn.microsoft.com/embedded/prodinfo/prodoverview/ce.net/ for information about Windows CE .NET.

- Purpose-built server services

  - Core server support
  - File transfer protocol (FTP) server
  - Remote access/point-to-point tunneling protocol (PPTP) server
  - File and print server support

- Security.

Figure 7.4 illustrates the five terminal software components: hardware abstraction layer (HAL), kernel, OS services, .NET compact framework (runtime environment), and applications and services development. The first three components (HAL, kernel, and OS services) correspond to the OS kernel defined in Figure 7.1, the .NET compact framework corresponds to the runtime environment, and the applications and services development component correspond to the middleware layer. The application layer is not depicted in Figure 7.4.

The HAL provides functionality to simplify the portability of the OS across heterogeneous devices. This layer includes the memory map, interrupt management, and a bus map.

The kernel provides the core OS kernel functionality, including virtual memory management, exception handling, process and thread management, scheduling, executable loading, synchronization, initialization, process switching, and memory mapped file management.

OS services provide several key OS components with functionality, such as communication and networking support, multimedia, graphics, device management, and object store and registry (persistent and nonpersistent data manager). Windows provides tools that allow developers to customize the OS services they want to deploy as part of the operating system kernel.

The applications and services development provides middleware services as well as frameworks to assist in the development of applications. This component provides services to access

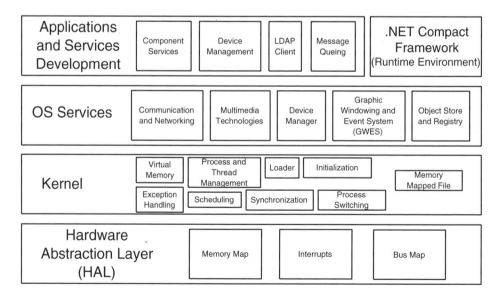

Figure 7.4 Windows CE .NET terminal software platform

directory services (Lightweight Directory Access Protocol (LDAP) client) and leverages message queuing, as well as component frameworks to standardize application development.

Finally, the .NET Compact Framework is a runtime environment developed by Microsoft that supports the development of safe, efficient, and portable applications. Section 7.3.2 provides more details about the .NET framework.

7.2.4 QUALCOMM BREW

Binary Runtime Environment for Wireless (BREW)[4] is a platform from QUALCOMM that offers a method to run software applications on a mobile device. Although BREW it is defined as a runtime environment, do not get confused with our previous runtime environment definition (for example, Java or .NET). BREW is a full terminal software platform.

BREW is a complete solution that includes both technical and business elements. It provides native support for C and C++ and can be extended with additional runtime systems, such as the Java Virtual Machine (JVM), thereby supporting additional languages. BREW supports OTA (over the air) application downloading and application management and provides a set of personal information management applications, such as calendar, contacts, e-mail, and instant messaging. BREW also provides support libraries for multimedia applications that are optimized to the device's hardware. Finally, BREW provides a unified billing mechanism.

BREW does not follow the generic terminal software platform (depicted in Figure 7.1) and therefore is different from the three previous platforms (Symbian, Palm, and Microsoft). BREW advocates for a model that relies on an execution environment that interacts directly with the chip and provides a common set of interfaces. Nevertheless, BREW still provides common functionality, such as memory and process management, though it is encapsulated in a monolithic image.

Figure 7.5 depicts BREW's architecture. At the lowest level, there is the ASIC software. The BREW platform runs directly on top of the chipset and provides support to run native applications, such as e-mail, photo sharing, position location, and push to talk. BREW allows installing extensions that support the execution of additional applications, such as browsers, video players, and Java software (Java VM extension).

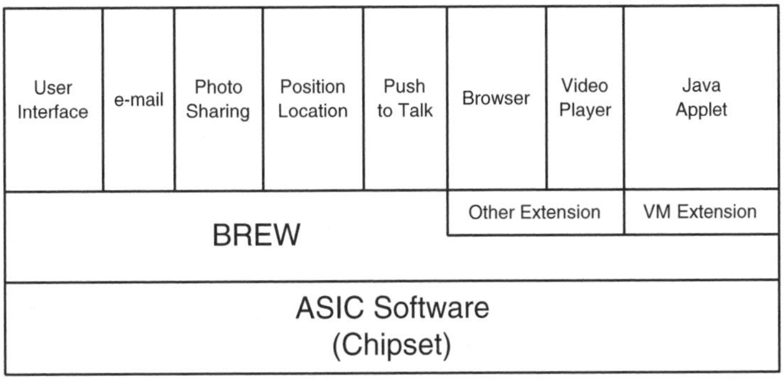

Figure 7.5 BREW client architecture

---

[4]See http://www.qualcomm.com/brew/ for detail about BREW.

7.2.5 Software Platform Comparison

In this section, we compare Symbian OS, Palm OS, and Windows CE platforms .NET. BREW is not discussed here, because its approach and infrastructure are different from the previous three OSs. In order to compare the three systems, it is important to understand the evolution of each OS. Symbian OS is specifically customized to smartphones, that is, cellular phones that provide advanced functionality, such as calendar, tasks, and e-mail. Palm OS was developed for the Palm personal digital assistant devices from scratch, taking into account the requirements and limitations of PDAs, and has been widely accepted, which is proved by the extensive amount of existing software. Finally, Windows CE was originally built for handheld devices, provides a look and feel similar to the desktop counterpart, and provides functionality to synchronize personal data and e-mails with the PC applications. Both Palm OS and Windows CE have been extended with telephony services, which are required for mobile terminals.

Symbian OS was the first to provide advanced functionality for cell phones and it is widely used, especially in Europe. Palm OS and Windows CE .NET have evolved from the PDA world and therefore provide an extensive collection of software titles.

Windows CE .NET provides a powerful programming environment that includes threads, memory management, and networking. Furthermore, the Windows CE API is compatible with existing Windows operating systems, such as Windows XP. As a result, porting applications across operating systems is simple. Windows CE .NET provides support for remote device management, including remote application configuration (Windows Mobile Start Service). One of the key strengths of Windows CE .NET is its seamless integration with existing PC-based productivity applications, such as Microsoft Outlook.

At the time of writing this book, Cobalt (a new version of Palm OS) is about to be released. Cobalt provides a significant number of improvements, including multithreading and multitasking, graphic acceleration support, extensible multimedia framework, and security.

As part of the future functionality, the three operating systems (Symbian, Windows, and Palm) will provide advanced multimedia frameworks, advanced security features, configurable infrastructures, faster execution times, and low power consumption.

7.3 Runtime Environments

Runtime environments are virtual machines that leverage the sandboxing model to restrict resource utilization and provide execution security guarantees. Mobile terminals download third-party applications – potentially malicious – and execute them locally. Therefore, it is important to ensure that application execution does not compromise the security of or crash the device. The two best-known runtime environments are Sun's JVM and Microsoft's .NET. We describe the two runtime infrastructures next.

7.3.1 Sun Java

The most popular and deployed runtime environment is J2ME (Java 2 Micro Edition (Microsystems n.d.)), which is the Java platform for consumer and embedded devices such as mobile phones, PDAs, TV set-top boxes, and in-vehicle telematics systems. J2ME (Figure 7.6) provides Java functionality to embedded devices and includes a flexible user

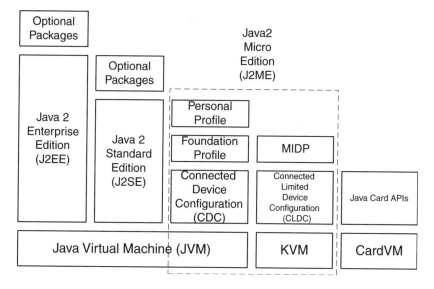

Figure 7.6 Java 2 platform

interface, a security model, a collection of built-in network protocols, and support for networked and disconnected applications. The architecture of J2ME consists of three layers: Java Virtual Machine (JVM), configuration, and profiles. The virtual machine sits on top of the hosting operating system; configuration sits on top of the virtual machine and defines the Java language and virtual machine features; profile is the topmost layer and addresses the specific demands of a certain vertical market segment or device family. J2ME provides a subset of the functionality offered by J2EE and J2SE, and it is customized to devices with limited resources.

The two most popular J2ME configurations are (Connected Limited Device Configuration) (CLDC) and CDC (Connected Device Configuration) (Lampsal n.d.). CLDC is the smaller of the two configurations, designed for devices with intermittent network connection, slow processors, and limited memory (CLDC requires at least 160 KB). CDC is designed for devices that have more memory (CDC requires 2 MB), faster processors, and greater bandwidth (for example, high-end PDAs). As illustrated in Figure 7.6, CDC uses the same JVM as J2EE and J2SE, while CLDC relies on a virtual machine (KVM) customized to devices with limited resources.

The JVM for CLDC (KVM) is compliant with JVM as specified by J2SE but does not include the following features:

- Support for Java Native Interface.

- Support for user-defined Java class loaders.

- Support for reflection.

- Support for thread groups and daemon threads.

- Weak references.

- Floating point support.

- Object finalization (invoking the finalize() method for cleaning up before garbage collection).

- Error handling (provides limited features).

- Standard class file verification, which is a highly expensive operation. CLDC relies on an external class verifier running on a server to ensure the class files are correct.

The JVM for CDC is called *compact virtual machine* (CVM). It is a virtual machine for devices with limited memory availability, which reduces the memory size of the libraries in 40%. The CVM provides additional features:

- Memory system (for example, small garbage collector, pluggable garbage collectors, and full separation of the virtual machine from the memory system)

- Portability

- Fast synchronization

- ROMable classes

- Native thread support (JNI support)

- Latest JVM support

- Stack usage (static analysis of the virtual machine code)

- Startup and shutdown

CDC provides the minimum required libraries from J2SE to support the JVM. Furthermore, CDC is a superset of CLDC and therefore supports all CLDC libraries as well.

At the time of writing this book, CLDC has only one profile customized to mobile information devices (PDAs and mobile phones): Mobile Information Profile Device (MIDP). MIDP defines properties specific to mobile information devices that extend or redefine certain aspects of the CLDC configuration. MIDP addresses the following issues:

- User interface support

- Event handling

- High-level application model

- Persistence support.

Finally, CDC defines three profiles: Foundation Profile, Personal Basis Profile, and Personal Profile. The foundation profile provides a J2SE class library and does not support GUIs. The personal basis profile, provides lightweight component support, support for Xlets, and supports the foundation profile APIs. Finally, the personal profile supports AWT, applets, and the APIs of the previous two profiles.

7.3.2 Microsoft .NET Compact Framework

Microsoft provides a runtime infrastructure called .NET compact framework, which provides a subset of the .NET framework functionality customized to mobile terminals. .NET compact framework is based on an intermediate language that is compiled on demand before execution. Figure 7.7 depicts the .NET framework. .NET compact framework is similar to the standard framework but removes certain services and libraries that are not considered essential for mobile devices, thereby reducing its size and resource requirements, and making it appropriate for mobile resource constrained devices.

The Common Language Runtime (CLR) provides a secure and robust execution environment. It provides functionality for memory management, multiple language support, exception management, threading, inheritance, and security. The CLR is composed of two components: the Common Type System (CTS), and the Execution Engine (EE). The CTS defines the types supported by the execution environment (everything is an object) and specifies rules for compound types, such as class, struct, enum, and interface. The EE compiles MSIL into native code before execution (just-in-time compilation), handles garbage collection, handles exceptions, enforces code access security (sandbox), and handles verification (managed versus unmanaged).

The Base Class Library (BCL) is similar to Java's system namespace, and it is used by all .NET applications. It provides support for IO, threading, database, text, graphics, console, sockets/web/mail, security, cryptography, COM, reflection, and assembly generation. The Data and XML component illustrated in Figure 7.7 provides additional support for database management and XML manipulation. The Web Services and User Interface components provide support to run and interact with web services and support for user interface management.

The .NET framework defines a common language specification (CLS) that is supported by all .NET languages. It is very similar to CORBA IDL (interface description language)

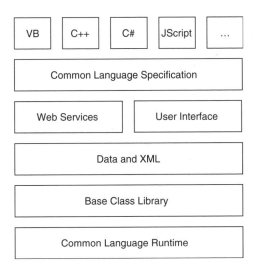

Figure 7.7 The .NET framework

types. CLS allows programmers to create software composed of component written in different languages.

7.3.3 Benefits of Runtime Environments for Mobile Handsets

Mobile handsets are highly heterogeneous in terms of hardware resources. As a result, traditional application development is a challenge that requires developers to take into account different hardware configurations. Java 2 Micro Edition and .NET Compact Framework allow developers to write applications that can be used in multiple devices. Only the runtime environment has to be customized to each hardware platform. This feature, combined with the ability to ensure security policies that restrict application access to hardware resources, makes J2ME and .NET Compact Framework a key component in the development of applications for mobile terminals. At the time of writing this book, J2ME is the most-used runtime infrastructure for mobile terminals. It is installed in millions of devices and allows users to download applications at runtime securely.

The original vision of Java and .NET was that of writing applications once and running them everywhere. The runtime is customized to different devices but the original code is unique. However, the size and resource requirements of these runtimes prevent most terminals from hosting them (although terminals' hardware features are improving fast). As a result, Java and .NET are modified so they can be executed in resource-limited devices. J2ME and .NET compact framework are the modified versions of J2EE/J2SE and .NET respectively, which remove features considered not essential for mobile terminals, and therefore require less resources to execute.

J2ME MIDP, for example, does not provide support for AWT and SWING graphic libraries. These libraries are large, and because of device heterogeneity, it is hard to support them. The MIDP profile defines a set of widgets that are natively mapped to each mobile terminal, thereby providing a common minimum set of GUI widgets. Furthermore, another key difference between J2EE/J2SE and J2ME is the lack of a *just-in-time* compiler, which compiles java bytecodes before executing them, generates native code, and thereby provides faster execution times.

One of the key differences between .NET compact framework and standard .NET is the lack of remoting functionality. Remoting is a framework that supports remote interaction. It is a fully customizable framework that can be customized to support different protocols. Furthermore, changes in the framework do not require any change in the applications. The .NET compact framework does not provide remoting, it only provides support for web service invocation. The remoting functionality requires significant resources (mostly in terms of memory) that are not available in mobile terminals.

7.4 Terminal Software Platform Security: Trusted Computing Group

The evolution of the terminal software platform gives developers the flexibility to create innovative software and the ability to customize the terminal in different ways. However, this flexibility brings many security concerns that must be addressed appropriately. As the number of APIs increases, so does the vulnerability of the software and hardware. The terminal software platform must provide mechanisms to enforce security and provide guarantees regarding functional integrity, privacy, and individual rights.

The Trusted Computing Group (TCG) (n.d.) is a not-for-profit organization that aims at defining open standards for hardware-enabled trusted computing and security technologies, including hardware building blocks and software interfaces, across multiple platforms, peripherals, and devices. Their primary goal is to help users protect their information assets (for example, data, passwords, and keys) from compromise due to external software attack and physical theft. The TCG has become the successor of the Trusted Computing Platform Alliance (TCPA) and has adopted and will extend their specifications. The TCG has over 50 members including AMD, HP, Intel, Microsoft, Sony, and Sun Microsystems.

TCG specifications address the increasing threat of software attack due to the sophistication of attack tools, vulnerabilities discovered in existing systems (mostly due to the complexity of these systems), and the mobility of users that increases the probability of digital and physical theft. According to the CERT (CERT Coordination Center) at the Software Engineering Institute operated by Carnegie Mellon University, the number of vulnerabilities has increased from 500 in 1999 to approximately 4100 in 2002, while the number of incidents has increased from 1000 in 1999 to approximately 80,000 in 2002. The increasing threat of software attack poses a risk for data stored on devices. TCG defines three types of theft:

1. Personal or enterprise data

2. Identity, which can be used to impersonate a user and access unauthorized data

3. Physical device.

7.4.1 TCG Specifications Overview

Figure 7.8 illustrates the TCG documentation roadmap, which includes an architectural overview, the main specification (which defines the trusted platform module), the trusted software stack specification, the platform-specific design guide, and a collection of platform-specific specifications (for example, PC and PDA).

Trusted Platform Module (TPM)

The Trusted Platform Module (TPM) specification is included in the TCG main specification. It is a hardware component that provides the following functionality (*Backgrounder* 2003):

- Asymmetric key functions for on-chip key pair generation using a hardware random number generator; private key signatures; and public key encryption and private key decryption of keys enable more secure storage of files and digital secrets.

- Secure storage of hash values representing platform configuration information in Platform Control Registers (PCRs) and secure reporting of these values, as authorized by the platform owner, in order to enable verifiable attestation of the platform configuration based on the chain of trust used in creating the HASH values. This includes creation of Attestation Identity Keys (AIKs) that cannot be used unless a PCR value is the same as it was when the AIK was created.

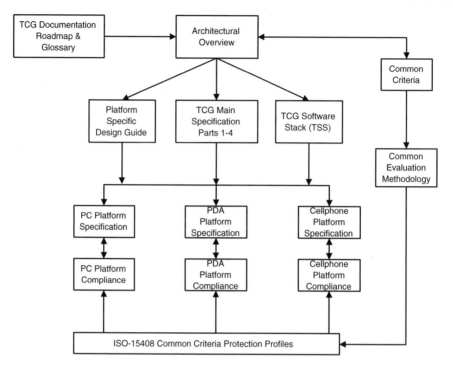

Figure 7.8 Trusted computing group document roadmap

- An Endorsement Key that can be used by an owner to anonymously establish that identity keys were generated in a TPM, thus enabling confirmation of the quality of the key without identifying which TPM generated the identity key.

- Initialization and management functions that allow the owner to turn functionality on and off, reset the chip, and take ownership, with strong controls to protect privacy.

TCG Software Stack (TSS)

The TCG Software Stack (TSS) provides a standard software interface for accessing the functionality of the TPM. This interface can be used to support existing interfaces, such as Microsoft's Crypto API (CAPI), the Common Data Security Architecture (CDSA), and Public Key Cryptography Standards #11(PKCS).

Platform-specific Design Guide

The TPM specification is platform independent. The platform-specific design guide ensures compatibility among implementations within each computing architecture. It includes the definition of the following components:

- The Core Root of Trust Management (CRTM)

- Trusted Building Block (TBB)

The TBB defines how the CRTM and the TPM are connected to a platform. For example, according to the TCG PC platform specific implementation, the CRTM is the BIOS or the BIOS boot block. The BIOS must load the hashes of preboot info into the TPM's PCRs, thus establishing an anchor for the chain of trust and the basis for platform integrity metrics.

Common Criteria Protection Profiles

These profiles refer back to the TCG specifications and are used to judge conformance with standard security properties and principles. The profiles provide certain security requirements, for example, environment, threats, objectives, and evaluation assurance level for platform subsystems (for example, TPM and TBB). Vendors or manufacturers can create a *security target* (ST) that describes their evaluated product or target of evaluation (TOE) that describes how these requirements are met and have this independently verified by a common criteria lab.

7.4.2 Trusted Computing Group and Mobile Devices

TCG is working on the specification of profiles for mobile devices, including cellular phones and PDAs. These profiles define the key building blocks to ensure a safe execution model that prevents unauthorized use of mobile devices, or unauthorized access to data stored in the devices. This work specifies the two components defined by the platform-specific design guide: the CRTM and the TBB. Furthermore, the TBB defines the interaction between the CRTM and the TPM components present in the mobile devices.

7.5 Terminal Software Platform Management: Over the Air Provisioning

As the functionality exported by the software platform increases, so does the complexity to configure it, personalize it, and customize it. Furthermore, increased functionality affects the size of the code and, therefore, the number of software errors. Configuration, personalization, and customization are complex and tedious tasks that cannot be given to the terminals' owners. The goal of these advanced software platforms is to provide new functionality to users, maximize their satisfaction, and avoid or minimize possible conflicts. Furthermore, software errors should be hidden from users and should be addressed by terminal manufacturers on the fly.

7.5.1 Open Mobile Alliance

The Open Mobile Alliance (OMA) group is a consortium with near 200 members, including mobile operators, device and network suppliers, information technology companies, and content and service providers. The mission of the OMA is to "facilitate global user adoption of mobile data services by specifying market-driven mobile service enablers that ensure service interoperability across devices, geographies, service providers, operators, and networks, while allowing businesses to compete through innovation and differentiation." (http://www.openmobilealliance.org). As of April of 2004, OMA has 17 working groups and committees, covering aspects from architecture, to data synchronization, and security.

The focus of this section is on one of these working groups: the Device Management Working Group (DMWG).

The goal of the DMWG is to specify protocols and mechanisms that achieve management of devices. Management includes setting initial configuration information in devices, subsequent installation and updates of persistent information in devices, retrieval of management information from devices, and processing events and alarms generated by devices. In the scope of device management, information includes (but is not limited to) configuration settings, operating parameters, software installation and parameters, software and firmware updates, application settings, and user preferences. Device management is a basic requirement for terminal software platforms.

The DMWG adopted the specifications on device management from the SyncML Device Management (DM) group and continues its work. SyncML DM leverages the SyncML Data Synchronization client software investment. SyncML DM is compact, efficient, designed with wireless applications in mind, extensible, and based on standards. The SyncML DM protocol supports the execution of management commands on management objects, which are the unit of management specified by SyncML DM. An example of a management object is a set of configuration parameters for a device, which accepts reading and writing parameter keys and values. Another example of management object is the runtime environment for software applications, which accepts actions, such as installing, upgrading, or uninstalling software.

As illustrated in Figure 7.9, the SyncML DM protocols consists of two parts:

1. Setup phase, which addresses authentication and device information exchange

2. Management phase, which includes interaction with management objects.

7.5.2 Over the Air Software Updating

Over the air software updating refers to the ability to modify the software running on a device. This modification includes replacement of software components (updates), as well as addition of new functionality to existing software (upgrades). OMA's DMWG does not

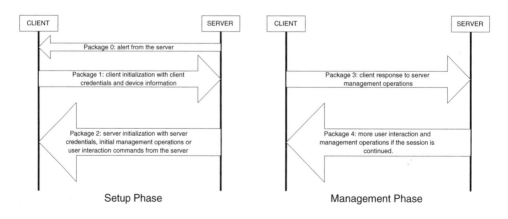

Figure 7.9 SyncML DM protocol phases

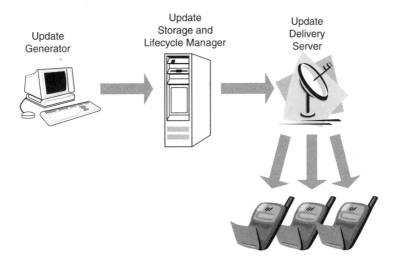

Figure 7.10 OTA software updating

support this functionality yet. However, they specifically refer to it in their documents (software and firmware updates). In this section, we present three companies that support terminal firmware upgrades.

The three companies that provide OTA software upgrading capabilities are BitFone (Bitfone n.d.), Redbend (Red Bend n.d.), and DoOnGo (*Intelligent wireless software manager* n.d.). Although they have proprietary solutions, the basic updating method is similar. Their approach replaces the binary image of mobile terminals over the air. They generate a new firmware image that fixes certain software problems or adds new functionality, calculate the binary differences with the original firmware image, and send the difference to the terminals. At the terminal side, there is a software component that receives the binary delta and replaces the existing firmware image, prior to user authorization.

Figure 7.10 illustrates the process. The update generator creates an update package with the binary delta between the original and the new software image. The update storage and lifecycle manager stores the existing updates and manages their lifecycle according to whether the updates have expired or are still required. The update delivery server receives the software updates and delivers them to the mobile terminals. Finally, the client update manager interacts with the update delivery server to download the updates, and implements a fault-tolerant algorithm to ensure failure-free software update.

7.6 Research Directions

Mobile terminals will play an important role in future computing environments. As the surrounding environment becomes digitally augmented, users need mechanisms to interact with it. Terminals become portals (Banavar et al. 2000) to the digital environment, bridging the gap between the physical and the digital world. The terminal software platform is a key component to enable XG terminals as advanced digital portals or advanced digital assistants.

However, while the technological components presented in previous sections are essential, they are not sufficient. Research is required to understand the computing paradigm of future computing environments and provide the appropriate services and technologies that will support it. As part of our research at DoCoMo Labs USA, we are studying future middleware services that will allow users to interact with distributed services seamlessly. These services rely on the functionality provided by the operating systems and runtime environments, and leverage security infrastructures, such as the one defined by the TCG. We are working on extending the basic middleware services provided by existing software platforms (synchronization, personal information management, and security) with functionality that will enable interaction with digitally augmented environments. Our goal is to provide the right supporting technologies and contribute to the evolution of the terminal software platform. We present a summary of our work.

7.6.1 Terminal Middleware Services

Integrating handheld devices into distributed systems requires functionality that allows these devices to interact with remote resources and allows remote resources access the resources the handheld exports. As a result, handheld devices become peers (clients and servers) of distributed environments. This integration requires a collection of middleware services for the handheld devices that take into account the challenges associated with these devices and provide functionality that adheres to their particular utilization pattern.

Figure 7.11 illustrates a collection of middleware services required to enable proactive interaction between handsets and external resources (such as printers, vending machines, displays, and handsets), handsets and ubiquitous computing environments (such as offices, airports, and homes), and handsets and web services (such as mapping services, location services, and business related services). These services are discovery, remote interaction, mobility, and disconnection. Discovery provides the functionality to advertise and find resources of interest. Remote interaction allows the software running in the handset to

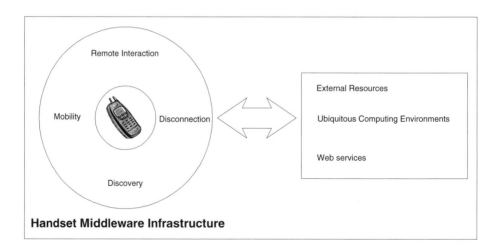

Figure 7.11 Middleware services for next-generation handsets

send and receive requests to and from remote resources. Mobility supports the migration of services and applications across devices. Disconnection enables the use of the handset software when there is no network connection available. The following sections describe the functionality provided by these services along with some related existing projects.

Discovery

There are two elements to discovery protocols: lookup and discovery (McGrath and Mickunas 2000). Lookup is passive and assumes the existence of a directory service that clients use to find appropriate resources on the basis of a query language. Discovery is spontaneous and assumes that the different entities in the system (such as clients, services, and directories) can find each other automatically. Services send information about their capabilities periodically, and clients and directories use these periodic broadcasts to find suitable services or to store information about existing services.

Consider the following example of discovery service. A user walks into a train station; the users' phone detects local services and notifies the user about a service to buy the ticket electronically. The phone also uses the discovery service to advertise a service to the train station that manages user's personal and banking information. The train station discovers the service and automatically initiates the process to sell the ticket by requesting permission to the user to initiate the transaction.

As an example of lookup, consider a user at home using the phone to interact with a lookup service to browse for services to control the lighting and music properties.

Existing services such as the LDAP (Wahl et al. 2000) and the Common Object Request Broker Architecture's (CORBA) Trader Service (Henning and Vinosky 1999) are examples of lookup services. They store a collection of entries that clients browse to retrieve information they require to complete their tasks. LDAP is a hierarchical directory service that defines a network protocol for accessing the information in the directory, a network information model defining the form and character of the information, a namespace on how information is referenced and organized, and an emerging distributed operation model detailing how data should be distributed and referenced. Furthermore, the protocol and the information model are both extensible. LDAP entries consist of attributes. Each attribute has a type and one or more values. Clients use the attribute values to resolve entries (Hodges 1997). CORBA's Trader Service stores entries consisting of attributes and a reference to a server object. Clients resolve entries by sending queries. These queries consist of attribute values that the trader uses to match the entries it stores.

Salutation[5], UPnP (Microsoft 2003a), SLP (Guttman et al. 1999), and Jini (Edwards 1999) are examples of currently used discovery services. They use multicast to advertise and discover services. Unlike lookup services (such as LDAP), discovery services do not rely on a centralized service where clients store and retrieve information. Each service running in the system broadcasts information about itself periodically. This model is appropriate for dynamic environments where services come and go frequently. New services automatically announce themselves. Services that are removed fail to broadcast their properties and reference, and therefore they leave the domain cleanly. Although discovery services do not rely on a centralized lookup service, a common approach implemented by services, such as SLP and Jini, provides an intermediate service that listens for broadcasts and maintains an

---

[5]See http://www.salutation.org for the Salutation web page.

internal list of available services. Unlike lookup services, these intermediate services can maintain its contents up to date.

Both lookup and discovery services are vital to users for locating and using remote services. They provide one of the essential components for expanding the capabilities of phones beyond their own resources. Lookup allows users to use their phone to browse and select existing functionality, while discovery notifies users proactively about services that are present remotely and have become available spontaneously.

Remote Interaction

Remote interaction is a fundamental middleware service that enables cell phones to access functionality from remote resources and enables remote resources to access functionality from the phones. A minimum set of functions necessary to enable remote interaction are:

- Remote resource connection establishment (for example, using a resource reference)

- Transmission and reception of data over the network

- Parameter marshaling and demarshaling

- Protocol management (e.g., header coding and decoding)

- Remote resource registration

- Remote resource functionality invocation (the ability to invoke the server's method specified by the client).

The implementation of these functional aspects depends on the specific details of each protocol. Examples of remote interaction services are CORBA (Henning and Vinosky 1999), SOAP (Bequet 2001), Java RMI (Oberg 2001), and Microsoft Remoting (Rammer 2002).

Although a standardization effort is under way, it is unlikely that a single standard will prevail. Different applications, such as real-time applications, require certain guarantees, and certain networks impose properties that affect the transmission assumptions. For example, a GPRS network has a high packet-loss rate and high delay. When connected to GPRS, the network management component reconfigures the underlying protocol with a customized flow-control algorithm. As a second example, when a user interacts with web services running on Internet servers, the communication middleware uses SOAP as the default wire protocol for compatibility issues; however, while interacting with other handsets, the communication middleware configures a customized binary protocol that reduces the size of the messages, therefore improving the throughput.

The remote interaction service described in this section addresses remote procedure call (RPC) functionality, which supports issuing and receiving requests remotely. However, there are additional remote interaction models, such as events and publish/subscribe. Each model is appropriate for different scenarios. For example, events are useful in environments where the number of entities generating information (suppliers) is smaller than the number of entities receiving the information (consumers). This model uses channels, where both consumers and suppliers register. When a supplier has new information, it sends it to the channel, which automatically forwards it to all the consumers. The main benefit of this model is the ability to send data to multiple listeners simultaneously. Furthermore, this model introduces an indirection level that improves fault tolerance. A faulty client can be replaced without affecting the consumers. Furthermore, consumers can join and leave channels without affecting the execution of the client. The second model, publish/subscribe, is a superset of the event model. In fact, the event-based model is referred to as a subject-based publish/subscribe model. Each channel is associated to a specific message type. Generic publish/subscribe systems, or content-based messaging systems, allow subscribers to register a query that describes the information they are interested in receiving. Publish/subscribe systems are composed of producers, consumers, and brokers. Brokers are responsible for routing information to the appropriate consumers on the basis of the queries they register. Furthermore, brokers are also responsible for handling client registration.

Although publish/subscribe systems and RPC are different models, publish/subscribe leverages RPC functionality, such as connection establishment, parameter marshaling, and protocol management. Furthermore, according to the open implementation model, we consider a remote interaction middleware infrastructure that can be dynamically adapted on the basis of the execution environment, that is, a middleware infrastructure whose functionality can be composed on demand. Therefore, it is possible to configure basic RPC functionality, or alternative configurations, such as publish/subscribe (Capra et al. 2002).

Mobility

Owing to the inherent mobile nature of cell phones, the software platform running on these devices must provide seamless application mobility. The form factor of cell phones makes them ideal for being carried around everywhere; however, their small dimensions make them unsuitable for interacting with certain applications for long periods of time. A possible solution to this issue is to support migration of applications from the handset to other devices and from other devices to the handsets (Ford and Lepreau 1994; Song et al. 2002). For example, consider a user that joins a remote slideshow presentation using a cell phone while traveling on the train. When the user reaches the office, he or she decides to continue with the slideshow using the office's PC and transfers the application from the handset to the PC. This action requires support from a service capable of suspending the application in the handset, starting a compatible application in the PC, transferring the state from the handset to the PC, and resuming the application in the PC. However, because of device heterogeneity, transferring the application state might require changing the format of the state. For example, the handset device may not have a native slideshow viewer application and, therefore, receives the slides as JPEG images. Furthermore, these images are scaled down to fit the size of the display of the handset. However, when the application is transferred to the PC, the PC has a native slideshow viewer application and therefore requests the slides in original format. Furthermore, the PC has a large display, and therefore the slides are modified to the original size.

Several projects address the issue of mobility from different perspectives. In fact, mobility has been studied for many years starting with Operating Systems, which provided support for thread (Ford and Lepreau 1994), process (Douglis and Ousterhout 1991), and task migration (Milojicic et al. 1992). The original goal was to support load balancing, improve reliability, and improve RPC response time. With the advent of wired and wireless networks, the Internet, and the proliferation of portable devices, users demand mechanisms to migrate applications and data across devices. A discussion of projects that deal with application mobility will demonstrate the different perspective that researchers have taken.

The one.world project (Grimm et al. 2002) provides support for application mobility (migration) through a hybrid of process migration and application-dependent mobility. It addresses three requirements: migration has to be visible to applications, needs to integrate persistent storage, and has to be easy to control and centralized in the source code. The one.world architecture utilizes Virtual Machine–based languages, such as Java and Microsoft Common Language Runtime. It organizes applications in environments, which contain all data of the applications, both runtime state and persistent storage. Transfer of application state involves copying of the entire environment, which mirrors process migration. Runtime state is transferred via a checkpointed byte stream of the running application. Applications that have external dependencies are required to resolve and reconfigure when the environment is moved, thus paralleling application-dependent mobility.

one.world relies on a one-to-one migration model. This model assumes that applications can move across devices as a single unit. That is, the whole application migrates from one device to another. However, there is another model, called *many-to-many*, that is relevant for handsets. This model assumes that applications can be dynamically partitioned, and therefore, it is possible to migrate individual components of the applications to different devices. The resulting application is spread across multiple devices.

Consider the following example: an invited speaker is reviewing the slides from the cell phone on the way to the conference room. When he reaches the conference room, he accesses the lookup service of the conference room through the phone to locate the main display server. Then, he moves the slide viewer to the display and keeps the controller on the phone so that he can control the slides. Many-to-many application mobility implies that an application can move from one device to many (prior application partitioning), that application components running on many devices can move to one device (dynamic application assembly), and that application components running on many devices can move to many devices (dynamic application partitioning and/or assembly).

The Gaia Application Framework (Roman and Campbell 2003) provides support for many-to-many application mobility. Gaia applications are built on top of three basic components called *Model* (application logic), *Presentation* (application output), and *Controller* (application input). The application framework provides a mobility service that allows the moving of presentations and controllers to multiple devices, effectively implementing many-to-many mobility.

Disconnection

Network disconnection of mobile terminals happens for various reasons, which can be classified into two categories: involuntary disconnection and voluntary disconnection.

One of the causes of involuntary disconnection is the insufficiency in wireless carriers' service coverage. For instance, voice or data service users may experience problems in sparsely populated areas such as parks with mountains. Wireless technology itself is far from perfect. Obstacles, such as large buildings, can totally block a wireless signal or make the signal so weak that it in effect appears as a disconnection to applications. In addition, mobile hosts moving at very fast speeds (such as a laptop with a 3G data service card in a fast moving vehicle) may experience a high bit-error rate and may not be able to communicate using the wireless connection. The other causes of involuntary disconnection include areas with such high concentration of mobile hosts that the overloaded wireless network disconnects users, or when a server for mobile clients fails.

Voluntary disconnection refers to disconnection caused by the user voluntarily. A user may voluntarily disconnect for various reasons, such as:

- Monetary concern (when wireless service is charged by the amount of data transmitted)

- Power consumption concern

- Security reasons (military personnel might disable wireless transmission to keep radio silence)

- When combined with caching techniques, voluntary disconnection may hide application interaction latency.

Phases of Disconnected Computing

Disconnected computing can be divided into three stages: the hoarding stage, the emulation stage, and the reintegration stage (Satyanarayanan 2002).

In the hoarding stage, the system starts to cache data that can be used during disconnected mode. In the case of possible involuntary disconnection, the system must continuously cache objects that are likely to be accessed later. Some systems, such as SEER and Rover, use automatic profiling (Joseph et al. 1997; Kuenning and Popek 1997; Tait et al. 1995) for creating such hoard databases, whereas others like Coda prefer to allow the user some level of control on creating a database of important objects (Ebling and Satyanarayanan 1998; Kistler and Satyanarayanan 1991). Bayou (Terry et al. 1995) on the other hand, maintains a fully replicated database on the mobile client for disconnected operation.

During emulation stage, the system must intercept all invocations to remote objects and service them locally from the cache. These operations are recorded to a stable log (Gruber et al. 1994; Joseph et al. 1997; Kistler and Satyanarayanan 1991; Terry et al. 1995; Valente et al. 2001). On a cache miss, several approaches can be taken. An error can be returned to the application or the request can be asynchronously processed with the application informed when it is successfully transmitted. A cache miss cannot happen in a system (like Bayou) that maintains the replicated database locally (Terry et al. 1995). In a system like Rover, security is another issue that crops up during the emulation stage. Rover relies on relocating data and code to the client as a precursor to execution. Hence, it is necessary that the execution environment is sufficiently isolated to prevent a malicious object from disrupting the rest of the system (Wahbe et al. 1993). Another solution is to have the object certified by a central certification authority, or to have the object demonstrate proof of compliance with the security policy of the client system (Necula 1997).

The trickiest part of disconnected computing is the reintegration stage. Once connection has been reestablished, any updates to cached data must be propagated to the other replicas of the data. Because of the variability of disconnection, it is usually not feasible to maintain an exclusive lock on the data, creating the possibility of a possible update conflict. In this case, a conflict-resolution procedure must be invoked to bring the data back into a consistent view. Reintegration problem is not just caused because of update conflicts. Applications might require very strict bounds on the *staleness* of the data. Thus, on reconnection it might be necessary to get a fresh version of the data and re-execute the operations on the object. It should be realized though that this re-execution approach is not always viable (given limited processing power, battery power, and connection time for mobile devices). Most systems that deal with file-based access punt on the update conflict issue. It is a known fact that the amount of write sharing in a file system is very low. Assuming this, systems, such as Coda and Ficus, do not stress on the efficiency in conflict resolution. In contrast to this approach, Bayou and Rover allow the application to dictate its conflict-resolution strategy. Bayou goes even further by embedding a conflict-resolution protocol with each write operation.

Several exemplary systems that deal with disconnection (such as Coda, Bayou, and Rover) are worth discussing in detail.

Coda

The evolution of Coda is documented by Satyanarayanan (Satyanarayanan 2002). Coda itself was an offshoot of research into the Andrew File System (AFS) developed at CMU. In trying to address the availability issues in AFS, Coda was given a feature set that was indispensable for mobile computing. Today, Coda is a mature file system that seamlessly supports disconnected and weakly connected operation. Coda has a client–server architecture. Clients communicate with servers using RPC (Satyanarayanan et al. 1990). Clients running a process called *Venus* use the local disk as a cache store, where files are cached in their entirety. A cached file is provided a callback (similar to a lease (Adya et al. 2002)) that is used to maintain consistency during a strongly connected operation.

Replication of server-side data occurs at the granularity of volumes called a *volume storage group* (VSG). Along with that, the client system keeps track of accessible volumes in a structure called *accessible volume storage group* (AVSG). The client is responsible for propagating changes back to the AVSG, thereby reducing server load for managing consistency. Since Coda follows the AFS semantics of open-close consistency, propagation of updated data only happens on a close. To handle disconnected operation, Coda utilizes an LRU caching scheme to hoard data. A user application can also be used to specify files that the user is likely to use during disconnection (especially if the disconnection is expected or voluntary). The information from the usage pattern is stored in the *Hoard Profile*. Once disconnection has occurred, Venus fulfills all client requests from the local cache. In the case of a cache miss, a failure code is returned to the client. All updates during the disconnected mode are logged and upon reconnection the updated files are reintegrated into the VSG. In case of an irresolvable conflict, the conflict is reflected back to the user for a solution.

Bayou

Bayou is a replicated data storage service developed at Xerox PARC in the mid 1990s. Key features of Bayou included support for disconnected operation (similar to Coda), weak consistency, update anywhere semantics, and no support for legacy applications. Unlike Coda (Satyanarayanan et al. 1990), Bayou does not maintain any support for legacy applications preferring to rely on application support for weak consistency to allow for better availability of data. In Bayou, each data storage unit is called a *data collection* and is fully replicated at a number of hosts (*servers*). It should be noted that Bayou does not limit the replication of the database to fixed workstations; instead, the database can be placed on mobile hosts.

Two basic operations are supported in the Bayou API: *read* for querying the database, and *write* for updating the database. A client can proceed as long as it has access to a single server (as can be the case during client disconnection where the server is a local database). Updates are propagated to other replicas using an antientropy protocol (Demers et al. 1987), but at no time is data access restricted by the underlying system. Thus, Bayou provides read-any/write-any semantics. Bayou also provides a *session abstraction* and provides for various *session guarantees* to allow clients to observe a consistent view of their *own* data (Terry et al. 1994). Each Bayou write contains an update procedure, a conflict detector, and a conflict-resolution procedure (Terry et al. 1995). Each write "packet" is identified by a globally unique identifier assigned by the accepting server. The storage system itself is an ordered collection of writes and the resulting data.

All operations are performed locally (including conflict detection and resolution), which makes local writes immediately available to any clients performing a read at the server. Updates are slowly propagated (given sufficient connectivity in the network) to all replicas. Bayou aims only for eventual consistency. That is, if no more updates are made and the network does not partition permanently, the replicated database will *eventually* converge to a globally consistent view. The philosophy behind Bayou's design is to exploit application-specific knowledge of the required level of data consistency to provide a highly available data store. Instead of following traditional approaches like Coda (Kistler and Satyanarayanan 1991), Bayou always uses the application to resolve conflicts (Terry et al. 1998). For stable storage, Bayou uses a flexible in-memory database (Petersen et al. 1997).

Rover

The Rover toolkit was developed at MIT to address the idea of disconnected computing using an object-centric environment. Unlike Bayou or Coda, which can be classified as systems, Rover is a toolkit that applications can be built on. The Rover toolkit provides mobile application with the abstraction of relocatable objects called *relocatable dynamic objects* (RDOs) and queued remote procedure calls (QRPCs). An RDO allows the application to move its functionality between the server and the mobile host. QRPCs allow clients to operate in disconnected mode by logging all remote procedure calls and replaying them over the network when there is connectivity. RDOs are safely invoked in a controlled environment for safety reasons.

Rover employs the check-in, check-out model of data sharing similar to source control systems (applications check-out objects for execution and check-in objects to update the shared state). When an object request is initiated by the application, Rover tries to locate the object in its local cache. Failing that, the object is fetched lazily from the server and all processing is done locally. If the fetch operation is delayed, the RPC operations are recorded in a local stable log that is replayed whenever the object is successfully fetched. Applications can register a callback routine to stay informed about that status of the fetch operation. When an object is updated, the Rover toolkit sends the method call (in the form of a QRPC) to the server.

Each object has a primary copy at a *home* server. The server executes the QRPC on the primary copy, checking for sharing conflicts. Each object can implement its own consistency mechanism, but the Rover server does check for update conflicts using a version vector scheme. Once an object is updated, the server returns a reply to the client, which can then mark the update as committed, instead of tentative. All communication in Rover is asynchronous because of QRPC. Control is returned to the application as soon as the request is logged on stable storage. The request is sent to the server whenever communication becomes possible. In contrast to traditional RPCs, QRPCs do not experience failure in the case of network disconnection; instead, the network appears slow (Joseph et al. 1995).

Thus, application progress and network connectivity are decoupled. Logged requests can be reordered on the basis of user-defined priority to better utilize intermittent connectivity. Furthermore, the request and reply need not be sent over the same physical link, thus better utilizing asymmetric communication mechanisms.

Other Systems

Ficus is a distributed file system with optimistic replication for higher availability developed at UCLA (Guy et al. 1990). Ficus tries to address the overhead associated with pessimistic locking in a distributed environment. Ficus concentrates on efficiently resolving file system conflicts during normal operation, with minimal user intervention. Fluid replication is a scheme to enhance the scalability of services by creating and relocating replicas when they are needed to maintain some level of quality. Fluid replication does not only seek to address disconnected computing but it also uses optimistic replication to maintain consistency among replicas. The bulk of the effort behind fluid replication is fast reconciliation (Cox and Noble 2001) and in deciding where and when to create replicas (Noble and Fleis 1999). Central to the idea of fluid replication was the selection of a correct consistency semantic for each application. This idea is furthered by Yu in TACT (Yu and Vahdat 2000). TACT is a middleware library for enforcing consistency bounds between distributed replicas. Instead of explicitly defining a fixed set of consistency types (as in Fluid Replication), a triplet of consistency metrics is defined. The level of consistency between replicas is measured as a function of *numerical error*, *order errors*, and *staleness*. The idea that different applications have different requirements for consistency has been brought up in other replicated systems like Bayou (Terry et al. 1998), but TACT addresses it in much more detail by analyzing the different requirements for each application based on their evaluation metrics. Such metrics include numeric error, order error, and staleness.

7.6.2 Mervlets: Leveraging the Web Model

Our previous work in enabling mobile devices for ubiquitous computing has focused on dealing with the complexities caused by the heterogeneity and dynamics in mobile environment. Heterogeneity in a mobile computing system comes from the divergence in device capability, the differences in network connections, the differences in application behaviors, and the differences in client preferences. Heterogeneity in a mobile computing environment requires middleware to take into account the preferences of users, applications, application servers, and the capabilities of the user devices and application servers, and it requires mobile middleware to be highly customizable and reconfigurable. Dynamics in a mobile computing system come from the dynamics in the environment (mainly network dynamics), the dynamics in user behavior, and the dynamics in application behavior. Such Dynamics require mobile middleware to agilely adapt to changes in the environment, in the application, and in user behaviors.

Initial research at DoCoMo Labs USA targeted such heterogeneity and dynamics issues in the context of web-based applications. Specifically, we designed an environment for the customizable, adaptive execution of Web applications on handheld devices. The environment supports adaptive on-device replication of service object, the dynamic user-interface binding for the applications, and the adaptable fault tolerance of the applications. The rest of this section describes this environment in detail, and in later sections, we will discuss how to extend this into a generic infrastructure to apply to broader application domains.

The Agile Operating Environment

Agile Computing Environment (AOE) is a middleware platform developed at DoCoMo USA Labs (Islam et al. 2004). It supports web-based applications where users request services through a browser from their user client device. It specifically targets applications that involve dynamic, personalized content. A typical AOE application consists of a cluster of *Mervlets* that can dynamically generate web pages. A Mervlet is similar to a Servlet except that it can be replicated and executed on client devices, its user interface can be dynamically attached, and it may recover from faults in the client device, network, or server. Each Mervlet within an AOE application implements the Mervlet interface, which consists of methods that can be used to help in replicating the Mervlet and synchronizing state among replicas. A developer of each Mervlet can also provide implementations of Adaptation Helpers that can help the AOE runtime to make adaptation decisions. An application developer can optionally create user interface widgets used specifically for a particular device. These widgets can be dynamically bound to the application. For each application, there is an application preference file that is specific for the particular deployment, and will be combined with client and server preferences to form the runtime preference of the application instance.

The AOE runtime is the environment for the execution of AOE applications. The client and the server each run an instance of the runtime (Figure 7.12). HTTP requests sent from the browser are intercepted by the AOE runtime on the client device (termed *Client AOE*). The client AOE can take one of two actions:

- It can pass the request to the AOE runtime on the server side (Server AOE) using any transport (such as a Reconfigurable Messaging System or RMS), in which case the Server AOE receives and serves the request and sends response to the Client AOE using AOE messages.

- It can serve the request locally when the requested page is locally available or when client AOE decides to execute the requested Mervlet locally.

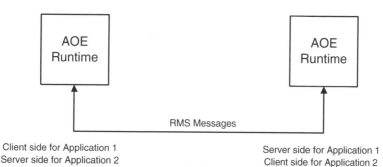

Figure 7.12 Symmetric AOE model

The response, which is usually the presentation of the application described in languages such as HTML, is optionally re-bound with the user interface library (UIL) deployed on the client device and finally returned to the browser in HTTP format. During the process, the client and server AOEs cooperatively assure that once the Client AOE receives an HTTP request from the browser, the request will be served according to a reliability guarantee, such as processing the request once and only once. Such reliability assurances are provided by the *Adaptive Reliability Manager* inside the AOE runtime, while the functions of deciding where to serve the request and dynamically rebinding user interface of the application are supported by *Replication Manager* and *UI Composer*, respectively (Figure 7.13). The details of the Replication Manager, the UI composer, and the Reliability Manager are described in later sections.

A key feature of AOE runtime system is that basic system facilities are adaptable. The three key facilities (UI Composer, Replication Manager, and Reconfigurable Messaging System) are made adaptable using three adaptation managers (the UI Adapter, the Replication Adapter, and the Adaptive Reliability Manager, respectively), which make adaptation decisions based on input from the Preference Manager and the Capability Profiler. The adaptation coordinator coordinates the individual adaptation managers of each facility:

- The Adaptation Coordinator is responsible for resolving conflicts between adaptation decisions made by the three Adaptation Managers (i.e., the UI Adapter, the Replication Adapter and the Adaptive Reliability Manager (ARM)). The adaptation managers consult the Adaptation Coordinator whether they should proceed before making any adaptation decisions.

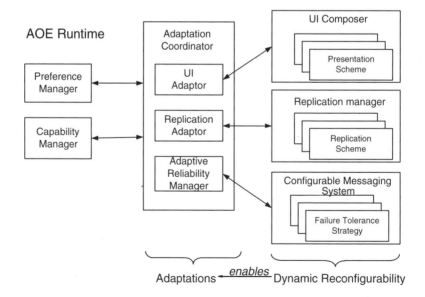

Figure 7.13 Layered components of AOE runtime

- The Preference Manager provides a means for devices, servers, and applications to jointly configure and reconfigure an AOE runtime, to allow devices and servers to control the behaviors of applications, and to allow the exchange of preferences between two AOE runtimes.

- The Capability Profiler of AOE runtime is a repository for keeping device and network capabilities, as well as some application characteristic data. Some of such capability or characteristic data (such as installed memory) are static, while others (such as current CPU load) are dynamic.

Mervlet On-device Adaptive Replication

Mobile users may have a poor experience because of longer response latency, lower throughput, wider variation in responses, and disconnection. While existing web caching and prefetching techniques are applicable to mobile web applications for static and some of the dynamically generated web pages, these technologies do not improve the users' experience for highly dynamic, highly personalized web content that is generated by server-side code units (such as servlets) and widely used in mobile web applications. AOE handles such highly dynamic/personalized web content by allowing the replication of server-side code units (or Mervlets) onto the client device.

The process of Mervlet replication can be divided into four phases: the selection phase, the populating phase, the invocation phase and the synchronization phase.

In the *selection phase*, before sending an HTTP request to the server, an AOE-enabled client will check local device preferences to see if the device allows Mervlet replication. If it does, the client will insert a field into the header of the HTTP request to indicate its willingness to be considered as site for replicating the service. Upon receiving such request, the server will first serve the request and generate a usual response. It then checks whether the application allows itself to be replicated; if so, it will send application and server preference and capability information to the client, again in the form of HTTP header fields. The profile of the application includes hints on the application's usage of memory, storage, CPU, and the application's consistency requirement, and data update frequency. Also included in the header is a private ID created by the server for the client.

In the *populating phase*, the client uses the private ID received from server earlier to identify proper session states to be downloaded, in addition to classes, immutable data, and shared mutable data.

In the *invocation phase*, when a client receives a request from user or other applications, the Client AOE checks preference and profiling information (both those derived locally and received from the server) to decide whether to serve the request locally or remotely. If the decision is to serve the request remotely, the service request will be forwarded to the server with no additional header fields.

In the *synchronization phase*, state modifications made in the invocation phase are synchronized between client and server AOEs to maintain a consistency level that satisfies the requirement of the application. Note that only application global states need to be synchronized at all times among replicas. Session states only need to be synchronized when changing from invoking client replica to server replica, or vice versa.

The selection and populating of a replica is customizable and adaptable in that devices, servers, and applications can define their own triggers for selecting a device as a replication site and populate the site, and the dynamic capabilities provided by the AOE runtime are used to automatically evaluate the predicates for the trigger. Adaptation in replica invocation is supported by per-request replica selection for invocation. That is, for each request received by the client, the client-side and server-side AOE runtime will collaboratively decide which replica to use for that particular request.

Dynamic On-device UI Binding for Mervlets

One of the challenges in developing applications for a ubiquitous environment is adapting the user interface for different types of devices. Traditionally the problem is solved using a server proxy that adapts content on its way to a device. The presentation parts of these applications can be written using XML and XLST style sheets. As there are several different types of devices, these types of solutions are cumbersome to update and maintain. Also, these solutions do not utilize the available resources in the new smart devices, in which the application can fully reside.

AOE requires applications to write their presentation in tags that help translate XML for rendering. However, the tags are not statically bound to the application. Instead, they are constructed as dynamically attachable libraries that can be chained together at runtime in a specific order. This decoupling of the presentation from the application is critical in mobile systems, because the systems may decide to deploy the presentation generation for an application in the server or on the device, and decide how to generate the presentation at runtime. Finally, our scheme reduces the complexity of application development by taking the presentation component out of the application and into the runtime.

To enable dynamic presentation binding, we choose a model similar to Java Server Pages (JSP), called *Mervlet Server Pages* or MSPs. Unlike JSP library, MSPs do not embed presentations in the applications at development time, but instead, add a wrapper for the presentation to the program. At runtime, the UI Adapter attaches the appropriate library to enable the presentation of the application. Presentations are implemented by UILs. Each UIL has an interface (UIL Interface) and an implementation (UIL Implementation). Each implementation is tagged by a capability specification.

At runtime, when the generated Mervlet makes a call that involves presentation, the system invokes the proxy corresponding to the appropriate UIL Interface. The proxy asks the UI Adapter service to provide an implementation of the specified library interface that is suitable for the device capabilities. The UI Adapter uses the characteristics of the device to find the appropriate library to install and link to the application.

On-device Support for Adaptive Fault Tolerance of Mervlets

Support for fault tolerance is realized with RMS and Recoverable Mervlets.

The RMS provides configurable message delivery functionality in the Mervlet environment. The RMS failure-free strategy interface encapsulates the RMS functionality. An application only calls methods on the interface. The actual implementation to be used is set by the ARM based on user or application requirements. For example, the RMS can be configured as a point-to-point messaging service or to use a centralized messaging server (such as JMS).

At the application level, recoverable Mervlets are used to provide fault tolerance. During failure-free operation, the Mervlet engine invokes the methods on the recoverable Mervlets

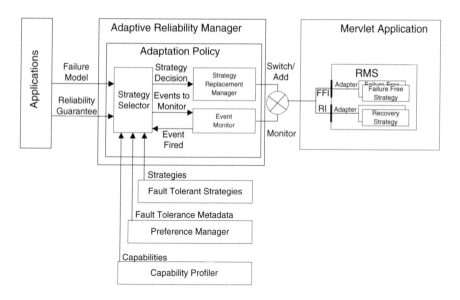

Figure 7.14 Reliability support in AOE

by getting the current failure-free strategy first and then calling the desired method (e.g., doPost and doGet) on the strategy. Recoverable Mervlets allow the same application to have different fault-tolerance mechanisms during different contexts. For example, the Web Mail application may be configured to be more reliable for corporate e-mail than personal e-mail.

Dynamic reconfigurability support in fault tolerance is achieved by allowing the two main components, the RMS and the Recoverable Mervlet, to have different failure-free and recovery strategies, which can be set dynamically by the ARM (shown in Figure 7.14). The separation between failure-free and recovery strategies helps in developing multiple recovery strategies corresponding to a failure-free strategy. For example, in case of RMS, one recovery strategy may prioritize the order in which messages are recovered, while another recovery strategy may not.

In our current implementation, the adaptability in fault-tolerance support is reflected in the ability to dynamically switch on and off server-side logging depending on current server load. Under high server load, the ARM can reconfigure the RMS to stop logging on the server side. In some cases, this can result in marked improvement in the client perceived response time.

7.7   Conclusions

The evolution of handheld devices clearly indicates that they are becoming highly relevant in users' everyday activities. Voice transmission still plays a central role but machine-to-machine interaction is becoming important and it is poised to surpass voice transmission. This data transmission is triggered by digital services running on the phone as well as on the network that allow users to access data and functionality everywhere and at anytime.

This digital revolution requires a middleware infrastructure to orchestrate the services running on the handhelds, to interact with remote resources, to discover and announce data and functionality, to simplify the migration of functionality, and to simplify the development of applications. At DoCoMo Labs USA, we understand that the middleware has to be designed to take into account the issues that are specific to handheld devices and that make them different from traditional servers and workstation computers. Examples of these issues are mobility, limited resources, fault tolerance, and security.

DoCoMo Labs USA also understands that software running on handheld devices must be built in such a way that it can be dynamically modified and inspected without stopping its execution. Systems built according to this requirement are known as *reflective systems*. They allow inspecting of their internal state, reasoning about their execution, and introducing changes whenever required. Our goal is to provide an infrastructure to construct systems that can be fully assembled at runtime and that explicitly externalize their state, logic, and architecture. We refer to these systems as *completely reconfigurable systems*.

8

Multimedia Coding Technologies and Applications

Minoru Etoh, Frank Bossen, Wai Chu, and Khosrow Lashkari

8.1 Introduction

As the bandwidth provided by next-generation (XG) mobile networks will increase, the quality of media communication, such as audiovisual streaming, will improve. However, a huge bandwidth gap (by one or two orders of magnitude) always exists between wireless and wired networks, as explained in Chapter 1. This bandwidth gap demands that coding technologies achieve compact representations of media data over wireless networks. Considering the heterogeneity of radio access networks, we cannot presume availability of high-bandwidth connectivity at all times. Figure 8.1 illustrates the importance of media coding technologies and radio access technologies. These are complementary and orthogonal approaches for improving media quality over mobile networks. Thus, media coding technologies are essential even in the XG mobile network environment, as discussed in Chapter 1.

Speech communication has been the dominant application in the first three generations of mobile networks. 8-kHz sampling has been used for telephony with the adaptive multirate (AMR) (3GPP 1999d) speech codec (encoder and decoder) that is used in 3G networks. The 8-kHz restriction ensures the interoperability with the legacy wired telephony network. If this restriction is removed and peer-to-peer communications with higher audio sampling is adopted, new media types, such as wideband speech and real-time audio, will become more widespread. Figure 8.2 illustrates existing speech and audio coding technologies with

Next Generation Mobile Systems. Edited by Dr. M. Etoh
© 2005 John Wiley & Sons, Ltd

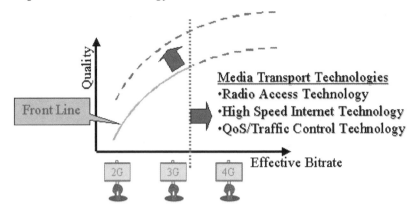

Figure 8.1 Essential coding technologies

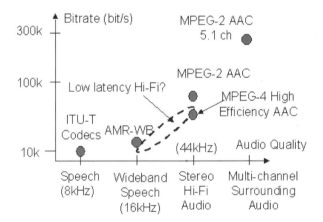

Figure 8.2 Speech and audio codecs with regard to bitrate

regard to usage and bitrate, where adaptive multirate wideband (AMR-WB) (ITU-T 2002) is shown as an example of wideband speech communication, and MPEG-2 of broadcast and storage media. Given 44-kHz sampling and a new type of codec that is suitable for real-time communication, low-latency hi-fi telephony can be achieved and convey more realistic sounds between users.

Video media requires a higher bandwidth in comparison with speech and audio. In the last decade, video compression technologies have evolved in the series of MPEG-1, MPEG-2, MPEG-4, and H.264, which will be discussed in the following sections. Given

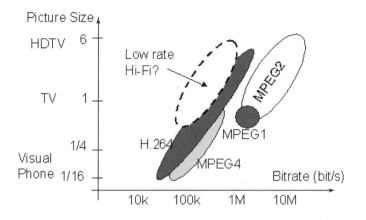

Figure 8.3 Video codecs with regard to bitrate

a bandwidth of several megabits per second (Mbps), these codecs can transmit broadcast-quality video. Because of the bandwidth gap (even in XG), however, it is important to have a codec that provides better coding efficiency. Figure 8.3 summarizes the typical existing codecs and the low-rate hi-fi video codec that is required by mobile applications.

This chapter covers the technological progress of the last 10 years and the research directed toward more advanced coding technologies. Current technologies were designed to minimize implementation costs, such as the cost of memory, and also to be compatible with legacy hardware architectures. Moore's Law, which states that computing power doubles every 18 months, has been an important factor in codec evolution. As a result of this law, there have been significant advances in technology in the 10 years since the adoption of MPEG-2. Future coding technologies will need to incorporate advances in signal processing local spectral information (LSI) technologies. Additional computational complexity is the principle driving codec evolution. This chapter also covers mobile applications enabled by the recent progress of coding technologies. These are the TV phone, multimedia messaging services already realized in 3G, and future media-streaming services.

8.2   Speech and Audio Coding Technologies

In speech and audio coding, digitized speech or audio signals are represented with as few bits as possible, while maintaining a reasonable level of perceptual quality. This is accomplished by removing the redundancies and the irrelevancies from the signal. Although the objectives of speech and audio coding are similar, they have evolved along very different paths.

Most speech coding standards are developed to handle narrowband speech, that is, digitized speech with a sampling frequency of 8 kHz. Narrowband speech provides toll quality suitable for general-purpose communication and is interoperable with legacy wired telephony networks. Recent trends focus on wideband speech, which has a sampling frequency of 16 kHz. Wideband speech (50–7000 Hz) provides better quality and improved intelligibility required by more-demanding applications, such as teleconferencing and multimedia services. Modern speech codecs employ source-filter models to mimic the human sound production mechanism (glottis, mouth, and lips).

The goal in audio coding is to provide a perceptually transparent reproduction, meaning that trained listeners (so-called *golden ears)* cannot distinguish the original source material from the compressed audio. The goal is not to faithfully reproduce the signal waveform or its spectrum but to reproduce the information that is relevant to human auditory perception. Modern audio codecs employ psychoacoustic principles to model human auditory perception.

This section includes an overview of various standardized speech and audio codecs, an explanation of the relevant issues concerning the advancement of the field, and a description of the most-promising research directions.

8.2.1 Speech Coding Standards

A large number of speech coding standards have been developed over the past three decades. Generally speaking, speech codecs can be divided into three broad categories:

1. Waveform codecs using pulse code modulation (PCM), differential PCM (DPCM), or adaptive DPCM (ADPCM).

2. Parametric codecs using linear prediction coding (LPC) or mixed excitation linear prediction (MELP).

3. Hybrid codecs using variations of the code-excited linear prediction (CELP) algorithm.

This subsection describes the essence of these coding technologies, and the standards that are based on them. Figure 8.4 shows the landmark standards developed for speech coding.

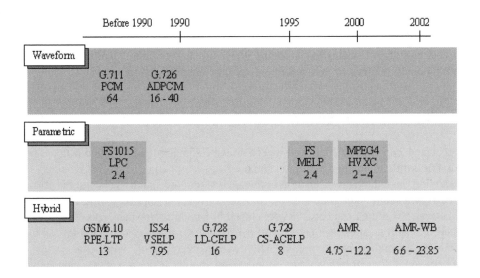

Figure 8.4 Evolution of speech coding standards

Waveform Codecs

Waveform codecs attempt to preserve the shape of the signal waveform and were widely used in early digital communication systems. Their operational bitrate is relatively high, which is necessary to maintain acceptable quality.

The fundamental scheme for waveform coding is PCM, which is a quantization process in which samples of the signals are quantized and represented using a fixed number of bits. This scheme has negligible complexity and delay, but a large number of bits is necessary to achieve good quality. Speech samples do not have uniform distribution, so it is advantageous to use nonuniform quantization. ITU-T G.711 (ITU-T 1988) is a nonuniform PCM standard recommended for encoding speech signals, where the nonlinear transfer characteristics of the quantizer are fully specified. It encodes narrowband speech at 64 kbps.

Most speech samples are highly correlated with their neighbors, that is, the sample value at a given instance is similar to the near past and the near future. Therefore, it is possible to make predictions and remove redundancies, thereby achieving compression. DPCM and ADPCM use prediction, where the prediction error is quantized and transmitted instead of the sample itself. Figure 8.5 shows the block diagrams of a DPCM encoder and decoder. ITU-T G.726 is an ADPCM standard, and incorporates a pole-zero predictor. Four operational bitrates are specified: 40, 32, 24, and 16 kbps (ITU-T 1990). The main difference between DPCM and ADPCM is that the latter uses adaptation, where the parameters of the quantizer are adjusted according to the properties of the signal. A commonly adapted element is the

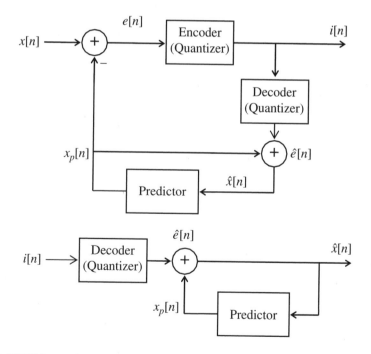

Figure 8.5 DPCM encoder (top) and decoder (bottom). Reproduced by permission of John Wiley & Sons, Inc.

predictor, where changes to its parameters can greatly increase its effectiveness, leading to substantial improvement in performance.

The previously described schemes are designed for narrowband signals. The ITU-T standardized a wideband codec known as G.722 (ITU-T 1986) in 1986. It uses subband coding, where the input signal is split into two bands and separately encoded using ADPCM. This codec can operate at bitrates of 48, 56, and 64 kbps and produces good quality for speech and general audio signals. G.722 operating at 64 kbps is often used as a reference for evaluating new codecs.

Parametric Codecs

In parametric codecs, a multiple-parameter model is used to generate speech signals. This type of codec makes no attempt to preserve the shape of the waveform, and quality of the synthetic speech is linked to the sophistication of the model. A very successful model is based on linear prediction (LP), where a time-varying filter is used. The coefficients of the filter are derived by an LP analysis procedure (Chu 2003).

The FS-1015 linear prediction coding (LPC) algorithm developed in the early 1980s (Tremain 1982) relies on a simple model for speech production (Figure 8.6) derived from practical observations of the properties of speech signals. Speech signals may be classified as voiced or unvoiced. Voiced signals possess a clear periodic structure in the time domain, while unvoiced signals are largely random. As a result, it is possible to use a two-state model to capture the dynamics of the underlying signal. The FS-1015 codec operates at 2.4 kbps, where the quality of the synthetic speech is considered low. The coefficients of the synthesis filter are recomputed within short time intervals, resulting in a time-varying filter. A major shortcoming of the LPC model is that misclassification of voiced and unvoiced signals can create annoying artifacts in the synthetic speech; in fact, under many circumstances, the speech signal cannot be strictly classified. Thus, many speech coding standards developed after FS-1015 avoid the two-state model to improve the naturalness of the synthetic speech.

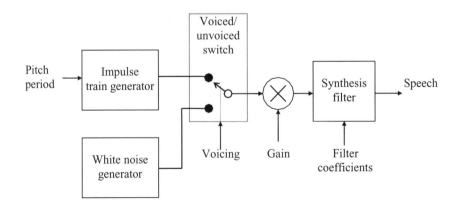

Figure 8.6 The LPC model of speech production. Reproduced by permission of John Wiley & Sons, Inc.

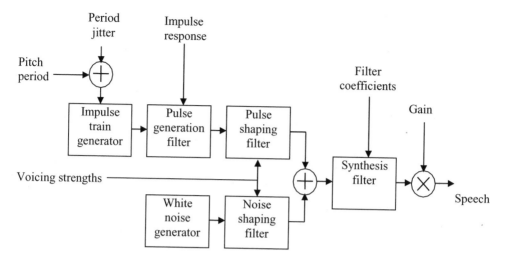

Figure 8.7 The MELP model of speech production. Reproduced by permission of John Wiley & Sons, Inc.

The MELP codec (McCree et al. 1997) emerged as an improvement to the basic LPC codec. In the MELP codec, many features were added to the speech production model (Figure 8.7), including subband mixture of voiced and unvoiced excitation, transmission of harmonic magnitudes for voiced signals, handling of transitions using aperiodic excitation, and additional filtering for signal enhancement. The MELP codec operates at the same 2.4-kbps bitrate as FS-1015. It incorporates many technological advances, such as vector quantization. Its quality is much better than that of the LPC codec because the strict signal classification is avoided and is replaced by mixing noise and periodic excitation to obtain a mixed excitation (Chu 2003).

The harmonic vector-excitation codec (HVXC), which is part of the MPEG-4 standard (Nishiguchi and Edler 2002), was designed for narrowband speech and operates at either 2 or 4 kbps. This codec also supports a variable bitrate mode and can operate at bitrates below 2 kbps. The HVXC codec is based on the principles of linear prediction, and like the MELP codec, transmits the spectral shape of the excitation for voiced frames. For unvoiced frames, it employs a mechanism similar to CELP to find the best excitation.

Hybrid Codecs

Hybrid codecs combine features of waveform codecs and parametric codecs. They use a model to capture the dynamics of the signal, and attempt to match the synthetic signal to the original signal in the time domain. The code-excited linear prediction (CELP) algorithm is the best representative of this family of codecs, and many standardized codecs are based on it. Among the core techniques of a CELP codec are the use of long-term and short-term linear prediction models for speech synthesis, and the incorporation of an excitation codebook, containing the code to excite the synthesis filters. Figure 8.8 shows the block

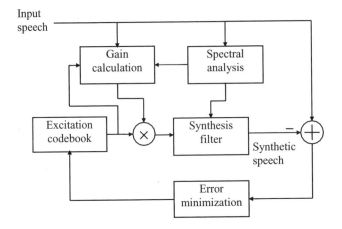

Figure 8.8 Block diagram showing the key components of a CELP encoder. Reproduced by permission of John Wiley & Sons, Inc.

diagram of a basic CELP encoder, where the excitation codebook is searched in a closed-loop fashion to locate the best excitation for the synthesis filter, with the coefficients of the synthesis filter found through an open-loop procedure.

The key components of a CELP bitstream are the gain, which contains the power information of the signal; the filter coefficients, which contain the local spectral information; an index to the excitation codebook, which contains information related to the excitation waveform; and the parameters of the long-term predictors, such as a pitch period and an adaptive codebook gain.

CELP codecs are best operated in the medium bitrate range of 5–15 kbps. They provide higher performance than most low-bitrate parametric codecs because the phase of the signal is partially preserved through the encoding of the excitation waveform. This technique allows a much better reproduction of plosive sounds, where strong transients exist.

Standardized CELP codecs for narrowband speech include the TIA IS54 vector-sum-excited linear prediction (VSELP) codec, the FS-1016 CELP codec, the ITU-T G.729 (ITU-T 1995) conjugate-structure algebraic CELP (ACELP) codec, and the AMR codec (3GPP 1999d). For wideband speech, the best representatives are the ITU-T G.722.2 AMR-WB codec (ITU-T 2002) and the MPEG-4 version of CELP (Nishiguchi and Edler 2002).

Recent trends in CELP codec design have focused on the development of multimode codecs. They take advantage of the dynamic nature of the speech signal and adapt to the time-varying network conditions. In multimode codecs, one of several distinct coding modes is selected. There are two methods for choosing the coding modes: source control, when it is based on the local properties of the input speech, and network control, when the switching obeys some external commands in response to network or channel conditions. An example of a source-controlled multimode codec is the TIA IS96 standard (Chu 2003), which dynamically selects one of four data rates every 20 ms, depending on speech activity. The AMR and AMR-WB standards, on the other hand, are network controlled. The AMR standard is a family of eight codecs operating at 12.2, 10.2, 7.95,

7.40, 6.70, 5.90, 5.15, and 4.75 kbps. The selectable mode vocoder (SMV) (3GPP2 2001) is both network controlled and source controlled. It is based on four codecs operating at 8.55, 4.0, 2.0, and 0.8 kbps and four network-controlled operating modes. Depending on the selected mode, a different rate-determination algorithm is used, leading to a different average bitrate.

In March 2004, the third-generation partnership project (3GPP) adopted AMR-WB+ as a codec for packet-switched streaming (PSS) audio services. AMR-WB+ is based on AMR-WB and further includes transform coded excitation (TCX) and parametric coding. It also uses a 80-ms superframe to increase coding efficiency. The coding delay is around 130 ms and therefore not suitable for real-time two-way communication applications.

Applications and Historical Context

The FS-1015 codec was developed for secure speech over narrowband very high frequency (VHF) channels for military communication. The main goal was speech intelligibility, not quality. MELP and FS-1016 were developed for the same purpose, but with emphasis on higher speech quality. G.711 is used for digitizing speech in backbone circuit-switched telephone networks. It is also a mandatory codec for H.323 packet-based multimedia communication systems. AMR is a mandatory codec for 3G wireless networks. For this codec, the speech bitrate varies in accordance with the distance from the base station, or to mitigate electromagnetic interference. AMR was developed for improved speech quality in cellular services. G.722 is used in videoconferencing systems and multimedia, where higher audio quality is required. AMR-WB was developed for wideband speech coding in 3G networks. The increased bandwidth of wideband speech (50–7000 Hz) provides more naturalness, presence, and intelligibility. G.729 provides near toll-quality performance under clean channel conditions and was developed for mobile voice applications that are interoperable with legacy public switched telephone networks (PSTN). It is also suitable for voice over Internet protocol (VoIP).

8.2.2 Principles of Audio Coding

Simply put, speech coding models the speaker's mouth and audio coding models the listener's ear. Modern audio codecs, such as MPEG-1 (ISO/IEC 1993b) and MPEG-2 (ISO/IEC 1997, 1999), use psychoacoustic models to achieve compression. As mentioned before, the goal of audio coding is to find a compact description of the signal while maintaining good perceptual quality. Unlike speech codecs that try to model the source of the sound (human sound production apparatus), audio codecs try to take advantage of the way the human auditory system perceives sound. In other words, they try to model the human hearing apparatus. No unified source model exists for audio signals. In general, audio codecs employ two main principles to accomplish their task: time/frequency analysis and psychoacoustics-based quantization. Figure 8.9 shows a block diagram of a generic audio encoder.

The encoder uses a frequency-domain representation of the signal to identify the parts of the spectrum that play major roles in the perception of sound, and eliminate the perceptually

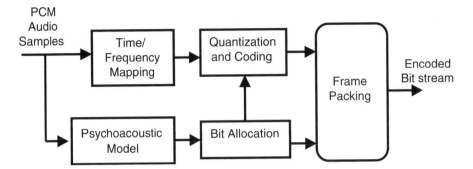

Figure 8.9 Generic block diagram of audio encoder

Figure 8.10 Generic block diagram of audio decoder

insignificant parts of the spectrum. Figure 8.10 shows the generic block diagram of the audio decoder. The following section describes the various components in these figures.

Time/Frequency Analysis

The time/frequency analysis module converts 2 ms to 50 ms long frames of PCM audio samples (depending on the standard) to equivalent representations in the frequency domain. The number of samples in the frame depends on the sampling frequency, which varies from 16 to 48 kHz depending on the application. For example, wideband speech uses a 16-kHz sampling frequency, CD quality music uses 44.1 kHz, and digital audio tape (DAT) uses 48 kHz. The purpose of this operation is to map the time-domain signal into a domain where the representation is more clustered and compact. As an example, a pure tone in the time domain extends over many time samples, while in the frequency domain, most of the information is concentrated in a few transform coefficients. The time/frequency analysis in modern codecs is implemented as a filter bank. The number of filters in the bank, their bandwidths, and their center frequencies depend on the coding scheme. For example, the MPEG-1 audio codec (ISO/IEC 1993b) uses 32 equally spaced subband filters. Coding efficiency depends on adequately matching the analysis filter bank to the characteristics of the input audio signal. Filter banks that emulate the analysis properties of the human auditory system, such as those that employ subbands resembling the ear's nonuniform critical bands, have been highly effective in coding nonstationary audio signals. Some codecs use time-varying filter banks that adjust to the signal characteristics. The modified discrete cosine transform (MDCT) is a very popular method to implement effective filter banks.

Modified Discrete Cosine Transform (MDCT)

The MDCT is a linear orthogonal lapped transform, based on the idea of time-domain aliasing cancellation (TDAC) (Princen and Bradley 1987). The MDCT offers two distinct advantages: (1) it has better energy compaction properties than the FFT, representing the majority of the energy in the sequence with just a few transform coefficients; and (2) it uses overlapped samples to mitigate the artifacts arising in block transforms at the frame boundaries. Figure 8.11 illustrates this process. Let $x(k), k = 0, \ldots, 2N - 1$, represent the audio signal and $w(k), k = 0, \ldots, 2N - 1$, a window function of length $2N$ samples. The MDCT (Ramstat 1991) is defined as:

$$X(m) = \sqrt{\frac{2}{N}} \sum_{k=0}^{2N-1} x(k)w(k) \cos\left[\frac{\pi(2m + 1)(2k + N + 1)}{4N}\right]. \tag{8.1}$$

Note that the MDCT uses $2N$ PCM samples to generate N transform values. The transform is invertible for a symmetric window $w(2N - 1 - k) = w(k)$, as long as the window function satisfies the Princen–Bradley condition:

$$w^2(k) + w^2(k + N) = 1. \tag{8.2}$$

Windows applied to the MDCT are different from windows used for other types of signal analysis, because they must fulfill the Princen–Bradley condition. One of the reasons for this difference is that MDCT windows are applied twice, once for the MDCT and once for the inverse MDCT (IMDCT). For MP3 and MPEG-2 AAC, the following sine window is used:

$$w(k) = \sin\left[\frac{\pi(2k + 1)}{4N}\right]. \tag{8.3}$$

Psychoacoustic Principles

Psychoacoustics (Zwicker and Fastl 1999) studies and tries to model the mechanisms by which the human auditory system processes and perceives sound. Two key properties of the auditory system, frequency masking and temporal masking, are the basis of most modern audio-compression schemes. Perceptual audio codecs use the frequency and temporal masking properties to remove the redundancies and irrelevancies from the original audio signal. This results in a lossy compression algorithm; that is, the reproduced audio is not a bit-exact copy of the original audio. However, perceptually lossless compression with compression factors of 6 to 1 or more is possible.

Figure 8.12 shows the frequency response of the human auditory system for pure tones in a quiet environment. The vertical axis in this figure is the threshold of hearing measured

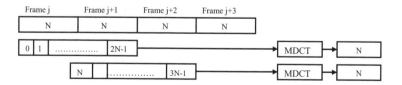

Figure 8.11 MDCT showing 50% overlap in successive frames

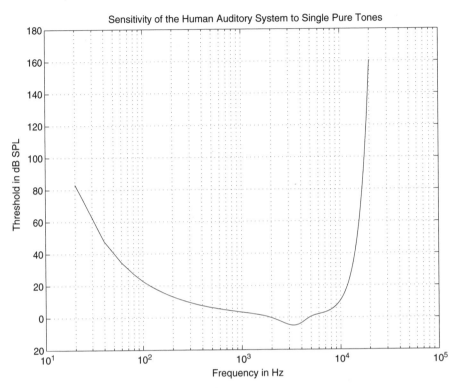

Figure 8.12 Sensitivity of human auditory system to single pure tones

in units of sound pressure level (SPL). SPL is a measure of sound pressure level in decibels relative to a 20-μPa reference in air. As seen here, the ear is most sensitive to frequencies around 3.5 kHz and not very sensitive to frequencies below 300 Hz or above 10 kHz. For a 2-kHz tone to be barely audible, its level must be at least 0 dB. A 100-Hz tone, on the other hand, must have a 22-dB level to be just audible, that is, its amplitude must be ten times higher than that of the 2-kHz tone. Audio codecs take advantage of this phenomenon by maintaining the quantization noise below this audible threshold.

Frequency Masking

The response of the auditory system is nonlinear and the perception of a given tone is affected by the presence of other tones. The auditory channels for different tones interfere with each other, giving rise to a complex auditory response called *frequency masking*.

Figure 8.13 illustrates the frequency-masking phenomenon when a 60-dB, 1-kHz tone is present. Superimposed on this figure are the masking threshold curves for 1-kHz and 4-kHz tones. The masking threshold curves intersect the threshold of the hearing curve at two points. The intersection point on the left is around 600 Hz and the intersection point on the right is around 4 kHz. This means that any tone in the masking band between 400 Hz and 4 kHz with SPL that falls below the 1-kHz masking curve will be overshadowed or masked by the 1-kHz tone and will not be audible. For example, a 2-kHz tone (shown in

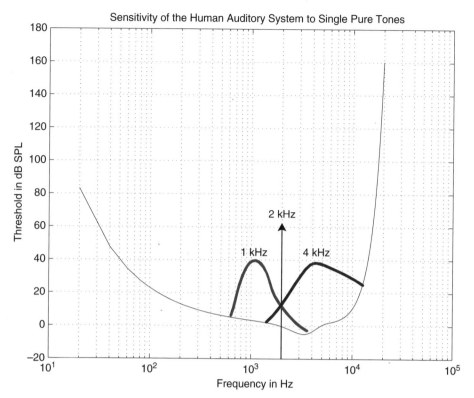

Figure 8.13 Frequency-masking phenomenon

Figure 8.13) will not be audible unless it is louder than 10 dB. In particular, the masking bandwidth depends on the frequency of the masking tone and its level. This is illustrated by the frequency masking curve for the tone at 4 kHz. As seen here, the masking bandwidth is larger for a 4-kHz tone than for a 1-kHz tone. If the masking tone is louder than 60 dB, the masking band will be wider; that is a wider range of frequencies around 1 kHz or 4 kHz will be masked. Similarly, if the 1-kHz tone is weaker than 60 dB, the masking band will be narrower. Thus, louder tones will mask more neighboring frequencies than softer tones, which makes intuitive sense. So, ignoring (i.e., not storing or not transmitting) the frequency components in the masking band whose levels fall below the masking curve does not cause any perceptual loss.

Temporal Masking

Temporal masking refers to a property of the human auditory system in which a second tone (test tone) is masked by the presence of a first tone (the masker). Here, the masker is a pure tone with a fixed level (for example, 60 dB). The tone is removed at time zero and a test tone is immediately applied to the ear. Figure 8.14 shows an example of the temporal masking curve. The horizontal axis shows the amplitude of the test tone in decibels. The vertical axis shows the response time corresponding to different levels of the test tone. It

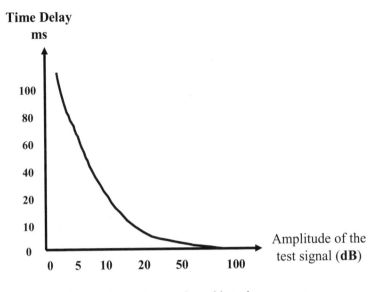

Figure 8.14 Temporal masking phenomenon

shows how long it takes for the auditory system to realize that there is a test tone. This delay time depends on the level of the test tone. The louder the test tone, the sooner the ear detects it. In other words, the ear thinks that the masking tone is still there, even though it has been removed.

8.2.3 Audio Coding Standards

Codec design is influenced by coding quality, application constraints (one-way versus two-way communication, playback, streaming, etc.), signal characteristics, implementation complexity, and resiliency to communication errors. For example, voice applications, such as telephony, are constrained by the requirements for natural two-way communication. This means that the maximum two-way delay should not exceed 150 ms. On the other hand, digital storage, broadcast, and streaming applications do not impose strict requirements on coding delay. This subsection reviews several audio coding standards. Figure 8.15 shows various speech and audio applications, the corresponding quality, and bitrates.

MPEG Audio Coding

The Moving Pictures Experts Group (MPEG) has produced international standards for high-quality and high-compression perceptual audio coding. The activities of this standardization body have culminated in a number of successful and popular coding standards. The MPEG-1 audio standard was completed in 1992. MPEG-2 BC is a backward-compatible extension to MPEG-1 and was finalized in 1994. MPEG-2 AAC is a more efficient audio coding standard. MPEG-4 Audio includes tools for general audio coding and was issued in 1999. These standards support audio encoding for a wide range of data rates. MPEG audio standards are used in many applications. Table 8.1 summarizes the applications, sampling frequencies,

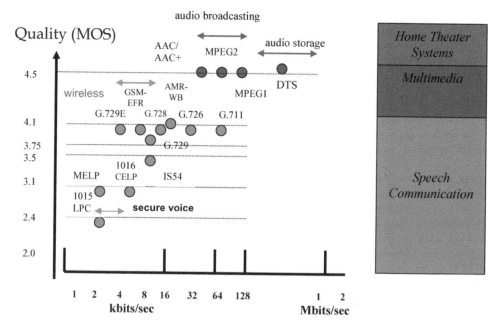

Figure 8.15 Applications, data rates, and codecs

Table 8.1 MPEG audio coding standards

Standard	Applications	Sampling	Bitrates
MPEG-1	Broadcasting, storage, multimedia, and telecommunications	32, 44.1, 48 kHz	32–320 kbps
MPEG-2 BC	Multichannel audio	16, 22.05, 24, 32, 44.1, 48 kHz	64 kbps/channel
MPEG-2 AAC	Digital television and high-quality audio	16, 22.05, 24, 32, 44.1, 48 kHz	48 kbps/channel
MPEG-4 AAC	Higher quality, lower latency	8–48 kHz	24–64 kbps/channel

and the bitrates for various MPEG audio coding standards. The following provides a brief overview of these standards.

MPEG-1

MPEG-1 Audio (ISO/IEC 1993b) is used in broadcasting, storage, multimedia, and telecommunications. It consists of three different codecs called *Layers I, II, and III* and supports bitrates from 32 to 320 kbps. The MPEG-1 audio coder takes advantage of the frequency-masking phenomenon described previously, in which parts of a signal are not audible because of the function of the human auditory system. Sampling rates of 32, 44.1, and 48 kHz are supported. Layer III (also known as MP3) is the highest complexity mode and is optimized for encoding high-quality stereo audio at around 128 kbps. It provides near CD-quality audio and is very popular because of its combination of high quality and high-compression ratio. MPEG-1 supports both fixed and variable bitrate coding.

MPEG-2 BC

MPEG-2 was developed for digital television. MPEG-2 BC is a backward-compatible extension to MPEG-1 and consists of two extensions: (1) coding at lower sampling frequencies (16, 22.05, and 24 kHz) and (2) multichannel coding including 5.1 surround sound and multilingual content of up to seven lingual components.

MPEG-2 AAC

MPEG-2 Advanced Audio Coding (AAC) is a second-generation audio codec suitable for generic stereo and multichannel signals (e.g., 5.1 audio). MPEG-2 AAC is not backward compatible with MPEG-1 and achieves transparent stereo quality (indistinguishable source from output) at 96 kbps. AAC consists of three profiles: AAC Main, AAC Low Complexity (AAC-LC), and AAC Scalable Sample Rate (AAC-SSR).

MPEG-4 Low-Delay AAC

MPEG-4 Low-Delay AAC (AAC-LD) has a maximum algorithmic delay of 20 ms and good quality for all types of audio signals, including speech and music, which makes it suitable for two-way communication. However, unlike speech codecs, the coding quality can be increased with bitrate, because the codec is not designed around a parametric model. The quality of AAC-LD at 32 kbps is reported to be similar to AAC at 24 kbps. At a bitrate of 64 kbps, AAC-LD provides better quality than MP3 at the same bitrate and comparable quality to that of AAC at 48 kbps.

MPEG-4 High Efficiency AAC

MPEG-4 High Efficiency AAC (MPEG-4 HE AAC) provides high-quality audio at low bitrates. It uses spectral band replication (SBR) to achieve excellent stereo quality at 48 kbps and high quality at 32 kbps. In SBR, the full-band audio spectrum is divided into a low-band and a complementary high-band section. The low-band section is encoded using the AAC core. The high-band section is not coded directly; instead, a small amount of information about this band is transmitted so that the decoder can reconstruct the full-band audio spectrum. Figure 8.16 illustrates this process.

MPEG-4 HE AAC takes advantage of two facts to achieve this level of quality. First, the psychoacoustic importance of the high frequencies in audio is usually relatively low. Second, there is a very high correlation between the lower and the higher frequencies of an audio spectrum.

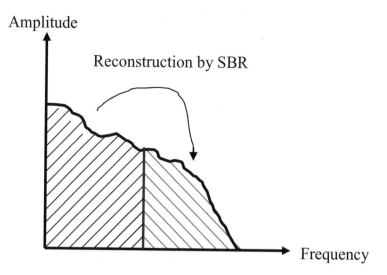

Figure 8.16 Spectral band replication in MPEG-4 HE ACC audio coder

Enhanced MPEG-4 HE AAC

Enhanced MPEG-4 HE AAC is an extension of MPEG-4 AAC and features a parametric stereo coding tool to further improve coding efficiency. The coding delay is around 130 ms and, therefore, this codec is not suitable for real-time two-way communication applications. In March 2004, the 3GPP agreed on making the enhanced MPEG-4 HE AAC codec optional for PSS audio services.

8.2.4   Speech and Audio Coding Issues

This subsection discusses the challenges for enabling mobile hi-fi communication over XG wireless networks and the issues of the existing codecs in meeting these challenges. A low-latency hi-fi codec is desirable for high-quality multimedia communication, as shown in the dashed oval in Figure 8.2.

Hi-fi communication consists of music and speech sampled at 44.1 kHz and requires high bitrates. Compression of multimedia content requires a unified codec that can handle both speech and audio signals. None of the speech and audio codecs discussed in the previous sections satisfy the requirements of low-latency hi-fi multimedia communication. The major limitation of most speech codecs is that they are highly optimized for speech signals and therefore lack the flexibility to represent general audio signals. On the other hand, many audio codecs are designed for music distribution and streaming applications, where high delay can be tolerated. Voice communication requires low latency, rendering most audio codecs unsuitable for speech coding. Although today's codecs provide a significant improvement in coding efficiency, their quality is limited at the data rates commonly seen in wireless networks. AMR-WB provides superior speech quality at 16 kbps and has low latency, but it cannot provide high-quality audio as its performance is optimized for speech sampled at 16 kHz, not 44.1 kHz. MPEG-4 HE AAC provides high-quality audio at

24 kbps/channel, but is suitable for broadcast applications, not low-latency communication. The low-delay version of AAC (AAC-LD) provides transparent quality at 64 kbps/channel. Even with the increases in bandwidth promised by XG, this rate is high and more efficient codecs will be required for XG networks.

The inherent capriciousness of wireless networks, and the fact that media is often transported over unreliable channels, may result in occasional loss of media packets. This makes resiliency to packet loss a desirable feature. One of the requirements of XG is seamless communication across heterogeneous networks, devices, and access technologies. To accommodate this heterogeneity, media streams have to adapt themselves to the bandwidth and delay constraints imposed by the various technologies. Multimode or scalable codecs can fulfill this requirement. Scalability is a feature that allows the decoder to operate with partial information from the encoder and is advantageous in heterogeneous and packet-based networks, such as the Internet, where variable delay conditions may limit the availability of a portion of the bitstream. The main advantage of scalability is that it eliminates transcoding.

Enhanced multimedia services can benefit from realistic virtual experiences involving 3D sound. Present codecs lack functionality for 3D audio. Finally, high-quality playback over small loudspeakers used in mobile devices is essential in delivering high-quality content.

8.2.5 Further Research

The following enabling technologies are needed to realize low-latency hi-fi mobile communication over XG networks.

- Unified speech and audio coding at 44.1 kHz

- Improved audio quality from small loudspeakers in mobile devices

- 3D audio functionalities on mobile devices.

Generally speaking, the increase in functionality and performance of future mobile generations will be at the cost of higher complexity. The effect of Moore's Law is expected to offset that increase. The following are the specific research directions to enable the technologies mentioned above.

Unified Speech and Audio Coding

Today's mobile devices typically use two codecs: one for speech and one for audio. A unified codec is highly desirable because it greatly simplifies implementation and is more robust under most real-world conditions.

Several approaches have been proposed for unified coding. One approach is to use separate speech and audio codecs and switch them according to the property of the signal. The MPEG-4 standard, for example, proposes the use of a signal classification mechanism in which a speech codec and an audio codec are switched according to the property of the signal. In particular, the HVXC standard can be used to handle speech while the harmonic and individual lines plus noise (HILN) standard is used to handle music (Herre and Purnhagen 2002; Nishiguchi and Edler 2002). Even though this combination provides reasonable quality under certain conditions, it is vulnerable to classification errors. Further research to make signal classification more robust is a possible approach toward unification. Another approach is to use the same signal model but to switch the excitation signal according to the signal type. The AMR-WB+ codec employs this approach by switching between the speech and the transform coded excitations (TCX). The problem is that to achieve high coding efficiency, it needs an algorithmic delay of 130 ms, which is not suitable for two-way communication. Reducing the coding delay while maintaining the quality at the same bitrate is a useful research direction. The improvement of quality without an increase in bitrate remains an important goal in media-coding research. Delay reduction is often in conflict with other desirable properties of a codec, such as low bitrate, low complexity, and good quality. Coding efficiency can be increased by better signal models and more-efficient quantization schemes. Signal models that are suitable for both speech and audio may provide the key to unified coding. For example, the MPEG-4 HILN model provides a framework for unified coding. In this model, three signal types (harmonics, individual lines, and noise) are used to represent both speech and audio. The signal is decomposed or separated into these three components. Each component is then modeled and quantized separately. Robust and reliable signal separation is needed for this scheme to work. Further research in signal classification and separation is a promising direction to make HILN successful. Explicit or hard signal classification has proved to be problematic in the past. Implicit or soft classification in which signal components are identified and sequentially removed is preferable.

Finally, sinusoidal coding, where the signal is analyzed as elementary sinusoids and separately represented through their frequency, amplitude, and phase might be a promising direction for unified coding. This signal model is very general and suits a wide range of real-world signals, including speech and music.

Scalability

Seamless communication across heterogeneous access technologies, networks, and devices is one of the goals of XG. Multimode and multistandard terminals are one way to deal with heterogeneity. This approach, however, requires multiple codecs and multiple standards. Scalability is a clear trend in speech coding and offers distinct advantages in this regard. Narrowband and wideband AMR and selectable mode vocoder (SMV) are examples of scalable codecs. MPEG-4 speech coding standards also support scalability (Nishiguchi and Edler 2002). Because of extra overhead, scalable codecs typically incur some loss of coding efficiency. Scalability may be achieved using embedded and multistage quantizers. To support scalability in a CELP codec, for example, one can use embedded quantizers to represent different parameters. In addition, the excitation signal can be represented using a multistage approach in which successive refinement is supported. Thorough evaluation of different approaches and their impact on system performance is needed to deploy a scalable codec working in an optimized manner.

3D Audio

XG promises to be user-centric. Realistic virtual experiences would greatly enhance the communication quality and tele-presence. Examples are virtual audio or video conferences where users can feel as if they are present in the room. To accomplish this, information about the 3D acoustic environment of the speaker must be gathered, and a method must be found to efficiently encode and transmit this information along with the audio bitstream.

On the receiver side, the 3D information must be recovered and the audio signals rendered and presented to the user. There are three specific challenges here. The first is speaker localization to identify the relative location of the speaker in an environment. In some applications, such as virtual teleconferencing, where speakers are stationary, head tracking may be sufficient to find the angle of the speaker relative to a reference (Johansson et al. 2003). The second challenge is compact representation of the 3D information. The 3D information of the sound is contained in the so-called echoic or wet head-related transfer function (HRTF). To spatialize a sound, that is to give it spatial dimension, the sound is filtered with the wet HRTF of the environment. Real environments exhibit complex acoustics because of reflections from obstacles, such as walls, floors, and ceilings. Efficient representation of complex echoic HRTFs is important in enabling 3D audio reproduction in bandwidth-limited mobile environments. The third challenge is low delay and efficient rendering. The decoder must be able to recover the transmitted 3D information and render high-quality 3D sound with low delay for two-way communication.

High-quality Audio

Codec technology has advanced to the point where analog I/O devices constitute a bottleneck in end-to-end quality. Because of size constraints, loudspeakers in mobile devices are very small. Small loudspeakers cannot reproduce the low frequencies present in speech and audio. They also have nonlinear characteristics. For example, the lowest possible frequency that can be faithfully reproduced by a typical 15-mm-diameter loudspeaker placed on a plate baffle is around 800 Hz. As a result, the bottleneck in quality is due to small loudspeakers, not coding technologies. Signal-processing techniques have been developed to model the loudspeaker nonlinearities. These models are used to find an equalizer or a predistortion filter to compensate for these nonlinearities (Frank et al. 1992). Several approaches have been employed. Volterra modeling is a general technique to model weak nonlinearities and produces promising results for small loudspeakers (Matthews 1991). The number of coefficients used in a Volterra model is on the order of a few thousand. Because of its large computational requirements, Volterra filtering may not be suitable for real-time operation. Wiener and Hammerstein models may be used for simpler models of nonlinearity. The Small–Thiele model provides a more compact representation with a small number of parameters. The main problem is to find a suitable inverse once the loudspeaker model has been identified. Work on precompensation techniques for small loudspeakers is essential for enabling high-quality multimedia playback over small mobile devices.

8.3   Video Coding Technologies

The deployment of video applications on mobile networks is much less advanced than voice and audio. This is not surprising because of the larger requirements for processing power, bandwidth, and memory for video applications. As processing power, bandwidth, and memory increase on mobile terminals, it is expected that video applications will become more common, much like what happened in the last ten years in the personal computer market. This section reviews the basics of video coding, the coding standards currently deployed, and those that will be deployed in a near future. The next section considers issues and solutions to make video applications ubiquitous in the mobile domain.

8.3.1   Principles of Video Coding

Video encoding is a process by which a sequence of frames is converted into a bitstream. The size of the bitstream is typically many times smaller than the data representing the frames. Frames are snapshots in time of a visual environment. Given a sufficiently high number of frames per second, an illusion of smooth motion can be rendered. The number of frames per second typically ranges from 10 frames/s for very low bitrate coding, to 60 frames/s for some high-definition applications. Rates of 24, 25, and 30 frames/s are common in the television and film industry, so a lot of content is available at those rates. While a low frame rate may be acceptable for some applications, others, such as sporting events, typically require frame rates of 50 and above.

Each frame consists of a rectangular array of pixels. The size of the array may vary between 176 by 144 for very low bitrate applications to 1920 by 1080 for high definition. In the mobile space, 176 by 144 is the most common frame size and is referred to as Quarter Common Interchange Format (QCIF). Each pixel may be represented by an RGB triplet that defines intensities of red, green, and blue. However, given the characteristics of the human visual system, a preferred representation for a frame consists of three arrays. These are a luma array that defines intensity for each pixel and two chroma arrays that define color. The chroma arrays typically have half the horizontal and half the vertical resolution of the luma array. This provides an instant compression factor of two without much visual degradation of the image because of the arrangement of photoreceptors in the human eye. Each sample within the luma and chroma arrays is represented by an 8-bit value, providing 256 different levels of intensity. Larger bit depths are possible but are generally only used for professional applications, such as content production and postproduction.

In typical video codecs, a frame may be coded in one of several modes. An I-frame (intraframe) is a frame that is coded without reference to any other frame. The coding of an I-frame is very much like the coding of a still image. A P-frame (predicted frame) is a frame that is coded with reference to a previously coded frame. Finally a B-frame (bidirectionally predicted frame) is a frame that is coded with reference to two previously coded frames, where those two frames may be referred to simultaneously during interpolation. B-frames generally contribute greatly to coding efficiency. However, their use results in a reordering of frames; that is, the display order and coding order of frames is different and additional delay is incurred. As a result, B-frames are generally unsuitable for conversational applications where low delay is a strong requirement.

Codecs typically partition a frame into coding units called macroblocks. A macroblock consists of a block of 16 by 16 luma samples and two blocks of collocated 8 by 8 chroma samples. A mode is associated with each macroblock. For example, in a P-frame, a macroblock may be coded in either intra or intermode. In the intramode, no reference is made to a previous frame. This is useful for areas of the frame without a corresponding feature in a previous frame, such as when an object appears. In the intermode, one or more motion vectors are associated with the macroblock. These motion vectors define a displacement with respect to a previous frame.

Figure 8.17 shows a simple block diagram of a decoder. Compressed data is parsed by an entropy decoder. Texture data is transmitted to an inverse quantizer followed by an inverse transform (e.g., an inverse DCT). Motion data is transmitted to a motion compensation unit. The motion compensation unit generates a predicted block using a frame stored in the frame

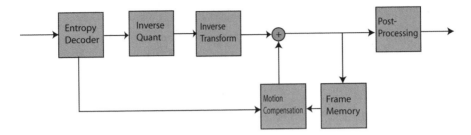

Figure 8.17 Decoder block diagram

buffer. The predicted block is added to the transformed data. Finally, an optionalpostprocess, such as a deblocking filter, is applied.

8.3.2 Video Coding Standards

The standardization of video coding standards began in the 1980s. Two bodies have led these standardization efforts, namely ITU-T SG16 Q.6 (Video Coding Experts Group or VCEG) who developed the H.26x series of standards and ISO/IEC SC29 WG11 (Moving Pictures Experts Group or MPEG) who developed the MPEG-x series of standards. Table 8.2 shows a timeline of the evolution of coding standards.

ITU-T led with the development of H.261 (ITU-T 1993) for $p \times 64$ kbps videoconferencing services. It was followed by MPEG-1 (ISO/IEC 1993a), which addressed compression for storage on a CD at 1.5 Mbps. Next, ITU-T and ISO/IEC jointly developed MPEG-2/H.262 for higher data rate applications. So far, MPEG-2 (ISO/IEC 2000) is one of the most successful standards with wide deployment in the digital television broadcasting and digital versatile disc (DVD) applications. The operating range is typically between 2 and 80 Mbps. ITU-T then addressed higher compression ratios with H.263 (ITU-T 1998) (including extensions) for PSTN videoconferencing at 28.8 kbps. On the basis of this H.263 work, MPEG standardized MPEG-4 (ISO/IEC 2001). ITU-T and ISO/IEC have recently produced a joint specification, called *H.264/MPEG-4 AVC (ISO/IEC 2003)*. This latest standard

Table 8.2 Timeline of evolution video coding standards

Year	Body	Standard	Application Domain
1989	ITU-T	H.261	p x 64 kbps videoconferencing
1991	ISO/IEC	MPEG-1	Stored media (e.g., video CD)
1994	ISO/IEC ITU-T	MPEG-2 H.262	Digital broadcasting and DVD
1997	ITU-T	H.263	Videoconferencing over PSTN
1999	ISO/IEC	MPEG-4	Mobile and Internet
2003	ITU-T ISO/IEC	H.264 MPEG-4 AVC	Mobile, Internet, broadcasting, HD-DVD

is expected to cover a wide range of applications from mobile communications at 64 kbps to high-definition broadcasting at 10 Mbps.

With respect to mobile networks and 3G in particular, MPEG-4 Simple Profile and H.263 Baseline are two standardized codecs that have been deployed as of mid-2003. 3GPP defines H.263 Baseline as a mandatory codec and MPEG-4 as an optional one. Both standards are very similar. 3GPP is likely to add H.264/MPEG-4 AVC as an advanced codec that provides higher coding efficiency in Release 6. The Association of Radio Industries and Businesses (ARIB) has further adopted the standard for delivering television channels to mobile devices. Although not a standardized codec, Windows Media Video 9 provides a good example of a state-of-the-art proprietary codec that may be used in mobile applications. The next sections discuss the technical details of these four specifications.

H.263 Baseline

H.263 is a flexible standard that provides many extensions that may be determined at negotiation time. The core of the algorithm is defined by a baseline profile that includes a minimum number of coding tools.

The H.263 decoder architecture matches the one described in Figure 8.17. Huffman coding is used for entropy coding. The inverse transform is the inverse discrete cosine transform of size 8 by 8. Even though this transform is mathematically well defined, limited precision is available in practical implementations. As a result, two different decoders may yield slightly different decoded frames. To mitigate this problem, an oddification technique is used.

In H.263, the precision of motion compensation is limited to a half pixel. Thus, motion vectors may take integer or half values. When noninteger values are present, a simple bilinear interpolation process is used, as in earlier standards, such as MPEG-1 and MPEG-2. A motion vector may apply either to an entire macroblock or to an 8 by 8 block within a macroblock. In the latter case, four motion vectors are coded with each macroblock. The selection of the number of motion vectors within a macroblock is done independently for each macroblock.

MPEG-4 Simple Profile

The MPEG-4 standard includes many profiles. However, only two are commonly used: Simple and Advanced Simple. Advanced Simple adds B-frames, interlaced tools, and quarter-pel motion compensation to the Simple Profile. In the mobile world, only the Simple Profile is used.

MPEG-4 Simple Profile is essentially H.263 baseline with a few additions, such as error resilience tools (i.e., tools that help a decoder cope with transmission errors). In MPEG-4, error resilience tools are mainly designed to cope with bit errors. These tools include data partitioning, resynchronization markers, and reversible variable-length codes (RVLC). In the data-partitioning mode, the coded data is separated into multiple partitions, such as motion and macroblock mode information, intracoefficients, and intercoefficients. Because the first partition is more helpful for reconstructing an approximation of a coded frame, it may be sent through the network with higher priority, or may be transmitted using stronger forward error-correction (FEC) codes. When only the first partition is available to a decoder, it is still able to produce a decoded frame that bears a strong resemblance to the original frame.

The use of RVLCs makes it possible to decode data in either a forward or backward direction. Thus, if an error occurs in the middle of a coded data unit, decoding of data before and after the error is possible. The reversibility feature imposes some constraints on the structure of the variable-length codes. Indeed, the variable-length codes are not complete, resulting in a decrease of coding efficiency.

H.264/MPEG-4 AVC

H.264/MPEG-4 AVC was developed jointly by VCEG and MPEG. Although it retains the basic hybrid motion-compensated structure of previous standards, it also includes several innovations. For example, the well-known 8 by 8 Discrete Cosine Transform (DCT) is replaced by an integer 4 by 4 transform. This transform approximates a DCT, and its inverse is precisely defined by a sequence of arithmetic operations. This guarantees that all decoders will reconstruct identical frames.

The improvement of prediction, both spatially and temporally, is the major component that drives the efficiency of MPEG-4 AVC. The smallest block that may be used for motion compensation is 4 by 4 pixels large. This enables a much more precise definition of motion boundaries. Furthermore, quarter-pixel interpolation improves on half-pixel interpolation. Prediction is further enhanced by the use of multiple reference frames. Although only one frame may be referenced at a time, different parts of a frame may use different reference frames for motion compensation.

Another significant improvement is the use of an in-loop deblocking filter. This filter has two effects: it improves the visual quality of decoded frames and improves the quality of motion-compensated prediction by removing some high frequencies in the reference frame. This in-loop filter does not preclude the use of an additional postfilter to further improve visual quality of the decoded frames.

The approach to handling channel errors is different in MPEG-4 AVC compared to MPEG-4 SP. MPEG-4 AVC was designed to cope with packet losses rather than bit errors. While the tools provided by MPEG-4 SP are useful in a circuit-switched network, MPEG-4 AVC will fare better in an IP-based network, such as XG. Flexible Macroblock Ordering (FMO) is one tool designed to cope with packet losses. For example, half the macroblocks in a checkerboard pattern may be transmitted within a same data packet and the other half in another data packet. When one of the data packets is unavailable to a decoder, good concealment remains possible. As for any error resilience tool, there is some degradation of coding efficiency, and in this case it is about 5%.

The standard defines three profiles: Baseline, Extended, and Main. Baseline is the simplest profile and is currently considered by 3GPP as a candidate for release 6. The Baseline profile does not provide support for B-frames.

Windows Media Video 9

As of April 2004, no detailed public disclosure of Windows Media Video 9 (WMV9) is available. Nevertheless, some information is available from a submission to 3GPP (Microsoft 2003b). WMV9 is a codec based on the same principles as its standardized counterparts. It features I-, P-, and B-frame types, quarter-pel motion compensation, integer transforms, and in-loop filtering, making it similar to MPEG-4 AVC in many ways.

8.3.3 Video Coding Issues

Although today's codecs provide a significant improvement in coding efficiency, the quality they provide at the data rates common on cellular networks remains limited. Indeed, the resolution of video (QCIF) and the frame rates (10–15 frames/s) are low. Even with the increases in bandwidth promised by XG, more efficient codecs will be required as the resolution of mobile terminal screens increases. Even if the network has the capacity to deliver video streams to mobile terminals, the economics remain an issue. Given that video streams are more voluminous than voice and audio streams, it is unclear whether a user would pay much more for video than for voice and audio. Therefore, it is important to be able to deliver video at relatively low data rates (e.g., 128 kbps) while maintaining high visual quality. Without acceptable quality, users are unlikely to use video applications. If the business model is changed to the model used for the Internet (i.e., flat fee, unlimited data transfers), it remains in the operator's interest to limit the amount of traffic by providing outstanding compression of video streams.

Currently deployed codecs are single-layer codecs that do not address scalability. Scalability is a desirable feature for applications, such as Multimedia Message Service (MMS), where the capabilities of a receiving device may not be known in advance, and where there may be multiple receiving devices with various capabilities. Scalability can also be a useful feature to address the heterogenity of XG networks and to adapt to channel conditions. The main advantage of a scalable codec is that transcoders are not needed when scaling the bitrate or resolution of a video stream. Scalable codecs have so far not been deployed in the market because of the loss of coding efficiency incurred by such codecs (a 50% overhead is not unusual). If scalable video codecs are to become viable, they need to offer coding efficiency comparable to that of single-layer codecs. Recent developments (MPEG 2003) suggest that this may be achievable.

8.3.4 Further Research

While standards, such as H.264/MPEG-4 AVC provide arithmetic coding as an optional tool, their core structure is still determined by Huffman coding. Macroblocks and the zigzag scan are two examples of coding tools that are a legacy of Huffman coding. By redesigning the structure around arithmetic coding, as was done in JPEG-2000 for still image coding, further significant gains may be within reach. The main difference between Huffman coding and arithmetic coding is that Huffman coding requires an integer number of bits for each coded element. There is no such restriction for arithmetic coding, making it more flexible and efficient. When probabilities of a coded element go well above one half, the inefficiency of Huffman coding is obvious. To avoid this, multiple elements are generally combined. For example, a single bit may indicate that a macroblock has zero motion and zero coded coefficients. One additional advantage of arithmetic coding is that adaptation to content is easier to implement. Although there are techniques for adaptive Huffman coding, their implementation is complex, so they are not widely used. This is not the case with arithmetic coding where adaptivity is typically built in.

Motion-compensated temporal filtering (MCTF) has been used to preprocess images, remove temporal noise, and increase coding efficiency. More recently, it has also been integrated into the coding loop to break with the traditional DPCM architecture (Ohm 1994).

The temporal integration provided by such an approach results in higher quality for those parts of the image that have temporal continuity. It is expected that this approach, including further refinements, will contribute to an increase of coding efficiency.

Perceptual visual quality remains poorly understood. Several objective metrics of the quality of a compressed video signal were designed (Süsstrunk and Winkler 2004). However, this has not led to any standard metric and the PSNR measure remains a widely used metric of quality. A better understanding of what makes an image look good should result in better encoders. Decoders could also benefit from better postprocessing algorithms that are designed to maximize quality with respect to a perceptual model.

Future advances in video coding, as discussed above, may result in a new standardized codec. Indeed, VCEG is planning to develop H.265 by about 2008 and MPEG is planning to develop a scalable video coder in the framework of MPEG-21.

8.4   Mobile Multimedia Applications

This section introduces mobile applications that have emerged as a result of the progress of wideband mobile networks and compression technologies. Speech is and will continue to be the traditional medium for communication, whether wired or wireless. Therefore, improved speech coding, such as low-latency hi-fi remains a significant objective. In addition to basic wireless telephony, DoCoMo is offering three multimedia applications: mobile TV phone, video clip download, and multimedia messaging services over its 3G network. Figure 8.18 summarizes the current mobile multimedia applications operated by DoCoMo as of early 2004.

8.4.1   Mobile TV Phone

With the rapid spread of mobile communication and the progress of standardization for 3G mobile networks, the ITU-T began studies on audiovisual terminals for mobile communication networks in 1995. Studies were made by extending the H.324 recommendation for the PSTN, and led to the development of H.324 Annex C in February 1998. When H.324 Annex C was standardized, functional enhancements were added to improve resilience against transmission errors caused by radio transmission. H.324 Annex C stipulates functional elements for providing audiovisual communication, as well as communication protocols that cover the entire flow of communication. For transmission methods that multiplex voice and image into one mobile communication channel and control messages exchanged in each communication phase, H.223 and H.245 are used.

On the basis of H.324 Annex C, the 3GPP Codec Working Group selected essential speech and video codecs and operation modes optimized for W-CDMA requirements, and prescribed the 3GPP standard 3G-324M in December 1999. In particular, a codec optimal for 3G was selected for this process without being restricted by the ITU-T standard. As a result, 3G-324M defines the specifications for audiovisual communication terminals for W-CDMA circuit-switched networks, optimally combining ITU-T recommendations H.324 Annex C and other international standards (3GPP 2002d, 2003e,g,h). Visual phones used in W-CDMA services are now compliant with 3G-324M. DoCoMo launched its visual phone service in 2001 for its 3G network, where it provides a 64-kbps circuit-switched connection

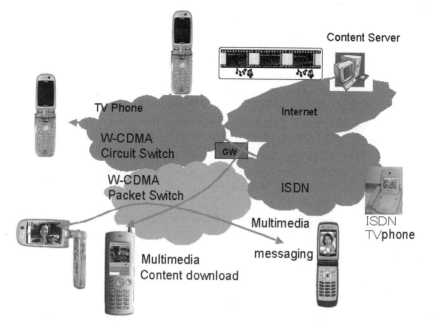

Figure 8.18 DoCoMo's multimedia services

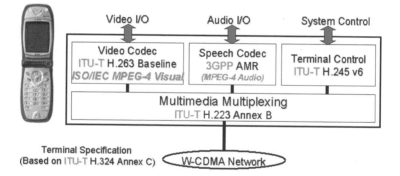

Figure 8.19 3G visual phone

that is compatible with N-ISDN. Figure 8.19 shows a 3G-324M terminal configuration. The 3G-324M standard is applied to speech and video codecs, the communication control unit, and the multimedia multiplexing unit. The speech codec requires AMR support. The video codec requires H.263 baseline and recommends MPEG-4 support. The support of H.223 Annex B, which offers improved error resilience, is a mandatory requirement for the multimedia multiplexing unit. Various media coding schemes can be used in 3G-324M by exchanging the terminal capability through the use of communication control procedures, and changing the codec setting upon the establishment of logical channels. 3G-324M defines a minimum essential codec to ensure interconnection between different terminals.

8.4.2 Multimedia Messaging Service

Mobile picture mail services have proved very popular in the 2.5G and 3G markets, as mentioned in Chapter 1. The picture mail service allows subscribers to transmit still images taken with compatible mobile phones featuring built-in digital cameras to virtually any device capable of receiving e-mail. As one of the 3G services, the mail service is extended in Japan to enable users to e-mail video clips as large as several hundreds of kilobytes taken either with the handset's built-in cameras or downloaded from sites. In this context, ISO/IEC and 3GPP standards have been adopted for Multimedia Messaging Services (MMS). The media types specified for MMS are text, AMR for speech, MPEG-4 AAC for audio, JPEG for still images, GIF and PNG for bitmap graphics, H.263 and MPEG-4 for video, and Scalable Vector Graphics (SVG). MMS specifications and usages are given in (3GPP 2002b,e, 2003d).

The generic stereo and multichannel (GSM) world, Europe in particular, and Japan use different aspects of MMS. In the GSM world, MMS is a store and forward messaging service that allows mobile subscribers to exchange multimedia messages with other mobile subscribers. As such, it can be seen as an evolution of SMS, with MMS supporting the transmission of additional media types, such as text, picture, audio, video, and combinations thereof. In Europe, MMS is an extension of SMS and is being used over GPRS networks. Because of limited bandwidth, a combination of AMR and JPEG pictures is typically used in GPRS. In Japan, operators started different picture mail services based on JPEG standards. These are not fully interoperable, and a gateway may be required to convert one format to another. In 3G, it was also anticipated that Japan's operators would start multimedia mail service without interoperability. To promote MMS as a successful application, interoperability among the operators is essential. For interoperability, it is better to use a common wrapper format of content as well as coding algorithms. With this view, DoCoMo has supported the 3GPP open standard and has opened their operational specification to the public. Figure 8.20 shows the history of the multimedia file format. The 3GPP standard-based file format is now being used for both DoCoMo's MMS and content download service.

8.4.3 Future Trends

Mobile applications evolve with the generations of mobile networks and compression technologies. Through the increase in available bandwidth, 3G and XG mobile systems offer new possibilities and services. One of these new services will be the ability to stream sound and movies to mobile devices. This section introduces media streaming that has not yet been commercialized, but it is anticipated that it would be done in the near future.

Streaming is a mechanism whereby media content can be rendered at the same time that it is being transmitted to a client over the data network. Streaming services are required whenever instant access to multimedia information needs to be integrated into an interactive media application. This differs from other multimedia services, such as MMS, where multimedia content is delivered to the user asynchronously by means of a message. 3G-324M provides a streaming mechanism that utilizes a N-ISDN compatible circuit-switched connection. Circuit-switched connections are most efficient for constant and continuous data streaming by definition, while they lack interactivity for web browsing.

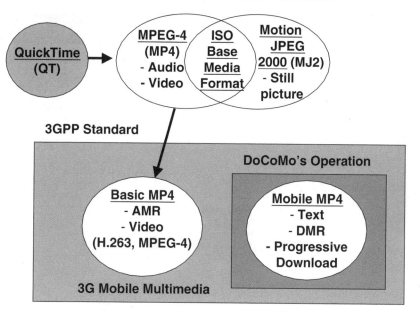

Figure 8.20 History of the MMS file format

Interactive applications that use streaming services include on-demand and live informa-tion delivery applications. Examples of on-demand applications are music, music video, and news-on-demand applications. Live delivery of radio and television programs are examples of live information delivery applications, which could, for example, make it possible to listen to a domestic radio station while abroad. A web server works with requests for information, delivers that information as quickly as possible, completes the transaction, disconnects, and goes on to other requests. A client connects to a web server only when it needs information.

A streaming service includes a TV or radio function by allowing the time series of media to be consumed as it is received, as well as still pictures and text. When consid-ering IP-transparent seamless connectivity, media streaming will certainly be based on a packet-switched connection, where a combination of IETF standards, such as the Real-time Transport Protocol (RTP), is adopted for 3G packet and XG networks. Figure 8.21 shows packet-based streaming protocol stacks standardized in 3GPP for W-CDMA networks. The details of the streaming specification are described in (3GPP 2002c,f, 2003f). The media types specified in this standard are very similar to 3G-324M and MMS.

Media transport technologies differ for each generation of mobile networks. Figure 8.22 summarizes the error-control technologies used in each generation. For real-time video con-ferencing over the circuit-switched network, DoCoMo uses H.324 Annex B for error control. H.324 Annex B is a sophisticated multiplexing scheme for bit errors. On the other hand, it cannot be used for packet streaming in which packet loss must be resolved. For robustness against packet losses, there are different techniques, such as forward error correction with data interleaving, and unequal error packetization (Stockhammer et al. 2003).

QoS support requires a different technology for each network generation (Figure 8.23). For 2G, voice and data networks are separated and each network realizes one specific QoS.

Content Stream Video/Speech/Audio/Image/Text					Session Control	Scene Des.	Menu	Security
H.263 Baseline (H.263, MPEG-4 Visual)	AMR, AMR-WB	MPEG-4 AAC-LC	JPEG	UTF-8 UCS2/ Unicode	SDP	SMIL	HTML/ WAP	Authentication Encryption Copy Control Renewal System etc...
					RTSP	HTTP		

RTP	TCP	Transport Layer Encryption & Authentication
UDP		
IP		
W-CDMA Packet, etc.		

Content Transport

Standardized	Not Discussed	Out of Scope

Figure 8.21 3GPP packet streaming

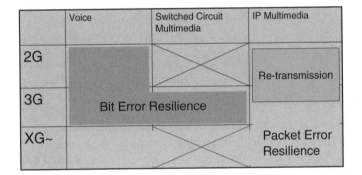

	Voice	Switched Circuit Multimedia	IP Multimedia
2G			Re-transmission
3G	Bit Error Resilience		
XG~			Packet Error Resilience

Figure 8.22 Mobile error-control technologies

One is for voice channels. For IP, every packet is reliably transmitted over a radio channel based on retransmission. For the 3G air interface, the QoS support is basically the same as in 2G, which means connections with different QoS requirements are recognized as separate connections, and the system controls each connection independently in order to provide different levels of service. On the network side, ATM provides the QoS framework.

In order to support various types of services, 3GPP defines four QoS classes: conversational, streaming, interactive, and background (3GPP 2003b). The quality of the synthetic (QoS) class used for each service type is determined by network operators. Typically, delay-sensitive services including circuit-switched service and VoIP are assigned to the conversational class, packet streaming services requiring a certain bit-rate guarantee are assigned to the streaming class, and best-effort and delay-insensitive services, such as web browsing and messaging, are in the interactive class or the background class. QoS class assignment is initiated by the terminal. The terminal requests the Network Layer to set up

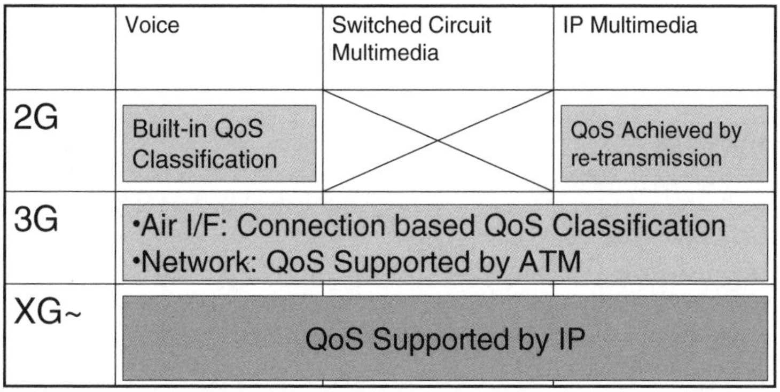

	Voice	Switched Circuit Multimedia	IP Multimedia
2G	Built-in QoS Classification		QoS Achieved by re-transmission
3G	•Air I/F: Connection based QoS Classification •Network: QoS Supported by ATM		
XG~	QoS Supported by IP		

Figure 8.23 Mobile network technologies for QoS control

network layer bearers. In GPRS and W-CDMA packet networks, this process is called *PDP context activation*. According to the 3GPP QoS model, different terminal applications may request differentiated QoS from the network according to the application needs, assuming that the network supports the QoS class. Typical radio network operation examples are as follows. According to the results in (3GPP 2001b), the quality of a video call using 3G-324M in circuit-switched service (AV bearer) shows unacceptable visual degradation at bit error rates (BER) of about 10^{-3}, and a BER of 10^{-4} or lower is required. According to (3GPP 2003a), a packet loss rate (PLR) of 10^{-3} at the transport layer is required for packet-switched conversational service (PSC). For these services, the required quality is satisfied by the transmission power control (TPC). The PSS, on the other hand, requires a PLR of 10^{-4} or lower (3GPP 2003c) at the transport layer. Since the streaming service tolerates a delay up to a few seconds, the required PLR is easily satisfied by the link-layer retransmission of RLC acknowledged mode. Moreover, because of the required compatibility with the ISDN bearer service, the unrestricted digital information (UDI) bearer service should satisfy a BER of 10^{-6}. Table 8.3 summarizes the required QoS in terms of link error rates.

As described in Chapter 3, the trend in emerging access networks beyond 3G is to use adaptive modulation and coding (AMC). In HSDPA, 1xEVDO and W-LAN, the main concept of network operation is *to maintain the link error rate to a small fixed amount by adjusting the dedicated bandwidth through the AMC*. Typically, the average BER requirement is set beforehand, depending on the class of application. The adaptive modulation is applied to maintain QoS, as evaluated directly by the signal-to-interference (SIR) power ratio or indirectly measured by BER. Accordingly, the Layer-1/2 transport control generally tends to provide two distinct states, quasi error free and burst errors during fading periods, while it leads to a large variation of bandwidth and delay. This trend will hold with various access networks in the next generation of systems.

In XG and beyond, QoS should be supported by the IP transport layer, where wireless QoS control should be in good accordance with the Internet QoS management framework, such as diffserv defined by the IETF. As the last hop of the wired IP network, a radio access network has its own unique set of complex characteristics depending on error conditions as

Table 8.3   QoS suggested by 3GPP standards

Service	Bit Error Rate	Transport Packet Loss Rate
UDI bearer	10^{-6}	N.A.
AV bearer	10^{-4} or better	N.A.
Packet-switched conversation	N.A.	10^{-3}
Packet-switched streaming	N.A.	10^{-4} or better

mentioned above. These characteristics should be associated with network-level QoS parameters, such as delay, jitter, error rate, and throughput, to meet end-to-end IP performance. This QoS support is a new challenge for the new generation (Etoh and Takeshi 2004).

9

Wireless Web Services

Henry Song, Dong Zhou, and Nayeem Islam

9.1   Introduction

A web service is an application or business logic that is accessible using standard Internet
protocols. Technically, a web service is defined by an interface that consists of a set of
operations offered by the service. This service definition resides in the network. Clients
can discover the definition and interact with the web service in a manner prescribed by its
definition, via XML-based messages conveyed by transport protocols.

Existing web services are intended to be served and consumed on servers and desktops
clients that have enough resources, such as processing speed and memory. As a result, the
overall architecture of web service, including its runtime modules, such as the XML parser,
is too heavyweight to run on a wireless mobile device. In a typical lifecycle of a web
services client application, a client device first sends a request (HTTP) to an application
server running remotely on the Internet. Upon receiving the request, an application workflow
is composed with multiple web service components. Then the application server executes the
workflow in a workflow engine, which may involve interaction with composing web service
components. Finally, the application server sends the response, in the form of an HTML
or a WML presentation, back to the client device. In this model, most of the computation
occurs on remote servers.

The explosion in the marketplace for wireless devices has opened new opportunities
for mobile applications. In the future, XG mobile devices will not be restricted to being
client devices that make requests to servers and receive responses from servers over the
network. XG mobile devices will become application servers that actively serve requests
from other mobile devices or desktops. For example, with camera-equipped mobile cell
phones, people can take pictures on their vacations, as well as author and publish their trip

Next Generation Mobile Systems.   Edited by Dr. M. Etoh
© 2005 John Wiley & Sons, Ltd

experience along with the pictures on their cell phones. This capability allows them to share their experience with family members and friends. In this scenario, cell phones become web service application servers that serve requests from other people.

Unfortunately, the technology necessary to support the previous vacation scenario is not quite available yet. To understand what is missing, one needs to understand the current web services architecture and the related web services technologies.

9.1.1   Emerging Web Services

Before web services, computing over the Internet was based on an information exchange through one-time and proprietary services. Often, data had to be converted to a mutually agreed format, and application logic had to be preintegrated when two companies wanted to interchange data.

Web services are quickly becoming the most prominent reusable distributed component technology with widespread acceptance and adoption from many industry players, including IBM, Microsoft, Sun, Oracle, and BEA systems. We believe that web services can potentially replace traditional distributed component technologies, such as OMG CORBA, Microsoft DCOM, and Java RMI.

The introduction of web services allows standard service interactions via the Internet in loosely coupled fashion. Web services have two key advantages over traditional, proprietary distributed services:

Interoperability: Web services build on standardized technologies that enable interactions among services with different implementations on different platforms.

Extensibility: Web services are based on XML (Recommendation 2002) technology, which allows services to extend in unlimited ways.

In a typical scenario, a service requester sends an HTTP (IETF n.d.) request to a service provider in the form of a SOAP (W3C n.d.b) message. The service provider receives the request, processes it, and returns the response back to the service requester, again via a SOAP message over HTTP. A well-known web service example is the stock-quote service. A client sends a request to the stock-quote service for the current price of a specific stock, and the stock-quote service sends the corresponding stock price back to the client.

9.1.2   Web Services Definition

A web service describes an interface that consists of a set of operations that can be accessed through XML messages over the network. The interface provides all details of a web service such that other web services can interact with this web service. The details include the web service location, the message format, and the transport protocol.

Web services separate interfaces from service implementations. The separation allows a web service to be independent of the hardware and software platforms on which it is implemented. In addition, a web service is independent of programming language in which it is written. This independence allows web service – based applications to be loosely coupled.

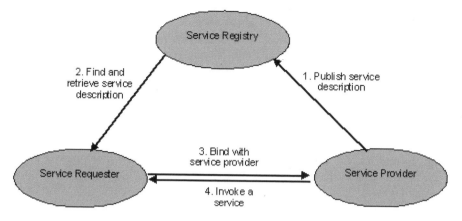

Figure 9.1 Web services entities and operations

9.1.3   Web Services Model

The web services architecture is based on interactions between three entities: the *service provider*, the *service registry*, and the *service requester*. The interactions among these three entities are *publishing*, *finding*, *binding*, and *invocation*. In a typical scenario, a service provider defines a service description for a web service and publishes it to a service registry. A service requester finds a particular web service in the service registry, retrieves the service description, and binds it with the service provider. After the binding, the service requester invokes and interacts with the service provider for the desired service. Figure 9.1 shows the interactions among these three entities.

Web Services Architecture Entities

As mentioned above, the web service model has three entities or actors that perform all the service operations.

A Service Provider is a person, an organization, or an application that owns a web service. The service provider provides the concrete implementation of the web service interface on a platform of its choice. An example of the service provider is Xignite.com, which provides a real-time stock-quote service on a Microsoft .Net platform.

A Service Registry is a searchable registry that hosts the published service description. A service requester looks up and finds the desired services in the service registry. The lookup and find can be done statically at the application development time or they can be done dynamically at the application execution time.

A Service Requester is a person, an organization, or an application that wishes to make use of a service provider's web service. The service requester finds and retrieves the service description from the service registry. After retrieval, it binds the service description with the service provider. Once the binding is complete, the service requester initiates an interaction with the service provider and exchanges messages.

Web Services Operations

For an application to use web services, four operations must take place: *publishing* service descriptions, *finding* service descriptions, *binding* services with service providers, and *invoking* the services. We discuss these operations in detail as follows.

Publishing: To allow service requesters to find and access a web service, a service provider needs to publish the description of the web services. Usually, the service provider publishes the service descriptions of the web services to a service registry.

Finding: To interact with a service provider, the service requester must retrieve a service description either directly from a service provider or query a service registry for the desired service. The find operation can be done in two different ways. The first way is to find and retrieve the service description statically at the application development time. The second way is to find and retrieve the service description dynamically at the application execution time.

Binding: Before the invocation of a web service, a service requester must perform a binding operation. The binding can be static or dynamic.

- Static binding allows a service requester to bind its service description to the implementation of the service description provided by a service provider at the application development time. This binding is fixed and cannot be changed at runtime.

- Dynamic binding allows a service provider to bind its service description to any implementation of the service description provided by different service providers at the runtime. After the binding operation, all abstract service type information on the service description interface is bound to the real service location, interaction type, interaction parameters, and interaction transport protocol.

Invoking: The service requester interacts with the service provider by invoking a request to the service provider. The interaction is usually performed in a request/response fashion. That is, the service requester makes a request to the service provider, and the service provider sends the response back to the service requester.

9.2 Web Services Architecture

The web services architecture consists of several layers: the service execution and processing layer, the service publication and discovery layer, the service description layer, the messaging layer, and the network layer. Web services architecture defines technologies for each layer. We first examine the architectural stack for web services, and then discuss the technology for each layer.

9.2.1 Web Services Stack

Web services architecture involves many layered technologies. These technologies, however, are also interrelated. Figure 9.2 shows the web services stack, which depicts web

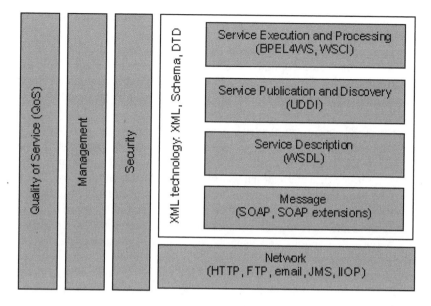

Figure 9.2 Web services stack

services technologies and their relations. This stack is a collection of standardized protocols and application programming interfaces (APIs) that let applications locate and utilize web services. Technologies involved in each stack layer are achieved by standard proposals, or standardized protocols and APIs. Standardization is the key to the ubiquitous deployment of web service architectures, and the ubiquitous deployment of the stack is the key to the success of web services.

In Figure 9.2, the upper layer technologies build upon the capabilities of the lower layers. The vertical bars represent requirements that must be addressed at each layer of the stack.

The foundation of the web services stack is the network. Web services must be network accessible. Service requesters invoke web services via the network, and web services that are publicly available on the network use common network protocols. One example is the HTTP protocol. Because of its ubiquity, it is currently the de facto standard network protocol for web services on the Internet. Other Internet protocols, such as SMTP (IETF n.d.) and FTP (W3C n.d.a), can also be supported.

The next layer is XML-based messaging. SOAP was chosen by the web service industry for XML messaging for several reasons:

- The SOAP protocol is the standard message delivery mechanism for exchanging documents and remote procedure calls (RPC) using XML.

- SOAP can be easily layered on HTTP messages with the SOAP envelope as the HTTP message payload.

- SOAP defines a standard mechanism to allow SOAP extensions in the SOAP messages.

The *service description* layer defines the standard to describe web services and their interfaces. WSDL (W3C n.d.c) is the de facto standard for XML-based service descriptions that support interoperable web services. WSDL defines the interfaces and the mechanics of service interaction.

Web services are defined to be network accessible via messages, and their interfaces are represented by service descriptions. The simplest web services stack would consist of using HTTP protocol for message transport, the SOAP protocol for the XML message exchange, and WSDL for service description. This is the interoperable and basic stack for all intraenterprise, interenterprise, and public web services.

To make a WSDL document available to service requesters, the web service provider must publish the WSDL. The simplest way for a service provider to publish the document is to send the WSDL document directly to a service requester (for example, via e-mail). Alternatively, a service provider can publish the WSDL document to a host. The host can be a private host within a specific domain or an enterprise, or it can be a host that is publicly accessible. After service providers publish their service descriptions, service requesters must discover these descriptions to use the services. Any method that allows service requesters to discover service descriptions and integrate them into applications is called a *service discovery*.

It is natural that a web service can be created by composing other web services. A web service composition can be done in several ways. A web service can be created to present a single web service interface by combining multiple web services together. Alternatively, a service requester may call multiple web services during its processing lifetime. The top service execution and processing layer describes how service-to-service and requester-to-service communications flows are performed.

In addition, for web services to be successful in the real world, infrastructures including security, management, and quality of service must be provided. These issues have to be addressed in each layer of the stack. The solutions at each layer can be independent of each other.

9.3   Web Service Technologies for Small Wireless Devices

Wireless web service is increasingly becoming the *de facto* standard for accessing services via the Internet. The push for web services on small devices requires the development of web services technologies that are designed specifically for mobile devices. These technologies were not developed for desktops or servers and were ported to small devices as afterthoughts. The goal of these technologies is to create a micro application server that interacts with web service application servers as well as web service application clients.

With the increasing capability of XG mobile devices in hardware, such as the increasing processing speed and the increasing amount of memory storage, and in software, such as web service server capabilities, we witness a wireless web service stack formulating and becoming mature. This wireless web services stack differs from normal web services stack in that it is designed and optimized to operate on mobile devices. In this section, we introduce existing technologies that are designed from the ground up for web services on small devices. Figure 9.3 shows a sample web service stack and its corresponding technologies for each layer.

In the wireless web service stack, one particular technology developed for the network layer is the kHTTP (kHTTP n.d.). kHTTP is a small HTTP server and provides a set of

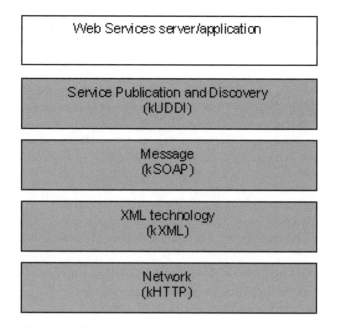

Figure 9.3   Sample web service stack on mobile devices

APIs for the J2ME (Sun n.d.a) platform. With technologies, such as kHTTP, J2ME/MIDP devices can be web service providers that serve web service requests.

Parsing XML documents into structured information for upper layer applications is a resource-intensive operation. On mobile devices where resources, such as memory and processing power, are limited, kXML (kObjects n.d.b) is designed to provide an XML parser that uses as little resource as possible. It is suitable for mobile platforms, such as J2ME.

kSOAP (kObjects n.d.a) is an SOAP API suitable for mobile platforms, such as J2ME, and it is based on kXML. kSOAP is a subset of the SOAP 1.1 specification. Because of the memory constraints on mobile devices, not all SOAP features can be included.

kUDDI (KUDDI n.d.) project is an effort to bring UDDI client APIs to mobile devices. The final goal of the project is to implement a subset of UDDI client APIs on mobile devices, such that mobile devices can discover and retrieve web service interfaces from UDDI registries.

With kXML and kSOAP, mobile devices can become web service clients that can make web services invocations to servers via the Internet. Coupled with a kHTTP server and kUDDI, mobile devices can also become web service servers that publish their web service interfaces to UDDI registries using kUDDI APIs. Clients can access the services offered by mobile devices by retrieving web service interfaces from UDDI registries and invoking the corresponding web services on mobile devices. This capability can change how mobile devices can be used in our daily lives. For example, camera-equipped cell phones can be web service servers that allow other people to access the pictures taken and stored on the mobile devices. Unlike today, when cell phone users have to upload their pictures to their

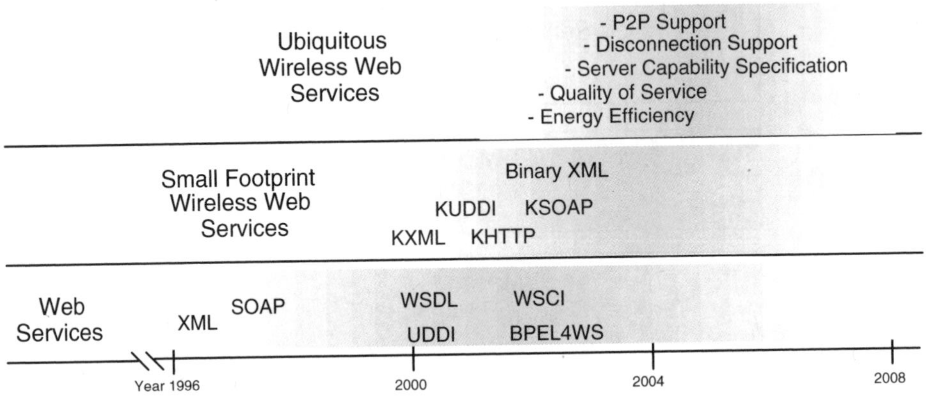

Figure 9.4 Current activities and future directions for wireless web services

desktop PCs or picture-hosting websites, soon pictures taken by cell phone users will be directly accessible via web service invocations to cell phones.

Figure 9.4 shows current activities and future directions in wireless web services area, along with activities in the general web services field. It divides activities in wireless web services into two generation: the Small Footprint Era, and the Ubiquitous Era. While current activities in this field have mainly focused on achieving small-footprint implementations, we project that future work in this area will target supporting ubiquitous access of wireless web services. Some of the research directions include

- energy efficiency, as smaller footprint does not necessarily translate to less energy consumption;

- specifications for wireless web service server capability on mobile devices, so that mobile device manufacturers and system software developers can meet the minimal requirements for running web services on wireless devices;

- peer-to-peer support, so that a mobile device can provide service to, and get service from, peer devices in an ad hoc network;

- disconnection support, so that mobile devices can cache web services to deal with disconnection and weak connection;

- quality of service support, so that predictable level of service quality can be guaranteed in a less predictable mobile environment.

The rest of this section describes in detail each of the web service technologies and their small-footprint implementations on mobile devices wherever applicable. While kXML, kUDDI, kHTTP and kSOAP are important web service technology implementations for mobile devices, there are also parallel works in each respective areas. We will cover some of them to give the reader a broader view of the field.

9.3.1 Communication Technologies

Web Service Communication Technologies

Web services are communicated via the network layer. This layer encompasses a wide variety of network protocols, such as HTTP, SMTP, FTP, Remote Method Invocation (RMI), and so on. The network protocol used in web services depends upon the web services' application requirements.

For those web services that are accessible on the Internet, the network protocol choice may favor the ubiquity of the network protocols, such as HTTP. Alternatively, other network protocols can be used for message exchange over the Internet, provided that service requesters and service providers agree upon the use of these network protocols. On the basis of the application requirements, such as security, performance, and reliability, other protocols (such as SSL) can be used.

The web services stack does not specify which network protocol to use; the message layer developers are responsible for the specifications using message-to-network binding. This network protocol specification is hidden from the application and service developers.

HTTP for Small Wireless Devices

kHTTP is a small HTTP server for resource-constrained mobile devices. It contains a small memory footprint runtime module and a set of programming APIs. Although the target platform of kHTTP is J2ME platform, its interfaces can be ported to any mobile platform.

kHTTP can operate in two modes: the normal mode and the proxy mode. In the normal mode, a kHTTP server, like any normal HTTP server, is accessible by a direct HTTP request via the network. In the proxy mode, a kHTTP server is communicated via a HTTP proxy server, which sits between the kHTTP server and the client.

Figure 9.5 shows the kHTTP server in the normal mode. The normal mode makes use of a class called *StreamingConnectionNotifier* in the CLDC connection framework that may not be available to many MIDP 1.0 implementations, because MIDP 1.0 specification only specifies *HTTPConnection* as the mandatory network connection. The MIDP 2.0 specification (JSR 118) (JCP n.d.a) has responded to these needs by adding support for socket and datagram; thus, providing mobile applications with more-capable networking interfaces that are available with the CLDC connection framework. However, such an addition is optional,

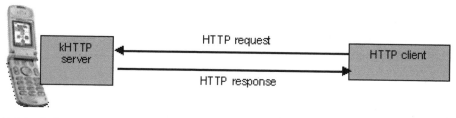

Mobile device

Figure 9.5 kHTTP server in normal mode

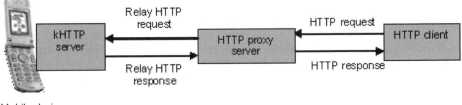

Figure 9.6 kHTTP server in proxy mode

which means a J2ME 2.0 implementer can select the support for such interfaces on the basis
of the target market segments and devices. Also, if a mobile device is on a private network
(such as one using NAT) or behind a firewall, the mobile device cannot be accessed by
direct HTTP requests.

To support a HTTP server in those devices, kHTTP can operate in the proxy mode, in
which a HTTP proxy sits between the mobile device and its clients. The HTTP proxy server
defines and implements an [HTTP Server Proxy] protocol that replays HTTP requests and
responses between a client and the kHTTP server. Figure 9.6 shows kHTTP server in the
proxy mode.

To conserve memory consumption on mobile devices, kHTTP implements a subset of
the HTTP server, including receiving requests and sending responses. It does not support
any memory-intensive functions, such as load balancing or name-based virtual hosting.

PocketHTTP is another HTTP client implementation targeting small devices (Fell
2004). It is an HTTP/1.1 client COM component running on Microsoft Windows platforms, including PocketPC. It is part of the PocketSOAP project (see Section 9.3.2).
PocketHTTP supports features, such as chunked encoding and persistent connections,
SSL support including SSL via proxy servers, compression support including the ability to compress the request body, server authentication, proxy server, session cookies,
and redirects. More information on PocketHTTP can be found at its project homepage
(http://www.pocketsoap.com/pocketHTTP/).

9.3.2 The Base Technology – XML

XML and Web Services

XML is a simple approach for marking up content with tags to convey information. The
tags delimit the content and they are expressed in a natural language to provide a human-
readable XML syntax. An XML document is in an ordinary text document, making it a
widely accepted solution for exchanging information between different platforms. Unlike
markup languages that have tags with fixed semantics such as HTML, XML allows users
to assign arbitrary names and semantics to tags. These features make XML easy to program
and extend.

XML is the backbone of the web service architecture. It provides the extensibility,
platform independence, and language neutrality that are keys to the loosely coupled and
standards-based application interoperability. This interoperability is the essence of the web
services value proposition.

XML for Small Wireless Devices

The kXML project provides an XML pull parser that is suitable for all Java platforms, including J2ME. Because of kXML's small footprint size, it is especially suitable for Java applications, such as Java MIDP applications on mobile devices. KXML features include:

- XML namespace support

- A relaxed mode for parsing HTML or other SGML formats

- A small memory footprint

- A pull-based parser for parsing XML structure

- XML writing support including namespace handling

- An optional kDOM

- Optional WAP support (WBXML/WML).

The parser generates different events for different elements:

- It generates a START_DOCUMENT event when it first encounters the start of an XML document.

- For each XML element, it generates a START_TAG event when it encounters a starting tag of an XML document.

- It generates a TEXT event when it reads the textual content of an XML element.

- For each XML element, it generates an END_TAG event when it reaches the end of an XML element.

- When it reaches the end of the document, it generates an END_DOCUMENT event.

Because the parser does not keep track of the XML document elements and their content (it passes these information to applications), the memory consumption of the pull-based parser is minimal and is suitable for memory-constrained mobile devices.

There have been considerable interests in using binary XML on mobile devices. As early as 1999, Ericsson, IBM, Motorola, and Phone.com submitted the WBXML note to the World Wide Web Consortium (W3C) on WAP binary XML content format (Martin and Jano n.d.). In 2003, W3C organized a workshop specifically targeting the binary interchange of XML information, whose participants included major industry players within and out of the mobile communication business (W3C 2003b). One of the interesting systems presented in this workshop is esXML and its canonical API: esDOM (Williams 2004).

esXML (or Efficiency Structured XML) and esDOM target XML and network application efficiency problems in data parsing, serialization, binding, allocation, and programming verbosity. It is a new portable structure that retains all XML features. Its representation is preparsed; its design reduces overheads in reading, traversing, transforming, and generates documents or data. It introduces new semantics including efficient pointers, copy-on-write layering of changes to a base document, and direct representation of binary content, such as images.

9.3.3   Messages – SOAP and its Extensions

SOAP and Web Services

SOAP provides a simple and lightweight mechanism for exchanging structured and typed information between server requesters and service providers in a decentralized and distributed environment. SOAP message syntax is based on XML. In fact, a SOAP message itself is an XML document. SOAP consists of three parts: the SOAP envelope, the SOAP encoding rules, and the SOAP RPC.

- The SOAP envelope defines an overall framework for expressing what is in a message, who should handle it, and whether it is optional or mandatory.

- The SOAP encoding rules define a serialization mechanism that is used to exchange instances of application-defined data types.

- The SOAP RPC defines a convention that is used to represent RPC and responses.

A SOAP message can be used in combination with or be re-enveloped by a variety of network protocols, such as HTTP, FTP, and RMI.

The basic requirement for a node in the network to participate in SOAP message exchange is the ability to build and parse a SOAP message and the ability to communicate over the network. Applications that communicate with web services using SOAP can be executed in four basic steps as shown in Figure 9.7:

- An application acting as a service requester constructs a SOAP message. The SOAP message is the request to the service provider to invoke a web service. The XML document in the SOAP message body can be a SOAP RPC request or a document message describing the operation of the web service according to WSDL. The application passes the SOAP message and the network address of the service provider to the SOAP runtime. The SOAP runtime then interacts with the underlying network protocol to send the SOAP message to the service provider.

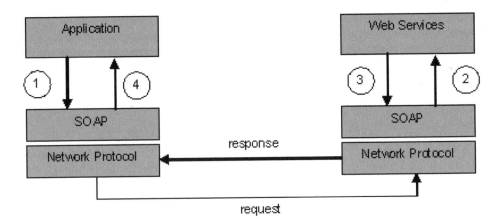

Figure 9.7   XML/SOAP message

- After the service provider's SOAP runtime receives the SOAP message, the SOAP runtime converts the XML message in the SOAP body to programming language-specific objects that can be understood by the service provider and passes it to the service provider.

- The service provider then processes the request message and creates a response. The response is encapsulated in a SOAP message. The service provider passes the SOAP message response to the SOAP runtime with the service requester's network address. The SOAP runtime coordinates with the network protocol to send the SOAP message response back to the service requester.

- After the SOAP message response is received by the service requester's SOAP runtime in the service requester, the SOAP runtime converts the XML message in the SOAP body to objects that are understood by the service requester application and passes them up to the application.

SOAP message exchanges between the service requester and service provider can be synchronous or asynchronous. They can also be performed in different modes: one-way messaging, notification, or publish/subscribe.

SOAP provides mechanisms that can be used to extend SOAP's capabilities. Examples of SOAP extensions include WS-Reliability (Sun n.d.b) and WS-security (IBM n.d.). The SOAP extensibility model provides two mechanisms to add extensions: the SOAP processing model and the SOAP protocol binding framework. The former describes the behavior of a single SOAP node with respect to the processing of a SOAP message. The latter mediates the interaction of sending and receiving SOAP messages between SOAP layer and network protocol.

The SOAP processing model enables SOAP nodes to extend their capabilities by including such features in the SOAP header blocks of the SOAP envelope. Such header blocks are intended for any SOAP nodes that can recognize such features when processing SOAP messages.

SOAP protocol binding operates between two adjacent SOAP nodes along a SOAP path. There is no requirement that the same network protocol to be used for all hops along a SOAP message path. This allows a SOAP node to specify the network protocol to be used between hops flexibly. There are also cases that a single network protocol is used along the entire SOAP message path.

SOAP for Small Wireless Devices

kSOAP is an SOAP API for J2ME based on kXML. The feature set of kSOAP is a subset of the SOAP 1.1 features. kSOAP supports basic functions for handling SOAP envelopes, such as reading/writing the content of SOAP headers and reading/writing the content of SOAP body. kSOAP also supports serialization/deserialization of SOAP messages and supports for transports of SOAP messages via HTTP network.

PocketSOAP is a SOAP client COM component for the Pocket PC platform (it also has a Win32 version that supports other Microsoft Windows family platforms)[1]. PocketSOAP is open source. It uses PocketHTTP as default transport although other transports can be

---

[1]For more information on PocketSOAP, see http://www.pocketsoap.com

easily added as well. It uses Expat XML Parser to parse SOAP response messages. The main features of PocketSOAP include support for Section 5 encoding, support for attachments (both DIME and SOAP with attachment), support for both 1999 and 2001Schema versions, and support for integration with the Win32 development environment (such as serialization of persistent VB objects).

9.3.4 Web Services Discovery (UDDI)

UDDI and Web Services

The discovery of a web service can take place at different times in the overall web service lifecycle. At one extreme, web service discovery can be purely static. The service requester already knows, or has an agreement with, a specific service provider. The service requester can search for that service provider from a registry and integrate with the provided service at the application development time. At another extreme, web service discovery can be purely dynamic. The service requester has only a list of service requirements. The service requester dynamically searches for a service provider that can fulfill its requirements and compose with the service just prior to the actual service execution. The Universal Description, Discovery and Integration (UDDI) (UDDI n.d.) of web services specification accommodates both types of web service discovery. UDDI defines the XML-based interfaces such that service requesters can find a desired service from service registries.

UDDI is a group of web-based registries that expose information about a web service interface published by service providers. UDDI registries are distributed over the Internet and administrated by different organizations. By accessing the public UDDI registries, one can search for information about web services. The information can help a service requester determine "who, what, where, and how" to interact with a particular web service. The contact information in UDDI specifies *who*. The web service classification based on industrial codes and products specifies *what*. The registered web service's address, such as a URL, where a service can be accessed, specifies *where*. Finally, registered web service's service interface reference specifies *how*. This reference is called a *tModel* in the UDDI specification.

A UDDI entry in a UDDI registry starts with a businessEntity element. A businessEntity element contains information about a business, including basic business information, such as the name of the business. A businessEntity element is a collection of businessService elements. A businessService element represents a web service. Each businessService element contains technical and descriptive information about the web service. A businessService element contains a collection of bindingTemplate elements. A bindingTemplate element describes how the businessService element uses various tModels. Figure 9.8 shows the structures of elements in a UDDI entry.

Technical Model (tModel) A tModel is a data structure that represents a service type registered in a UDDI registry. Each registered web service is categorized on the basis of the service type. This categorization enables service requesters to find a web service or service provider by searching the service types.

Each tModel consists of a name, a service description, and a Universal Unique Identifier (UUID). The tModel name corresponds to a service type, for example, uddi-org:http. The tModel description provides more information about the service, for example, *an http or web browser-based web service*. The tModel UUID, also known as the tModelKey, is a series

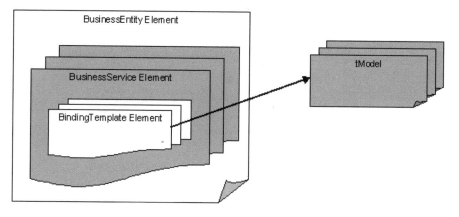

Figure 9.8 UDDI entry structure

of alphanumeric characters for identifying the service type, for example, uuid:68DE9E80-AD09-469D-8A37-088422BFBC36. The above example is a tModel, which describes a web service that is invoked through a web browser and the HTTP protocol.

Mapping WSDL and UDDI The mapping of WSDL and UDDI maps each WSDL elements to a separate UDDI entity, and therefore accurately represents the WSDL structure and semantics. portType and binding elements in WSDL map to a tModel in UDDI. WSDL service element maps to a UDDI businessService element. WSDL port element maps to bindingTemplate element. Figure 9.9 shows the mapping between WSDL elements and elements in UDDI version 2.

A web service description can be published in a variety of ways and can be published to several service registries. The simplest way of publishing a service description is with a direct publish. A direct publish means that a service provider sends a service description directly to a service requester. This can be done via e-mail, FTP, or media distribution (such as a CD-ROM). A direct publish occurs after a service provider and a service requester have established a partnership over the Internet. A more dynamic publishing way is to publish service description on one or multiple service registries. Several types of service registries can be used depending on how a web service is used.

Internal UDDI registry – Web services that are intended for use within an organization should be published to an internal registry. The scope of this internal registry can be departmental or enterprise level. The service registry is behind organization's firewall and is not accessible from the outside. Figure 9.10 shows the diagram of internal registry.

Portal UDDI registry – Web services can be published to a portal registry for external partners. A portal registry is outside a service provider's firewall and between firewalls. The portal registry only contains service descriptions that an organization wishes to provide to its partners. Usually, a role-based controls system is installed in the portal

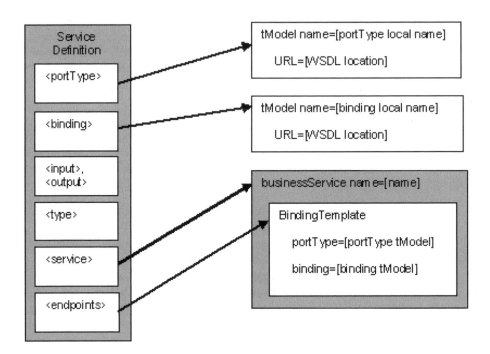

Figure 9.9 WSDL to UDDI v2 binding

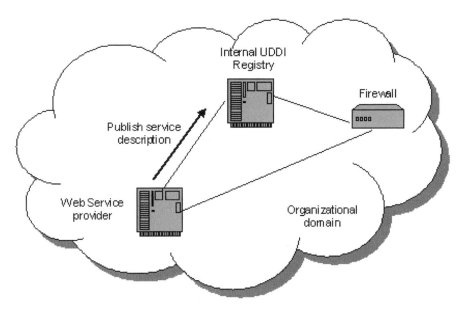

Figure 9.10 Internal UDDI Registry

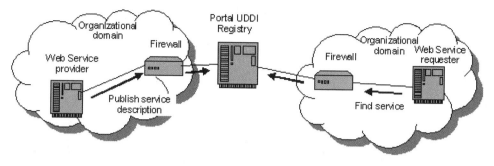

Figure 9.11 Portal UDDI Registry

registry to limit the visibility of services to authorized partners. Figure 9.11 shows the diagram of portal registry.

Partner UDDI registry – If web services from a service provider are intended for a particular partner, the web services can be published to its partner's registry. The partner registry is behind the partner's firewall. The partner registry contains only tested, legitimate service descriptions. Figure 9.12 shows the partner registry.

Public UDDI registry – A service description can be published to one or more public registries to compete in an open market. The public registry can be hosted by organizations, such as standardization bodies or consortiums, or by organizations that build their business around content hosting. The public registry is accessible by any service requester who searches for services.

UDDI for Small Wireless Device

kUDDI is an effort to port UDDI4J (JCP n.d.b), a heavy Java API library for interactions with UDDI registries, to the J2ME platform. In contrast to UDDI4J, which allows SOAP messages to bind with different network transport protocols, kUDDI is based on a simple kSOAP implementation that only allows HTTP connections. kUDDI does not support a

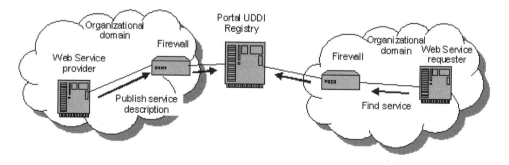

Figure 9.12 Partner registry

HTTP proxy server that can redirect requests to a HTTP server via an HTTP proxy. However, kUDDI is not short of functionality. kUDDI supports the complete API interfaces that UDDI 2.0 specification provides, including inquiry functions and publishing functions.

9.3.5 Web Services Description (WSDL)

To allow interoperability across heterogeneous systems, web service requesters and providers need to have a mechanism that describes precise message structure and data types. WSDL is an obvious choice today as the means to provide such precise description of web services. WSDL allows service requesters and providers to specify the inputs, outputs, and procedures of the web service binding and invocation.

Technically, WSDL is an XML document that describes web services as a set of endpoints operating on a message containing documents or RPC-oriented messages. The operations and messages are described abstractly and then bound to a concrete network protocol and message format to define an endpoint. WSDL can be extended to describe endpoints and their messages, regardless of what message formats or network protocols are used. Figure 9.13 shows the format of a WSDL document.

A WSDL document contains a set of service definitions. A service definition is defined by six major elements.

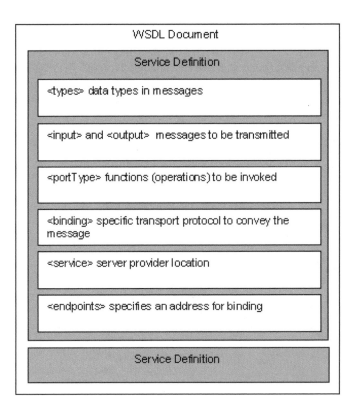

Figure 9.13 WSDL document version 2.0 overview

- The <types> element provides data type definitions for data in the exchanged message. The data types are usually defined with XML Schema to achieve maximum platform neutrality.

- <input> and <output> elements represent the required incoming and outgoing messages for an operation. The <input> and <output> elements respectively contain an optional message attribute that represents the content of the message, and an optional message reference attribute that identifies the role of the associated message in a message exchange pattern of a given operation.

- The <portType> element defines a set of operations. Each operation refers to an input message or an output message. The <portType> element can be thought of as a function library or a class in conventional programming languages.

- The <binding> element describes the protocol and the data format for operations and messages defined in a <portType>. The <binding> element contains two attributes: the name and the type. The name attribute defines the name of the binding, and the type attribute depicts the port where the binding occurs.

- The <service> element contains a collection of <endpoints> elements and an interface. The <service> element defines a conceptual web service. The <endpoints> element specifies the location of the implementations of the web service, and the interface defines the set of operations that the web service supports.

- The <endpoints> element specifies the location of the web service implementation, that is, the address for binding. For example, the location can be a network address or URL.

The WSDL document contains sufficient information to describe how service requesters can invoke and interact with a web service. The service interfaces defined in a WSDL are platform independent. A web service provider can implement the web service interfaces defined in the WSDL document on platforms of its choice.

The publication of web services includes the creation and publishing of web service descriptions. The service description can be created by hand coding or combining existing service interfaces.

9.3.6 Web Services Execution and Process

Beyond the description of individual messages and the process of discovering services, a web service has other process descriptions that allow the execution or integration of web services. They include the process for describing multipart and stateful sequences of messages, and the process for aggregating processes into higher-level processes.

Web services use a loosely coupled integration model to allow flexible integration of heterogeneous systems from different domains. Web services include B2C, B2B, and enterprise application integration among service requesters and service providers. Web services interaction is more than a simple request – response transaction with standard network protocols and message exchanges. Web services can fulfill their full potential when service requesters and service providers are able to integrate their complex interactions using a standard process integration model. The model in WSDL is a stateless model of synchronous and asynchronous interactions. Models for complex web service interactions require a sequence of message exchanges among service providers and service requesters. The message exchanges can be performed in a synchronous or asynchronous fashion with state information and a long duration period.

To define such interactions, a formal description of message exchange protocols among multiple parties is needed. The definition of such message exchange protocols requires specification of the message exchange behavior of each party that is involved in the protocol without revealing their internal implementation. There are two reasons to separate public interfaces of services from internal implementation. The first one is that a service may not want to reveal its internal logic and data management. The second is that separating public interfaces from private implementations provides the flexibility to change internal implementations without affecting the public interfaces.

There are two competing efforts to standardize protocols for message exchanges among web service requesters and providers. They are BPEL4WS and WSCI.

Web Service Composition and Work Flow (BPEL4WS)

The Business Process Execution Language for Web services (BPEL4WS) (Andrews et al. 2003) represents the merging of WSFL (WSFL n.d.) and XLANG (Thatte n.d.), which currently are the most-dominating flow languages that describe business processes. There is a push to make BPEL4WS the basis of a standard for web service composition. BPEL4WS combines the graph-oriented processes defined in WSFL and structural constructs for processes in XLANG into a cohesive package that supports the implementation of business process. In addition to being an implementation language, BPEL4WS is used to describe the interfaces of business processes.

As an executable process language, the role of BPEL4WS is to define a new web service by composing a set of existing services. The interfaces of the composite service are described as a collection of WSDL portTypes. The composition (called *process*) indicates how the service interface fits into the overall execution of the composition. Figure 9.14 shows a view of the BPEL4WS process.

The BPEL4WS process is composed of various process steps. Each step is called an *activity*. There is a collection of primitive activities: invoking an operation on a web service, waiting for a message to be received, generating the response of a web service interface, copying data from one place to another, handling errors, and terminating processes. These primitives can be combined into more complex flows using BPEL4WS language structures, such as <sequence>, <switch>, and <pick>.

BPEL4WS processes consist of making interactions with other web services with invocations to other services or with invocations by other services. BPEL4WS calls these services *partners*. A partner is either a service that invokes the BPEL4WS process or a service that is invoked by a BPEL4WS process.

BPEL4WS uses service link types to define partners. A partner is defined by giving it a name, indicating the name of a service link type, and identifying the role that the process will play from that service link type and the role that the partner will play. The partner role is indicated by <receive>, <reply>, and <invoke> tags.

BPEL4WS uses <throw> and <catch> language constructs to handle and recover from errors in the processes.

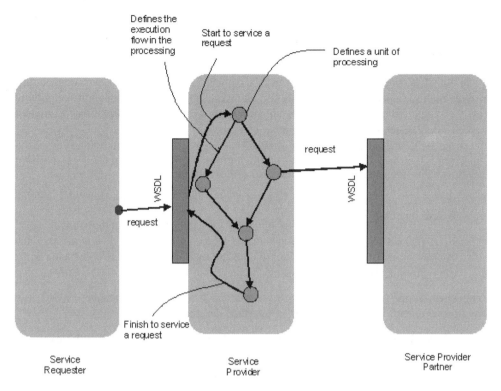

Figure 9.14 Process-based view of BPEL4WS

Web services that are implemented as BPEL4WS processes have an instanced lifecycle model. It means that a client of these services always interacts with a specific instance of these services. Instances of BPEL4WS are created implicitly when messages from clients arrive for the service.

Web Service Choreography Interface (WSCI)

A competing effort in web services processing submitted by Sun, BEA, and SAP is the Web Service Choreography Interface (WSCI) (Arkin et al. n.d.). This specification was submitted to W3C in August 2002. In January 2003, W3C formed a new Working Group, called *Web Services Choreography Working Group*, to target the specification.

WSCI addresses web service choreography from two aspects. First, WSCI builds upon the WSDL portType to describe the flow of messages that are exchanged in a process. This flow is from the point of view of the web service's own interface. In addition, WSCI describes the external *observable behavior* of a web service in the expression of sequential and logical dependencies of exchanging messages. WSCI also describes the semantics of message correlations, message groups as transactions, and exception handling in the <interface> tag.

Second, WSCI defines language constructs that allow composition of two or more WSCI definitions into a collaborative process that involves multiple web services. WSCI calls this the *global model*. The WSCI global model allows one web service interface to link other web services' interfaces and to specify the direction of message flow between those web services interfaces. This global model provides a message-oriented view of the overall web services process.

In WSCI, a process is a conversation that a client establishes with a web service, and invokes multiple operations on the web service interfaces. WSCI provides a description of how these multiple operations are to be organized for a successful conversation. This helps a client to understand how to interact with web services.

There are several critical concepts in WSCI:

Interface: The interface describes a scenario of how to interact with a web service.

Activity: Activities describe the behaviors of a web service. An activity is mapped to a WSDL operation.

Process: A process is a part of a behavior that is labeled with a name. It is a reusable unit that can be invoked from anywhere within an interface. A process is instantiated by receiving a message or by spawning the process itself.

Properties: Properties are analogous to variables in programming languages. WSCI uses property to reference a value in an interface definition.

Context: A context defines an environment in which a set of activities can be executed. An activity is defined in exactly one context.

Message correlation: A web service may be involved in multiple conversations at the same time. Message correlation describes how conversations are organized and which properties can be exchanged among messages to retain the semantic consistency of conversations.

Exceptions: Exceptional behavior may occur during a conversation. WSCI describes exceptions as a part of the context definition. Examples of exception are occurrence of a fault or time out.

Transaction: A transaction describes the transactional properties of an activity in a context. All activities in transactional context execute in an all-or-nothing manner.

WSCI describes the behavior of a web service in activities. Activities can be atomic or complex. An atomic activity is the basic unit of behavior of a web service, such as sending a message. A complex activity is composed of other activities. WSCI supports activities in sequential, parallel, loop and conditional execution.

BPEL4WS versus WSCI

BPEL4WS and WSCI approach web services choreography with different models. BPEL4WS is from an executable process model, and WSCI is from individual web service-centric choreography model. However, both approaches follow a bottom-up approach where choreography is built from individual web service interfaces that are defined in WSDL.

It is very desirable that these two specifications can converge into a single and coherent specification that all vendors can uniformly embrace.

9.3.7 J2ME Web Services Specification – JSR 172

JSR 172 specification (JCP 2003) is part of the Java Community Process (JCP)[2]. The specification is JCP's response to the considerable interest in the developer community in extending enterprise services out to J2ME (Java 2 Micro Edition) clients. It intends to provide two new capabilities to the J2ME platform:

- Parsing XML data

- Access to remote SOAP/XML-based web services.

On XML data parsing side, the specification selects a strict subset of JAXP 1.2 functionality to be supported. On the web services side, it selects a subset of the JAX-RPC 1.1 functionality, providing access to web services from J2ME but without server capabilities.

The criteria in the selection of API subsets are that they should meet platform size requirements (i.e., they should ensure that the API fits within the footprint requirements of the target devices) and that they should meet platform performance requirements (i.e., they should ensure that the API can be implemented within runtime memory and processing requirements for the target devices). As a result, the specification explicitly excludes DOM and XSLT support, and does not consider server capabilities.

9.3.8 Research on Wireless Web Services

As described in the previous section, there are a number of web service technologies that are designed from the ground up for the wireless web service. These technologies form a foundation for further development and deployment of web service applications on small devices.

Looking into the future, we foresee continuous push for web services on mobile devices. For web services to be successful in the wireless and mobile space, the unique problems, such as energy efficiency, server capability specification, disconnected computing, peer-to-peer support, and quality of service, need to be addressed. We discuss them briefly in the following.

Energy-efficient Wireless Web Service Implementation

While current effort in enabling web services on mobile devices has largely focused on providing small-footprint implementations, it is worthwhile to note that smaller footprint does not necessarily translate into lower energy consumption. For example, a Java virtual machine (JVM) with a smaller base of boot classes offers smaller footprint, but may need to consume more energy to complete the same task than a JVM with a larger base of boot classes if boot classes are precompiled rather than interpreted. As another example, a simple web browser without features, such as on-device caching support, certainly offers smaller footprint, but it may need to spend more energy on fetching otherwise cachable pages.

Energy efficiency has been heavily researched in OS, networking and hardware architecture areas. However, we believe that there are unique research issues in energy efficiency for distributed applications in general, and web services in particular. We further believe that

---

[2]For information on JCP, see http://jcp.org.

frameworks similar to that described in (Zhou et al. 2003) can be used to, at least partially, address the issue.

Server Capability Specification

A server capability specification for wireless devices does not seem to be a pressing issue as of today. Such a specification, however, is a must if our vision of cell phones' publishing services is to be accomplished. The need for a server capability specification is not obvious because web services are defined by their interfaces rather than their implementations, which run within servers and are invisible to the clients. Indeed, web services are functioning well on today's desktops without a server capability specification. However, there are two important distinctions between the wired desktop world and the wireless mobile device world:

- The difference in available resources: Cell phones in the future will undoubtedly have plenty of computing power and memory resources. But because of their much smaller form factors, they will always be at a disadvantage compared with powerful desktops, which are needed today, and most likely in the future, to run those resource-demanding enterprise web servers. Much more importantly, it is even harder for cell phones to compete with desktops in resources, such as energy and effective network bandwidth.

- The difference in server administration: Today's web services are largely deployed and managed by experienced professionals, who participate in the selection of hardware and software platforms, the fine-tuning of the services, and the safe-guarding of the servers. It is unrealistic to expect similar expertise from an ordinary cell phone user.

  The wireless web service industry thus needs a server capability specification which, directly or indirectly, establishes minimal resource requirements for hardware manufacturers and network carriers, details architectural solutions for dealing with energy and bandwidth constrains, defines easier procedures for service configuration and deployment, and provides better protection from malicious attacks. For example, part of this specification could be a Java server platform tailored for wireless mobile devices, and another part of the specification could be on the protocol between a mobile web server and its reverse proxy (see (Wessles and Claffy n.d.), (Squid-Cache n.d.)).

Disconnection

Unlike wired connections, web services cannot assume that wireless connectivity will be present all the time. Wireless networks can suffer from network disconnections caused by physical obstacles, such as in tunnels, or by moving outside the wireless network coverage areas. This is particularly problematic during a secure web service transaction. This is a very important topic in the wireless research, and to read more, please refer to Chapter 7.

Wireless Web Services in Ad hoc Networks

Wireless web services in ad hoc networks is an interesting research area because disconnections in wireless networks can create partitioned ad hoc groups and because many valuable wireless web services will be location-based services, which coincides with the spatial locality of wireless ad hoc networks.

We expect research in this area will leverage and expand related research in peer-to-peer systems, and corresponding development will mingle industry efforts in both peer-to-peer computing and web services (see (Oreilly 2001) and (Gartner n.d.) for early attempts). One of the research focuses will be on the publishing and discovering of services in ad hoc networks, and the security, resource management, and usage accounting issues involved in the process.

Quality of Service (QoS)

A wireless network is much less predictable than a wired network. Signal noises and fading in a wireless network often causes fluctuating link bandwidth. Mobile/wireless networks need to provide end-to-end QoS support for web services and applications. This is challenging in the presence of scarce and variable wireless bandwidth, bursts of wireless channel errors, and user mobility. Providing QoS in mobile/wireless network is an active research area. We provide a few samples of research work in this area.

In recent years, researchers have proposed architectures to support QoS in mobile/wireless networks. The Mobiware (see (Balachandran et al. 1997) and (Angin et al. 1998)) project developed a QoS-aware middleware that is capable of supporting adaptive multimedia applications. Srivastava et al. (Srivastava and Misha 1997) proposed a novel architecture for QoS support in a mobile network. They argued for a simple wireless link layer and more-sophisticated applications with QoS negotiation and adaptation. TIMELY (Bharghavan et al. 1998) is an adaptive resource management architecture that provides resource reservation and resource adaptation by coordination between different layers of a network. TOMTEN (Silva et al. 1999) is a framework for managing resources in mobile network. Yasuda et al. (Yasuda et al. 2001) discussed an end-to-edge QoS framework that consists of a mechanism for resource reservation, QoS translation, and QoS negotiation.

9.4   Web Services and the Open Mobile Alliance

The explosive market growth of mobile devices prompts new service enablers based on open standards, such as MMS, Java, and XHTML, to provide new services for mobile users. It is a new source of growth for the mobile industry. To ensure the continuing successful mobile services, it is important to minimize the fragmentation of service platforms and seamless interoperability.

Owing to this reason, Open Mobile Alliance (OMA) (OMA n.d.) was formed in June 2002 and includes approximately 300 companies. The member companies represent the world's leading mobile operators, device and network suppliers, information technology companies, application developers, and content providers. The mission of OMA is to grow the market for the entire mobile industry by removing the barriers of global user adoption, ensuring seamless application interoperability, and allowing business to compete through innovation and differentiation.

9.4.1   Web Services in OMA

A Mobile Web Service (MWS) Working Group was formed within OMA. The goal of MWS is to develop a specification that defines the application of web services within the OMA architecture and to ensure that the specification provides for the application of web services that is converged with the work of external activities. MWS is working on the specifications to integrate services offered by mobile network operators with the end-user services provided by value-added service providers (VASPs).

Mobile network operators have established relationship with their subscribers: operators know who they are and their current location, and bill them for the services that they use. The operators can package these information into web services and publish their service descriptions to one or more service registries. VASP applications can find the network service and offer services to their customers. For example, a mobile user may wish to find the nearest post office. He/she uses his/her mobile phone to access a VASP's "post-office finder" service. To provider the service to the mobile user, the VASP needs to know the mobile user's current location. VASP finds the location service provided by a mobile operator from one of the web service registries. Then, VASP makes a request to the mobile operator's location server and receives the response. After that, VASP looks up its post-office location database and provides the final information to the mobile user. Additionally, VASP can also request the mobile network operator to charge the user for the service on the VASP's behavior and send a predetermined share to the VASP.

9.4.2   Location-aware Messaging Service (LMS)

A location-aware messaging service (LMS) in a mobile network is a service provided to the mobile subscriber based on his/her current geographic location. This position can be obtained by user entry, such as user-provided address or a GPS receiver that he/she carries with him/her, or the use of a function built into the mobile network that uses the known geographic coordinates of the base stations. The mobile network operator can wrap such functions into web services interfaces and publish the service description of such services to one of the service registries. A value-added service provider that provides location-aware service can retrieve user's current location by binding the mobile network operator's location service and provides value-added messaging service. For example, a VASP that provides location-aware instant messaging can use user's current location and the information about an on-going discount sale in a near-by store.

9.5   Conclusion

Web services architecture is promising for future interoperable applications over the network. Web services architecture consists of several layers and defines technologies for each layer. However, existing web service technologies are defined for desktops and servers, and are too heavy for mobile devices with limited resources. Several attempts are underway to design web services technologies for mobile devices. These technologies are designed from the ground up for mobile devices as opposed to porting existing technologies to mobile devices as afterthoughts. The success of the above attempts allows XG mobile devices to act as both clients and application servers so that web services can be solely run on XG mobile devices.

Industrial leaders in mobile network and mobile computing areas have formed OMA on standardizing mobile service architectures and interfaces to enable new services for mobile users. One working group in the alliance is working on the web services specifications to ensure the integration and interoperability of mobile network operators and various VASPs.

Part IV

Security

The previous three parts of this book have provided an overview of XG research directions and architecture, mobile network technologies, and middleware and applications. This part deals with a critical and fundamental issue that is often given a great deal of lip service, but not sufficient effort and attention, in system design – security.

We believe that security will gain paramount importance in the operation of XG systems and applications. This is for two reasons. The first is that the dramatic rise in the number of threats to networks in general, particularly the Internet, ranging from spam to viruses, scams, and outright sabotage. There is every cause to believe that these threats will spread to mobile wireless networks, and XG mobile networks will become a target as they become popular and ubiquitous. In fact, wireless spam is already a significant problem in 3G mobile networks, and operators such as DoCoMo are moving aggressively to curtail it.

The second reason is that XG mobile networks will, by design, be far more open in terms of underlying technology than previous generations. This technological openness will exist at several layers of the system, and is driven by diverse factors.

Firstly, unlike previous generations of networks, XG systems will feature terminal devices, including mobile phone handsets, which are open and extensible. This is driven by the desire to grow the functionality of the terminal beyond that of its original design, in order to accommodate new applications, as well as by the need to reduce the cost and nuisance of maintenance and software updates. Needless to say, this openness and flexibility raises the possibility of security breaches in a key part of the system, one that the user regards as valuable and personal property and which is likely to contain sensitive information.

At the radio access layer, the desire to integrate heterogeneous devices and wireless technologies in order to provide the highest data rates and ubiquitous service is likely to mean that radio access networks of diverse security capabilities will become part of the system. Further, some of these access networks may be owned and operated by the customer (e.g., home wireless LANs) or third parties (e.g., enterprise wireless networks or specialized ad hoc networks).

At the core network layer, the desire for an IP-based network is driven by the need to integrate seamlessly with the Internet, the attractive economies of scale of standardized IP equipment leading to lower capital expenditure, as well as lower operating expenses due to simpler management processes and the availability of a relatively large pool of trained operations personnel. However, all these factors also lower some of the technical and nontechnical security barriers that are present in existing cellular networks.

Next Generation Mobile Systems. Edited by Dr. M. Etoh
© 2005 John Wiley & Sons, Ltd

At the applications layer, the entire design of the XG mobile network is driven by the need for openness and programmability in order to facilitate the rapid deployment of innovative new services. Clearly, allowing third parties, many of whom may not be telecommunications or even networking experts, to develop applications also raises the possibility of inadvertent as well as malicious security faults. Further, the more complex and sophisticated applications become, the more likely that they will become vulnerable to security breaches.

Finally, we believe that aside from protecting the network itself, security will become a key differentiator in the eyes of end users as well as third-party service and content providers. In fact, it is possible that security itself, in the form of trust brokerage or credentials services, may become a revenue-generating service in its own right.

In this part, we describe three key topics of security research that are applicable to threats at various layers in the XG mobile network, namely, cryptography, AAA, and secure code.

We begin with Chapter 10, which discusses the fundamental issues in cryptography as they apply to network systems. Chapter 10 first covers some cryptographic techniques that are currently used in 2G and 3G systems. This includes secret-key cryptography, as used in GSM and 3GPP systems, public-key cryptography, as used in SSL, and proposal for a general public-key infrastructure (PKI). The chapter then defines the formal notion of a secure cryptosystem, in terms of provable security, using public-key encryption and signing as examples. The last part of the chapter summarizes important research directions in applying cryptography for securing XG mobile systems, including cryptographic techniques for coping with heterogeneity, the development of efficient cryptographic primitives required in the resource-constrained mobile wireless environment, and the development of trusted or tamper-resistant devices.

Chapter 11 addresses a key practical issue in network security, namely, Authentication, Authorization, and Accounting (AAA). After defining the terms and the evolution of AAA and the special needs of AAA in mobile networks, the chapter introduces a common framework for discussing AAA research. It then describes current AAA technologies, particularly protocols such as RADIUS, Diameter, PANA, EAP, WLAN protocols, and 3GPP as well as 3GPP2 approaches. This forms a solid basis for the last part of the chapter, which discusses AAA research relevant to XG networks, focusing on the challenging and interesting area of AAA for access networks with ad hoc extensions, followed by AAA for WLAN handover optimization required for seamless mobility over heterogeneous networks. Finally, initial work toward a unified AAA approach is summarized.

Chapter 12 focuses on the security implications of downloading software to open and extensible terminal platforms. The chapter describes how a security manager is required to inspect and monitor downloaded code, to perform various checks such as conformance to published interface specifications, safe use of software data structures such as stacks and arrays, as well as valid and bounded usage of resources such as CPU and memory. The chapter then reviews security managers from the research literature, starting with the Java 2 manager, a standard dynamic monitor. Eight selective dynamic monitors, which are more flexible than standard dynamic monitors, are then discussed. Finally, static managers, which verify code before execution, are presented. Each manager is discussed along various dimensions and compared to other managers where appropriate.

The three chapters discussed in Part IV thus cover, in each of their areas, existing security technologies and present a survey of research directions toward security in XG mobile systems. These form a basis for much of the exciting research work being done in this important area.

10

Cryptographic Algorithms and Protocols for XG

Craig Gentry and Zulfikar Ramzan

10.1 Introduction

Cryptography is usually described as providing certain functionalities that are designed to make communications "secure," the typical functionalities being:

Confidentiality: Ensure that only the intended recipient of a message can read its content

Integrity: Allow the message recipient to verify that the message has not been modified in transit

Authentication: Allow the message recipient to verify that some specified entity (e.g., the message sender) "approved" of the message's content

Nonrepudiation: Prevent some specified entity (e.g., the message sender) from later denying that it approved the message

All of these functionalities are very important, and they essentially embody how cryptography is currently used in practice, but they certainly do not encompass all of the functionalities that cryptography can provide. Moreover, while these basic functionalities have sufficed for 3G, the large scale, pervasive, and heterogeneous nature of XG (as we envision it) will drive the demand for significantly more flexible and advanced cryptographic techniques.

Next Generation Mobile Systems. Edited by Dr. M. Etoh
© 2005 John Wiley & Sons, Ltd

10.1.1   The Challenge of Securing XG

Cryptography has a very broad objective: to design schemes that (*whatever their intended functionalities*) are robust against malicious attempts to make the schemes deviate from those functionalities (Goldreich 1999). *Encryption* and *digital signatures* are well-known examples of cryptosystems – they help make the confidentiality and authenticity of wireless communications robust against malicious attacks, such as eavesdropping and spoofing – but they are just specific examples of cryptosystems within the broad objective, designed to protect a few specific functionalities. It is really no exaggeration to say that purview of cryptography is every bit as broad as the universe of systems that it seeks to protect and the universe of malicious attacks against these systems that an adversary may devise.

Given cryptography's broad objective, the challenge of securing XG is that it will offer a rich variety of functionalities (applications, network services, personalized aspects, etc.) over increasingly heterogeneous devices and contexts. As the offered functionalities become more complex, so does the cryptographic problem of making these functionalities robust against attacks. Unfortunately, to solve this problem (efficiently, anyway), it is often not enough to simply kludge together simple atomic cryptosystems in the right way; instead, cryptosystems often need to be tailored for individual services and applications.

Consider a simple example. One of the hallmarks of XG will be heterogeneity – devices of differing capabilities (e.g., laptops, PDAs, cell phones, wrist watches, and kitchen appliances), differing connection characteristics (wired and wireless, lossless and lossy, cable modem and wireless LAN, etc.), and changes in context (such as network handoff due to user mobility). To handle this heterogeneity and enhance performance in other ways, one might consider placing value-added proxies near user devices to transcode data before transmitting it to users, so that the data is customized for a user's specific device and connection characteristics, or to facilitate the user's changes in context. For example, one might enable a network proxy to adapt a content provider's multimedia stream to the particular display capabilities of a user's cell phone in real time. However, neither the content provider nor the user is likely to trust such a proxy completely; both would like some assurance that the proxy has not modified the stream too much. In other words, we need a cryptosystem that allows limited transcoding but still provides robustness against malicious attempts (by the proxy or anyone else) to modify the data beyond what is permitted by the content provider. Traditional digital signature schemes, in accordance with the principle of end-to-end security, do not allow any modification of the message after the content provider has signed it; if a single bit of data is modified, the signature verification check will fail. Making this limited transcoding functionality robust against attacks, while preserving the ability of the proxy to operate in real time, requires a cryptosystem specifically tailored for the application.

As a slightly more complex example, consider one of the most commonly used applications on the Internet: a search engine. What robust functionalities might we want a search engine to have? First, we may want to ensure that the user's queries and the search engine's responses are authentic, unmodified, and confidential with respect to third parties. (These functionalities can essentially be secured using conventional encryption and signature schemes.) Second, we may want to make the search engine robust against denial-of-service attacks. Third, we may want to ensure that the search engine does a thorough search of the contents in its database. Fourth, we may want to ensure the privacy of the user's query in various ways, even against the search engine. For example, the user may want to hide its identity and location; on the other hand, the user may want to reveal enough verifiable information about its identity and location to prove to the search engine that it is "authorized" to make the query or to receive a query response that contains location-relevant information. Also, the user may want to ensure that the search engine does not publish the content of its search query. Many other functionalities may also be desired. Can cryptography make these functionalities robust against malicious attacks? For example, how can we prevent a malicious search engine from disclosing the content of users' searches (e.g., to interested advertisers)? Amazingly, there are existing cryptosystems for *private information retrieval* (PIR) that allow a user to get useful search results without even revealing the contents of its search to the search engine! (Instead, at a high level, the user reveals only an encrypted form of its search query that the search engine cannot decipher.) However, although these PIR cryptosystems are much more efficient than the trivial solution in which the search engine simply sends its entire database to the user, many more advances need to be made – particularly in allowing more complex private searches – to make these schemes efficient and versatile enough for practical use. A similar efficiency/versatility problem applies to cryptosystems covering many other functionalities (when such cryptosystems exist at all).

Since the precise set of functionalities that will compose XG is obviously difficult to predict, it is not always clear what sort of cryptosystems we should be constructing. Even when it is clear what functionalities we want to protect, it is often not clear how to create practical cryptosystems that protect them. This is the challenge of securing XG. Even beyond XG, as long as the functionalities themselves are in flux, this situation is unlikely to change.

10.1.2 Chapter Overview

In the next few subsections, we cover some cryptography currently in use, including secret-key cryptography (exemplified by GSM and 3GPP), public-key cryptography (exemplified by SSL), and a few different proposals for public-key infrastructure (PKI). Using public-key encryption and signing as examples, we then discuss how cryptography formally defines what it means for a cryptosystem to be secure. Then, we discuss a variety of more advanced functionalities that cryptography can efficiently protect, as well as cryptographic research directions for XG.

10.2 Secret-key Cryptography

In a *secret-key* (or *symmetric*) cryptosystem, two (or more) users share the *same cryptographic key* k. In secret-key encryption, the encrypter uses k to encrypt its plaintext message m and generate a ciphertext c (i.e., $c = E_k(m)$ for a given encryption algorithm E), after which the decrypter uses k to decrypt c and recover the plaintext m (i.e., $m = D_k(c)$ for the decryption algorithm D). One user can also authenticate itself to another using secret-key cryptography by proving its knowledge of k through a challenge-response protocol – for example, the second user issues a random challenge string $RAND$, and the first user responds with $f(k, RAND)$, where f is a preestablished one-way function (such as a cryptographic hash function, which is hard to invert).

But, one may ask, how can the users agree on a secret key in the first place (unless they already have a secure channel)? One solution to this chicken-and-egg problem of secret-key distribution is to distribute the key out of band. For example, in GSM (discussed shortly), an operator distributes secret keys to its subscribers *manually* by giving them smartcards containing preloaded keys. Later, we will discuss a cryptographic solution to this problem, which is called *public-key cryptography*.

10.2.1   Some History

Secret-key encryption schemes have existed practically since the dawn of civilization. The earliest documented example of cryptography dates back to ancient Egypt some 4000-odd years ago when nonstandard hieroglyphs were used in an inscription. Since then, Julius Caesar used a simple cipher that involved shifting letters of the alphabet (e.g., *D* might be substituted for *A*, *E* might be substituted for *B*, *F* might be substituted for *C*, and so on) when communicating with government officials. While cracking Caeser's cipher might have been a challenge for his messenger, who was most likely illiterate to begin with, it would clearly not provide much security today. Eventually, people began to build devices that could perform more complicated mathematical calculations in much less time than one could by hand. One such machine, known as Enigma, was used by the Germans in World War II. Ultimately, with the aid of some captured encrypted text and about a month's worth of previously used Enigma keys, the cipher was broken by a Polish mathematician, Marian Rejewski. Further advances in attacking Enigma were made by a crew at Bletchley Park in England that included Alan Turing.

Although many of the encryption schemes invented before and during World War II were quite ingenious, the most lasting legacy of this period lies not in the schemes themselves but in the formation of an important principle elucidated in the nineteenth century that is often described as the basis of modern cryptography. Known as "Kerckhoff's Principle," it states that the security of a cryptographic algorithm should rest on the secrecy of a cryptographic key, not on the secrecy of the algorithm itself. The rationale of this principle is that an algorithm is more difficult to keep secret than a short cryptographic key, and, once the algorithm is revealed, it is more difficult to replace; moreover, subjecting an algorithm to public scrutiny often weeds out the inevitable flaws overlooked by the designers. (Many of the recent debacles involving standardized security protocols such as 802.11 WEP and GSM, the latter of which is discussed below, are the direct consequence of a failure to follow Kerckhoff's Principle and subject the proposed security protocols to public scrutiny during the standardization process.)

Over time, cryptosystem designers not only became more mathematically sophisticated, but they started to use digital computers as a tool in cipher design.[1] In the mid-1970s, a team at IBM in consultation with the United States National Security Agency designed the Data Encryption Standard (DES). To this day, the only "practical" method of attacking DES is to exhaustively try every possible cryptographic key until one that works is found. While this might have been costly when DES was invented, as computing became faster, cheaper, and more prevalent, such brute-force attacks were no longer out of the question. In fact, in 1998, a group of cryptographers built a device that can perform such an attack in a matter of days for approximately $250,000. An interim fix to this problem had already been presented much earlier in the form "Triple DES." Here, one iterates DES three times and uses two *different* keys in the order (k_1, k_2, k_1). Because of the increased key size, a brute-force attack

---

[1] Of course, the caveat is that anyone trying to break the resulting cryptosystem could perhaps also rely on having a digital computer to carry out any calculations needed in the attack.

is no longer presently feasible, and it is unlikely to be feasible for some time to come. Of course, repeatedly using DES with different keys was a serious performance hit, and it was clear that a successor to DES was long overdue.

In the United States, the National Institute of Standards and Technology (NIST) began an effort in the late 1990s to establish a replacement for DES. Many of the top block cipher designers in the world took part, submitting candidate algorithms and/or providing constructive comments on them. The result was the adoption of an algorithm called *Rijndael*, now often called *Advanced Encryption Standard* (AES), which was developed by Joan Daemen and Vincent Rijmen. However, other standardization bodies are free to propose the adoption of different block ciphers; indeed, the 3GPP standard (successor to GSM) adopts a block cipher called *KASUMI* for use in third-generation mobile phones.

10.2.2   GSM

The Global System for Mobile Communications (GSM) standard, implemented in most mobile phones worldwide as of 2003, is an interesting case study in the use of symmetric cryptography for encryption and authentication, as well as an object lesson in standardizing cryptosystems.

In terms of security, GSM's goal is to make the mobile phone system as secure as the public switched telephone network. To this end, GSM uses symmetric cryptography to enable unilateral authentication of mobile stations to base stations and to protect the confidentiality of subscribers' data and identity information. Of course, to get this functionality, the mobile station and the network must share a common secret key.

In GSM, secret-key distribution is handled as follows. A subscriber's mobile station consists of two items – a mobile handset (which is not, by itself, security enabled) and a smart card called a *subscriber identity module* (SIM). A subscriber obtains its SIM when it signs up for service. It stores the following numbers:

- A personal identification number (PIN), which unlocks the card

- An International Mobile Subscriber Identification (IMSI), which uniquely corresponds on to a mobile phone number

- A subscriber authentication key K_i, a 128-bit secret key also known to the subscriber's authentication center (AC) in its home network.

The subscriber should personalize its PIN upon obtaining its SIM, but the other numbers are fixed.

After the subscriber enters its PIN to unlock the SIM, a mobile station may authenticate itself to a base station as follows. First, the mobile station transmits its IMSI, which is then relayed to the subscriber's AC. The AC then generates several triplets, which are relayed back to the base station, such that each contains a random challenge *RAND*, an authentication tag *XRES*, and a cipher key K_c. The AC derives *XRES* and K_c by first mapping the subscriber's IMSI to its key K_i, and then setting *XRES* and K_c to be the two encryptions of *RAND* with the key K_i using the encryption algorithms A3 and A8 – that is,

$$XRES = \text{A3}_{K_i}(RAND) \quad \text{and} \quad K_c = \text{A8}_{K_i}(RAND).$$

The base station transmits the challenge *RAND* to the mobile station. Assuming the mobile station's SIM contains K_i, it can set is response *SRES* to be $A3_{K_i}(RAND)$. Finally, the base station completes authentication of the mobile station by confirming that *XRES* = *SRES*.

Once authentication is complete, the cipher key K_c is used to encrypt traffic between the handset and the base station using an algorithm called *A5*. Since A5 is used as a stream cipher – specifically, each packet is XORed with the pseudorandom sequence generated by K_c and the packet number – a bit error in the ciphertext only causes a corresponding bit error in the plaintext (as opposed to the "cascade" of decryption errors caused by some other techniques). See Figure 10.1 for an illustration.

To protect the confidentiality of the subscriber's *identity*, the network generates a temporary mobile subscriber identification (TMSI) for the subscriber, records the association between IMSI and TMSI in its database, and transmits TMSI to the mobile station encrypted with K_c. The network changes the mobile station's TMSI frequently, and the mobile station can use the most-recent TMSI when attempting access to that network.

The designers of GSM did not attempt to make it robust against active attacks on the system's components; thus, there are several ways in which a sophisticated adversary can "break" GSM security without having to break any of the symmetric cryptosystems. Even more troubling, the symmetric cryptosystems themselves – A3, A8 and A5 – have serious flaws. In 1998, Wagner et al. (Briceno et al. n.d.) found an attack on Comp128, the algorithm used as A3 and A8 by many operators. To extract a SIM's secret key K_i, an adversary needs about 150,000 chosen *RAND-SRES* pairs, which an adversary can easily get (in about 10 h for most SIMs) by loading the SIM into its laptop and using freely available phone emulation software. Comp128 was not subjected to public scrutiny during the standardization process; instead, the attacks were found after its details were leaked to the public.

The A5 series of algorithms, which includes A5/1 (used, e.g., in Europe) and A5/2 (a weaker version exported to many other countries), was also not subjected to public scrutiny, and was also cryptanalyzed (with varying success) after the algorithms' details were leaked

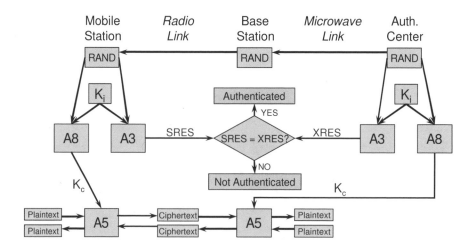

Figure 10.1 GSM authentication, cipher key generation and encryption

and reverse-engineered. In 2000, Biryukov et al. published an attack on A5/1 in which the adversary, after precomputing a large file of stream sequences and associated algorithm states, can extract K_c given a few seconds of encrypted known plaintext and a few minutes of computation on a PC. There are much more efficient attacks on A5/2 (Barkan et al. 2003), and when a mobile station supports both algorithms (as do European GSM phones), an adversary may be able to negotiate the mobile station down to first using A5/2 (and then switch to using A5/1 with the same (broken) key).

10.2.3   3GPP and Kerckhoff's Principle

The Third-Generation Partnership Project (3GPP) standard, formerly the Universal Mobile Telecommunications System (UMTS) standard, improves upon GSM in a variety of ways. For example, it makes the system robust against rogue base stations by requiring *mutual* authentication. Also, all user traffic and signaling data is encrypted over the air, preventing the interception of triplets. In general, an effort has been made to secure the system components against active attacks.

Moreover, the Comp128 and A5 secret-key cryptosystems have been replaced by a block cipher called *Kasumi*. Kasumi's design, which is public, is based on the "Misty" block cipher, which has held up to public scrutiny for years.

While it may seem that keeping a cryptosystem's design secret gives "added security," it seems difficult in practice to prevent leaks and reverse engineering. The more sensible approach, in accordance with Kerckhoff's principle, is to sacrifice the (probably unsustainable) secrecy of algorithm design to obtain instead an algorithm strengthened by public critique.

10.3   Public-key Cryptography

In a *public-key* (or *asymmetric*) cryptosystem, a user generates a public/private key pair (k_{pu}, k_{pr}). The user publishes the public key k_{pu}, making it available to everyone, but it keeps its private key private. Although the two keys in a public/private key pair have an inextricable mathematical relationship, it should be hard for third parties to compute the private key from the public one; otherwise, anyone could compute the user's private key, and the public-key cryptosystem would be insecure.

In a public-key encryption scheme, the encrypter uses k_{pu} to encrypt its plaintext message m and generate a ciphertext c (i.e., $c = E_{k_{pu}}(m)$ for a given encryption algorithm E) after which the decrypter uses k_{pr} to decrypt c and recover the plaintext m (i.e., $m = D_{k_{pr}}(c)$ for the decryption algorithm D). In other words, for fixed algorithms E and D and for *all* messages m, it should be the case that $D_{k_{pr}}(E_{k_{pu}}(m)) = m$. (Clearly, for this to be possible, k_{pu} and k_{pr} must have a strong mathematical relationship.) Third parties should not be able to decrypt c using the public key k_{pu}; only the private-key holder should be able to decrypt.

In a public-key signature scheme, the signer uses k_{pr} to generate its signature s on a message m (i.e., $s = S_{k_{pr}}(m)$ for a given signing algorithm S). To verify a signature, a verifier runs a verification algorithm on the signature that outputs a "1" or a "0," depending on whether the signature is valid or not. It should be the case that, for all messages m, $V(m, k_{pu}, S_{k_{pr}}(m)) = 1$ for the verification algorithm V – that is, all legitimately produced signatures should be confirmed valid. Third parties should have only a negligible probability of producing a forgery that passes the verification test; only the legitimate signer should be able to produce valid signatures.

How does public-key cryptography resolve the secret-key distribution problem? Consider public-key encryption: the encrypter and decrypter do not need a confidential channel to preestablish a secret key; instead, the encrypter can acquire the decrypter's public key over a public channel. Similarly, for signatures: the verifier does not need a preestablished secret key to verify a signer's signature, only its public key.

There are various ways of trying to imitate this functionality using secret-key cryptography. For example, recall that GSM base stations are able to authenticate subscribers without knowing the subscriber's authentication key by using triplets forwarded by the AC. However, this works only because the AC *does* know the subscriber's authentication key, and it *should* (as in 3GPP) transmit these triplets to the base stations confidentially; without this confidentiality, eavesdroppers could pass challenge-response protocols easily. Another example is Kerberos, in which two users that share different secret keys with a *trusted third party* (TTP) can use this TTP to liaise a secret-key agreement between themselves. However, this is not scalable, since the TTP must share a secret key with every user, and must always be on-line to distribute secret keys for every session between two or more users. Moreover, it requires a high degree of trust in the TTP, since the TTP (whether it is one entity or a federation of entities) knows all of the secret keys in the system.

Public-key cryptosystems are very useful for establishing secure communication in dynamic environments, such as the Internet, where one often interacts with new people. For example, Secure Socket Layer (SSL), which we discuss shortly, is a very successful application of public-key cryptography to secure websites (such as for credit-card transactions).

Secret-key cryptosystems work best where the communicating parties are essentially static, as in the (GSM) network access scenario. Public-key cryptosystems are typically much slower than secret-key cryptosystems, but this performance deficit can be minimized by using a hybrid asymmetric/symmetric approach – for example, using a public-key encryption algorithm to encrypt a symmetric key, and then using the symmetric key and a symmetric algorithm to encrypt the bulk of the message.

10.3.1   Some History and Well-known Schemes

Unlike secret-key cryptography, public-key cryptography is a thoroughly modern invention. The *notion* of public-key cryptography was first published in 1976 by Diffie and Hellman in their seminal paper "New Directions in Cryptography." In fact, they sketched both public-key encryption and public-key signatures, in an abstract format similar that given above, but they were unable to find actual schemes to instantiate these proposed functionalities. During this period, Merkle described a notion of cryptographic puzzles, which may be considered an early (but inefficient) form of public-key cryptography.

Despite not finding public-key encryption and signature schemes, Diffie and Hellman did address the *key-exchange* problem, in which two parties who have not previously met can agree on a secret key despite communicating over a channel over which anyone can eavesdrop. The Diffie – Hellman key agreement scheme is as follows. Let p be a large (e.g., 2048-bits) prime number and let g be a generator for the group $(\mathbb{Z}/p\mathbb{Z})^*$. (These "parameters" can be shared by everyone in the system.) Suppose Alice and Bob want to agree on a key. Alice first picks a random value x_A between 1 and $p-1$ and sends $z_A = g^{x_A} \bmod p$ to Bob. Similarly, Bob picks a value x_B and sends $z_B = g^{x_B} \bmod p$ to Alice. Now, given z_B, Alice can compute $y = z_B^{x_A} = g^{x_A x_B} \bmod p$. Bob can also compute $y = z_A^{x_B} = g^{x_A x_B} \bmod p$. At this point y is a common value known to both Alice and Bob, and they can use it to derive a shared secret key. On the other hand, an eavesdropper that wants to compute the key must somehow derive $g^{x_A x_B} \bmod p$ given only $(p, g, g^{x_A} \bmod p, g^{x_B} \bmod p)$.

To this day, there is no known efficient solution to this so-called "Diffie – Hellman Problem." If p is a very large number – such as, 2048-bits long – then even with state-of-the-art computing equipment and the best-known algorithms today, an eavesdropper would need millions of years to derive $g^{x_A x_B} \bmod p$ from $(p, g, g^{x_A} \bmod p, g^{x_B} \bmod p)$. However, it is entirely possible that an efficient "polynomial-time" algorithm for solving the Diffie – Hellman problem will be found, thereby breaking the Diffie – Hellman key agreement scheme. We do not yet know how to prove that the Diffie – Hellman problem is "hard"; in fact, if we would prove that there is no polynomial-time algorithm for solving it, we would have proven the complexity theory conjecture that $P \neq NP$. We only *believe* that the Diffie – Hellman problem is hard, since many cryptographers have tried and failed to construct an efficient algorithm for solving it.

The honor of finding the first practical public-key encryption and signature schemes went to Rivest, Shamir, and Adelman. Their scheme, known as RSA (formed from the first letters in each author's surname), is one of the most widely used cryptosystems in use today. In fact, it is the cryptographic technology behind SSL, which enables secure credit-card transaction capabilities found on most worldwide websites. RSA works essentially as follows. Let p, q be large (e.g., 1024-bit) prime numbers and let $n = pq$. Let e, d be such that $ed \equiv 1 \pmod{(p-1)(q-1)}$. The public portion of the key is (n, e). The private portion is (p, q, d). To encrypt a message m, one computes $c = m^e \bmod n$. To decrypt the ciphertext c, one computes $c^d \bmod n = m^{ed} \bmod n = m^{1+k(p-1)(q-1)} \bmod n = m \cdot m^{k(p-1)(q-1)} \bmod n = m$ The last equality follows from the previous since $(p-1)(q-1)$ is the order of the group $(\mathbb{Z}/n\mathbb{Z})^*$ and any element raised to a multiple of the order results in 1. For an adversary to determine m given c, he would have to be able to compute c^d given just (c, e, n). To this day, there is no known efficient algorithm for solving this so-called RSA problem, but as with the Diffie – Hellman problem, we cannot prove that no efficient algorithm exists. Currently, the fastest known methods for solving the RSA problem require that the adversary first factor n, though this does not rule out the possibility of a method that avoids factoring. If n is a very large number – for example, 2048-bits long – then even with state-of-the-art computing equipment and the best-known algorithms today, an adversary would need millions of years to factor n.

The RSA cryptosystem can also be used for digital signing. In this case, when Alice wants to sign a message, she computes $\sigma = m^d \bmod n$. Anyone knowing Alice's public key can verify the signature by checking the equality $\sigma^e \bmod n = m \bmod n$. The equality follows from the same argument as in the case of RSA encryption. (For simplicity, we avoid discussing how to encode the message properly before exponentiation during RSA encryption or signing.)

Interestingly, it was revealed in 1997 that researchers in the United Kingdom's Government Communications Headquarters (GCHQ) had invented schemes similar to RSA encryption and Diffie – Hellman key agreement in the early 1970s, but their results were classified.

Since the early history of public-key cryptography, faster and more bandwidth-efficient public-key cryptosystems have been invented, such as those using elliptic curves. Also, public-key cryptography has evolved beyond just encryption and signing to cover a wide array of features, some of which are discussed in Section 10.6.

10.3.2 Certification of Public Keys

As mentioned earlier, public-key cryptography addresses the chicken-and-egg problem of secret-key distribution, in that a public key can be transmitted across a public channel (rather than a confidential channel). This is a significant achievement, and it helps make cryptography practical and scalable in dynamic environments, such as the Internet.

However, there is a second (albeit, more tractable) key-distribution problem, for both symmetric and asymmetric cryptography – namely, key authentication. Suppose, for example, that Bob obtains Alice's public key, perhaps from a public-key directory or perhaps from someone claiming to be Alice. How does he know that this public key is really Alice's, and not someone else's? How can he be sure, when he encrypts a message for Alice under her putative public key, that an attacker that is the *true* owner of the public key (and knows the corresponding private key) will not intercept his transmission and decrypt his confidential message? A resourceful attacker may be able to corrupt a directory or fool Bob into believing that he is obtaining Alice's key directly from Alice.

For public-key cryptography, this problem can be mitigated by using a TTP called a *Certificate Authority* (CA) (Figure 10.2). The CA publishes its public key, so that it is accessible to everyone. After verifying (e.g., through some out-of-band mechanism) that Alice's public key really does belong to Alice, the CA binds Alice's identity and her public key together by using its private key to generate its public-key signature on a message that includes Alice's identity (e.g., her name of some other identifier) and her public key, together with any bookkeeping information the CA may want to include, such as a serial number and an expiration date. The CA's signature, together with the information it signed, is Alice's public-key certificate. The certificate attests to the CA's belief that Alice's putative public key is indeed her own, and that it likely will remain so until the embedded expiration date (unless the CA revokes Alice's certificate prematurely). Now, if Bob trusts Alice's CA, he can verify the CA's signature in Alice's public-key certificate using the CA's public key, and thereafter be confident that Alice's putative public key indeed belongs to Alice. The current ISO standard format for certificates is known as X.509. The general infrastructure for managing users' public keys and certificates is typically called "Public Key Infrastructure" (PKI), which we discuss in more detail in Section 10.4.

However, public-key certification induces another chicken-and-egg problem: how can Bob be sure that the CA's putative public key actually belongs to the CA? Of course, the CA's public key may be certified by a "higher-level" CA, but eventually this recursion must terminate with a "root CA"; who certifies the root CA? Essentially, there is no cryptographic solution to this problem; it must be handled out-of-band. For example, SSL handles this problem by directly embedding a root CA's public key in web browsers. Although no solution is very appealing, one should appreciate that using CAs makes the problem much more tractable; rather than authenticating (out-of-band) the public keys of *everyone he*

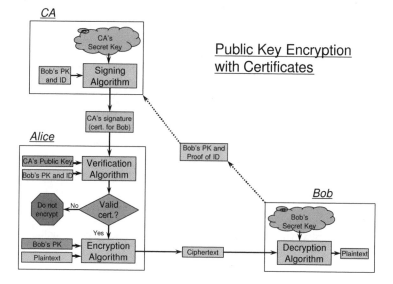

Figure 10.2 Alice encrypts to Bob after verifying his certificate

corresponds with, Bob instead only needs to authenticate (out-of-band) the public keys of a few root CAs.

One should also notice that public-key cryptography with CAs requires less trust in its TTPs than secret-key cryptography (e.g., Kerberos). In secret-key approaches, the TTP knows all of the secret keys. This means that the TTP can, for example, passively decrypt all user communications without even needing to perform an active attack; an attacker that compromises the TTP can also do this. In public-key cryptography with CAs, since the CA does not know the private keys corresponding to Alice's and Bob's public keys, it cannot decrypt their correspondence. Of course, a CA can mount an active attack in which it distributes erroneous certificates – for example, attesting that the CA's own public key belongs to Alice – and the CA can decrypt Bob's message if Bob relies on such a certificate, but the CA is more likely to be caught in such an active attack.

10.3.3 SSL

Secure Socket Layer (SSL) is an encryption system used by most web browsers to secure on-line transactions; connections using SSL begin with `https://` rather than `http://`. It protects http requests and responses against modification and eavesdropping by using public-key cryptography. Each website server has a public key that should be certified by one of the "root" CAs (either directly, or though a "chain" of lower-level CAs) whose public keys are stored in the user's browser.

Although details can vary, SSL works essentially as follows:

1. The client sends a CLIENT-HELLO message to the server, containing its name C, some cipher specifications SP_C, and a random challenge R_C.

2. The server sends back a SERVER-HELLO message, containing its name S, some cipher specifications SP_S, a random connection-ID R_S, its public key K_S, and its public-key certificate C_S.

3. The client verifies the server's certificate C_S to ensure that K_S actually belongs to S. Then, the client generates a pre-master-secret-key $PMSK$. From $PMSK$, the client derives three other symmetric keys, including a master-secret-key MSK, a client-write-key CWK and a server-write-key SWK – perhaps according to the formulas $MSK = H(C, S, PMSK, R_C, R_S)$, $CWK = H(C, PMSK, R_C, R_S)$ and $SWK = H(S, PMSK, R_C, R_S)$, for some cryptographic hash function H. To the server, it sends $PMSK$ (encrypted with K_S), as well as a FINISHED message including a message authentication code (MAC) or "keyed hash" of MSK and all past transmissions (encrypted with CWK).

4. The server sends back a FINISHED message with a MAC of MSK and all past transmissions (encrypted with SWK). Then, the server begins sending actual data.

In summary, the protocol flow is as follows:

$$C \rightarrow S : C, SP_C, R_C$$

$$S \rightarrow C : S, SP_S, R_S, K_S, C_S$$

$$C \rightarrow S : \{PMSK\}_{K_S}, \{\text{FINISHED}, \text{MAC}(MSK, everything)\}_{CWK}$$

$$S \rightarrow C : \{\text{FINISHED}, \text{MAC}(MSK, everything)\}_{SWK}, \{data\}_{SWK}$$

Currently, SSL typically uses RSA as the public-key encryption algorithm. As approximately a decade has passed since SSL was designed, and as elliptic curve cryptography (ECC) has continued to withstand public scrutiny since its invention in 1984, ECC would probably be a better choice if SSL were redesigned today.

Although SSL uses public-key certificates, it does not provide a good mechanism for revocation of these certificates. Certificate revocation may be necessary, for example, if a website server's private key is compromised. We explore this issue more fully in the next section.

10.4 Public-key Infrastructure

A certificate should include some type of expiration date, but a CA may want to "revoke" a certificate prior to this time for several reasons. For example, a person's position or affiliation (or other germane identifying information) might have changed. Also, perhaps a user's private key became compromised and thus can no longer be used. This problem is serious and has been an impediment to the widespread deployment of public-key cryptography.

A *certificate revocation list* (CRL), which is a signed and time-stamped list issued by the CA specifying which certificates have been revoked according to some identifier like a serial number, can be used to address this problem. These CRLs must be distributed periodically even if there are no changes to prevent illegitimate reuse of stale certificates. A CRL is analogous to the black lists issued by credit-card companies advising merchants which cards are no longer valid. One appealing aspect of this approach is its simplicity. However, the management of CRLs may be unwieldy with respect to communication, search, and verification costs. By encoding the list as the leaves of a *Merkle tree*, one can achieve some performance improvements. This idea was introduced by Kocher (Kocher 1998).

One can avoid using lists altogether by giving the certificate authority the capability of answering on-line validity queries about specific certificates. The *On-line Certificate Status Protocol* (OCSP) (Myers et al. 1999) espouses this approach. This approach has a major drawback. In particular, the CA's responses must be transmitted securely, which requires that each such response be digitally signed. This process is expensive, especially considering that the CA will potentially be responding to numerous queries concurrently. Moreover, the signatures themselves are long, so the communication costs of OCSP are also high. If we distribute the CA's capabilities, then its signing key would have to be distributed, which increases the risk of the signing key being compromised. On the other hand, if the CA is centralized, then the resulting scalability problem may be severe.

10.4.1 Hash-based Certification

The NOVOMODO scheme of Micali (Micali 1996, 1997, 2002) addresses many of these problems by using *hash chains* in conjunction with a single digital signature to amortize the digital-signature cost over many intervals. Also, hash chains decrease communication costs significantly.

The NOVOMODO scheme works essentially as follows. Suppose, for example, that certificates expire after one year (365 days) and that we wish to be able to revoke certificates with the granularity of one day. Suppose also that H is a hash function that maps n-bit strings to n-bit strings and has suitable security properties (e.g., it is one way on its iterates). For a given client, the CA chooses values y_0 and N_0 at random from $\{0, 1\}^n$, and computes $y_{365} = H^{365}(y_0)$ (called *the validity target*) and $N_1 = H(N_0)$ (called *the revocation target*), where $H^i(x) = H(H^{i-1}(x))$. The CA then signs the concatenation of the validity target, the revocation target, and other pertinent certificate data, such as the expiration date. On day i after the certificate's creation, the CA sends out $H^{365-i}(y_0)$ if the certificate is still valid, or N_0 otherwise. (Only the CA knows these values, because of the properties of H.) Given $H^{365-i}(y_0)$, third parties can verify certificate validity by confirming that $H^i(H^{365-i}(y_0)) = H^{365}(y_0) = y_{365}$ is the correct value of the validity target. Similarly, given N_0, third parties can confirm revocation by confirming that $H(N_0)$ equals the revocation target.

While a single hash function computation is much cheaper than a digital-signature computation, the above scheme may start to get expensive if there is a sizable gap between validity checks. Therefore, Naor and Nissim, as well as Gassko et al., pointed out that one could replace chains with Merkle trees. Currently, the most efficient such proposal in terms of communication and computational complexity is by Gentry and Ramzan (Elwailly et al. 2004), which uses so-called QuasiModo trees, a refinement of Merkle trees that more effectively utilizes the tree's internal nodes.

10.4.2 Certificate-based Encryption

Although the above hash-based certification schemes are very efficient computationally, there are compelling reasons for handling the certification of public encryption keys using a different mechanism. First, let us consider a digital-signature transaction, with Alice producing a signature for Bob. To verify Alice's signature, Bob needs not only Alice's public verification key but also the CA's certificate reconfirmation attesting that Alice's key is still valid (e.g., this reconfirmation is produced daily in NOVOMODO). Bob could, of course, obtain this reconfirmation from the CA, but since Alice is sending a signature to Bob anyway, she might as well also send her reconfirmation certificate. Thus, when dealing with public verification keys, the CA need not handle "pull" queries by third parties regarding the certificate status of its clients; instead, the CA has the option of using a potentially more efficient "push" system, pushing Alice's reconfirmation certificate directly to Alice.

For public encryption keys, the situation is different. When Bob encrypts a message to Alice, he must check her certificate status before encrypting to her. He could, of course, obtain Alice's reconfirmation certificate either from the CA or from Alice herself, but this adds an undesirable round of additional interaction. It would be preferable if Bob could encrypt his message without checking Alice's status beforehand, yet somehow be sure that Alice will not be able to decrypt unless she is still certified. This is the purpose of Gentry's certificate-based encryption (CBE) (Gentry 2003).

In CBE, Alice generates her own public – private key pair, as in a regular PKI. Also, Alice obtains her certificate from her CA as she would in a regular PKI: she gives the CA proof of authorization, and the CA returns its digital signature on Alice's public key. The difference is that the CA's certificate as well as its subsequent reconfirmation certificates also function as a secondary decryption key. To encrypt to Alice, Bob must know Alice's public key as well as the CA's public key (again, as in a regular PKI). He does *not* need to have Alice's reconfirmation certificate, but he must know what the CA signed (e.g., Alice's name, her public key, the expiration date of her certificate, etc.). This information, like her public key, is *long-lived* information (e.g., it may be valid for one year); it need not be *fresh* like (daily) certificate status information. Using this information, as well as the current "time period," he encrypts his message to Alice. Alice will be able to decrypt Bob's message only by using her two decryption keys: (1) her secret key and (2) a fresh reconfirmation certificate from her CA. Using CBE, the CA can enjoy the same infrastructure advantage that it has in the signature context: it need not handle third-party queries on certificate status.

The scheme of Gentry (Gentry 2003), like refinements to NOVOMODO, also describes how one uses a hierarchical structure to improve efficiency and scalability. Although the advantages of this approach with respect to infrastructure are not as significant outside of the encryption context, Al-Riyami and Paterson (Al-Riyami and Paterson 2003) describe certificate-based signature and key-agreement schemes (though they prefer the term "certificateless," opting to call the CA's "certificate" a "partial key").

10.4.3   Identity-based Cryptography

Like the schemes above, identity-based cryptography could also be considered an approach to certificate management; however, it is so different that it deserves a separate treatment. In fact, depending on how one defines "certificate," one could say that identity-based cryptography abolishes not only certificates but also public keys. Yet, identity-based cryptography retains many of the advantages of public-key cryptography and also has some significant advantages. Shamir (Shamir 1985) introduced the concept of identity-based cryptography in

1984, providing an identity-based signature scheme in the same article, but practical identity-based encryption schemes (Boneh and Franklin 2001; Cocks 2002) have been invented only recently.

Identity-based cryptosystems use a TTP called a *private-key generator* (PKG), which is somewhat analogous to a CA in a traditional PKI. As with a CA, clients of the PKG must provide proof of identity. Also like a CA, the PKG has a private – public key pair. However, instead of giving each client a signature, as a CA does in a PKI, the PKG provides an identity-based decryption key that corresponds to the client's identity, as given to the PKG by the client. This identity, which could be anything from a name to an e-mail address to an IP address, acts as the public key. Identity-based cryptography immensely simplifies public-key distribution. The PKG is the only entity whose public key is an arbitrary string; the public key of each client is some string, such as e-mail address, that a correspondent will likely know anyway. It also simplifies certificate management, since a PKG client will only be able to sign or decrypt if it has received a private key from the PKG; there is no need to check certification separately.

The primary disadvantage of identity-based cryptography is that it allows *key escrow* – specifically, the PKG, since it generates all private keys, can decrypt everyone's messages, and/or forge everyone's signatures. One can deploy safeguards against such violations – such as requiring the cooperation of multiple PKGs, using c08f012 cryptography, to generate identity-based private keys – but the implications are nonetheless worrisome. CBE (Gentry 2003) was designed to retain some of the benefits of identity-based encryption, while eliminating key escrow.

Another disadvantage of identity-based cryptography is that it is highly centralized. Traditional PKIs mitigate centralization by using a hierarchy of CAs, with a "root" CA at the top of the hierarchy, and where each CA certifies the identities and public keys of its children. This problem was overcome by hierarchical identity-based cryptography (Gentry and Silverberg 2002), in which only the root CA has a traditional public key; the public keys of all lower-level entities are a function of their identities and their positions in the hierarchy. For example, if Alice's e-mail address is alice@cs.univ.edu, then her hierarchical identity consists of the ID-tuple (edu,univ,cs,alice), which Bob can use in combination with the root's public key to encrypt to Alice. Like identity-based cryptography, its hierarchical extension suffers from key escrow: each of Alice's ancestors in the hierarchy (e.g., the system administrator in her computer science department) can decrypt her messages. However, there are ways of limiting escrow to Alice's ancestors at or below Alice and Bob's lowest-level common ancestor.

10.5 Proving that a Cryptosystem is Secure

What does it mean for a cryptosystem to be secure? In this section, we discuss the notion of *provable security*, by which one can formally *prove* (from clearly stated assumptions) that a cryptosystem is secure. The provable security paradigm is useful, because provably secure cryptosystems are less likely to have design flaws and are more likely to gain acceptance quickly and be standardized. Having described public-key signatures and encryption in the abstract, we will describe a concrete signature scheme based on the assumption that it is hard to factor large numbers, and show how to prove that this scheme is secure if this assumption is true.

10.5.1   The Provable Security Paradigm

As the name implies, any claim that a given cryptographic protocol is provably secure must be supported by a proof. However, much more is actually required. In particular, the provable security analysis typically involves the following components:

1. An *adversarial model* that accurately describes the adversary's capabilities and resources.

2. A *security definition* that describes what it means for an adversary to succeed in breaking the system. Along with such a security definition, one typically includes a *claim* that states that the cryptosystem meets the security definition.

3. A set of *underlying assumptions* that will be made in the analysis. Typically, an assumption will state that one or more "hard" problems cannot be solved within certain resource constraints (typically time or space).

4. A *proof* – typically called a *reduction* – that demonstrates that any series of steps for violating security, in accordance with the security definition and employing adversaries who fall within the adversarial model, can be leveraged (usually as a subroutine) to show that one of the assumptions is false (e.g., to solve the underlying hard problem with the adversary's specified capabilities and resources) (Figure 10.3).

For this provable security paradigm to be meaningful, the components discussed above should be as robust as possible while encompassing what might happen in real life. In particular, the adversarial model should give the adversary at least the amount of resources (in terms of computational power, ability to interact with the cryptographic systems, etc.) that a real-world adversary may have. Likewise, the security definition should at least account for those security breaches that have nontrivial negative real-world consequences. Finally, the underlying assumptions should be believable, and ideally have been well studied. For example, an often-used assumption is that factoring is a hard problem – that is, that if p and q are large enough (e.g., 1024-bit) primes, then recovering p and q from $n = pq$ takes more

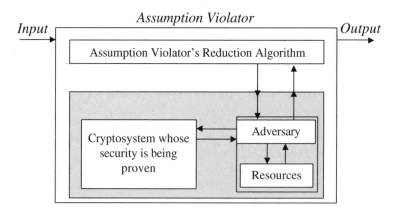

Figure 10.3   A cryptosystem breaker is used as a subroutine to prove an assumption wrong

computational resources than a real-world adversary can possibly muster. If the assumptions underlying the security proof turn out to be false, then the security proof may be completely useless.

The notion of a *reduction* is the same as that used in complexity theory, where, for example, one can prove that "Vertex Cover" is an NP-Complete problem by reducing 3-SAT (an NP-Complete problem) to Vertex Cover – that is, by demonstrating that a polynomial-time algorithm for solving Vertex Cover can be used as a subroutine in a polynomial-time algorithm to solve 3-SAT.

10.5.2   Example: The Rabin Signature Scheme

Here, we make the provable security paradigm more concrete with an example: the Rabin signature scheme. The Rabin signature scheme is similar to the RSA signature scheme, except that it is provably secure (in the so-called Random Oracle Model) assuming that it is hard to factor large numbers. Before discussing the Random Oracle Model and the details of Rabin signing, let us first consider what it means (in general) for a signature scheme to be secure.

Suppose that an adversary wants to forge Alice's signature – that is, compute Alice's signature on a message that she never legitimately signed. What resources can the adversary use in its attack? Of course, the adversary will have certain *computational* resources – such as, processing power, and memory – that it can apply to any hard computational problems it encounters. It will also have access to various types of pertinent information about Alice and her signing process – information that, collectively, we may call the adversary's view. For example, as part of its view, the adversary will (at a minimum) see Alice's public key, which contains information about her private key (though that information may be very difficult to extract. The adversary may also see signatures legitimately produced by Alice, since legitimately produced signatures are typically intended to be public. The adversary may even be able to influence what messages Alice signs – for example, if the adversary has engaged in transactions with Alice. The adversary may have even greater access to Alice's signing power if, for example, the adversary can get access to Alice's device for a certain period of time and then return it without being detected. (If the adversary is detected, Alice may request that the CA revoke her public-key certificate.) Is it possible to give the adversary this much power, and yet still prove that a signature scheme is secure against such an adversary?

In fact, in the provable security paradigm, this corresponds to the minimal acceptable notion of security, called security against "existential forgery under chosen-message attacks." To formalize this notion abstractly, it is modeled by the following interactive "game" played between a challenger and the adversary, which the adversary should have only a very small probability ε of winning if the scheme is secure:

- The challenger gives the adversary PK, the public signature verification key.

- The adversary adaptively chooses messages M_i to be signed by the challenger.

- The challenger sends back σ_i, a valid signature for (PK, M_i).

- After at most q such queries and responses, the adversary outputs an attempted signature σ for (PK, M).

The adversary wins if σ is a valid signature for (PK, M) and the adversary did not ask the challenger to sign M during the interactive phase. An adversary (t, q, ε) breaks the scheme, if the model adversary is limited to t computation, makes at most q "queries" (e.g., signature requests), and wins the above game with probability greater than ε.

This "game" may seem rather abstract, but it actually models reality quite well. If the prover uses a weaker security model, it must justify why a real-world adversary would be unable – for all settings in which the signature scheme is used – to collect a long-enough "transcript" of its target's signatures on messages of its choice.

Now, we give the details of Rabin and demonstrate how to apply the above framework to prove Rabin signing secure. As in the RSA signature scheme, a signer generates its key by choosing two large (e.g., 1024-bit) prime numbers p and q satisfying $p \equiv 3 \pmod 8$ and $q \equiv 7 \pmod 8$, and setting $n = pq$; the public key is n, while the private key is (p, q). To sign a message M, the signer does the following:

1. Computes $m = H(M) \in (\mathbb{Z}/n\mathbb{Z})^*$, where H is a cryptographic hash function

2. Computes $s \in (\mathbb{Z}/n\mathbb{Z})^*$ such that $s^2 \equiv cm \pmod n$ for $c \in \{1, 2, -1, -2\}$.

(The notation $(\mathbb{Z}/n\mathbb{Z})^*$ refers to set of numbers relatively prime to n.) Notice that the signature is essentially a (modular) square root of the hashed message. The reason that we need the "fudge factor" c is that not all numbers in $(\mathbb{Z}/n\mathbb{Z})^*$ are exact (modular) squares; however, given the special way we chose p and q, cm is a modular square for every $m \in (\mathbb{Z}/n\mathbb{Z})^*$ for exactly one $c \in \{1, 2, -1, -2\}$. On the flip side, every number that *is* a square modulo n has 4 (modular) square roots; the signer should choose one of these square roots randomly as the value of s, but (for reasons described shortly) it should always pick the same square root for a given value of m. To verify that the signature is correct, the verifier computes $m = H(M)$ and confirms that $s^2 \equiv cm \pmod n$ for some $c \in \{1, 2, -1, -2\}$.

Why, in our version of the Rabin signature scheme, should the signer always give the same value of s for a given m? As mentioned, cm (for some $c \in \{1, 2, -1, -2\}$) has four square roots modulo n. The reason is that cm has two square roots modulo p (because if $(s_p)^2 \equiv cm \pmod p$, then also $(-s_p)^2 \equiv cm \pmod p$), and cm similarly has two square roots $\pm s_q$ modulo q; thus, there are four possibilities $(\pm s_p \pmod p, \pm s_q \pmod q)$ for the value of the square root s modulo n. Suppose that the signer disclosed two modular square roots s_1 and s_2 of n such that $s_1 \equiv s_2 \pmod p$ and $s_1 \equiv -s_2 \pmod q$. Then, anybody could figure out p (and thereby factor n) simply by computing $p = \text{GCD}(s_1 - s_2, n)$. Similarly, values of s_1 and s_2 of n such that $s_1 \equiv -s_2 \pmod p$ and $s_1 \equiv s_2 \pmod q$ also give away the factorization of n.

Now, we finally get to the security proof of Rabin signing. In the security proof, we use a heuristic called the "*Random Oracle Model,*" in which we assume that the cryptographic hash function H behaves indistinguishably from a "random oracle." When we give an input to a "random oracle," it chooses a random bit string of the appropriate length, and sends this string back as its output; however, for identical inputs, it sends back identical outputs. A cryptographic hash function is, in some sense, similar; its outputs seem quite random, but it gives back the same output for identical inputs. Thus, it is a fairly good heuristic to pretend that a cryptographic hash function behaves indistinguishably from a random oracle from the perspective of the adversary (even though there are recent results on schemes that are uninstantiable in the random oracle model – that is, they are provably secure in the random oracle model, but are insecure for every specific instantiation of the hash function). In the random oracle model, we also make the simplification that the challenger gets to choose the hash function output. For Rabin signatures, as long as the hash function is unrelated to factoring, giving the challenger control of the hash function should not somehow enable the challenger to factor. Thus, if an adversary can forge and if the adversary's forgery allows the challenger to factor, it should not be because the challenger has control over the hash function; it should instead be because of some inherent weakness in the scheme.

So, in the random oracle model, the proof is as follows. The challenger has a number n that it wants to factor; it gives n to the adversary as its public verification key. At any time, the adversary can pick a message M_i and ask the challenger for the value of $m_i = H(M_i)$, computed according to the random oracle H controlled by the challenger. To respond, the challenger picks a random value $s_i \in (\mathbb{Z}/n\mathbb{Z})^*$ and a random value $c_i \in \{1, 2, -1, -2\}$ and sets $m_i = s_i^2/c_i \pmod{n}$. Also, at any time, the adversary can pick a message M_i and ask the challenger to sign it. To respond, the signer sends back s_i, which is a legitimate and verifiable signature on M_i. The hash output and signing responses have the same statistical distribution that one would expect in the "real world," as required. Eventually, after at most q such queries and responses, the adversary attempts to output an attempted signature s_i' for message M_i. If s_i' actually passes the verification test, then, since the adversary does not know that square root s_i of $c_i m_i$ that the challenger already knows, it may happen (in fact, with 50% probability) that $s_i' \neq \pm s_i \pmod{n}$. If so, then $GCD(s_i - s_i', n)$ is a nontrivial factor of n. Thus, the adversary's successful forgery will, with 50% probability, enable the challenger to factor n. Thus, factoring is only twice as hard as breaking the Rabin signature scheme (in the random oracle model). Q.E.D.

10.6 Advanced Functionalities and Future Directions

10.6.1 Electronic Cash and Other Privacy-preserving Protocols

As discussed above, public-key cryptosystems permit digital signatures. One interesting question that arises is whether we can use such a capability to create some form of digital currency. The basic idea is as follows. A bank creates a public – private key pair. When one of its customers makes a withdrawal from his account, the bank provides it with a digitally signed note that specifies the amount withdrawn, together with a serial number, and other miscellaneous information. This note is, in effect, a digital monetary instrument. The customer can present it to a merchant, who can then verify the bank's signature. Upon completing a transaction, the vendor can then remit the note to the bank, which will then credit the vendor the amount specified in the note.

At first glance, this scheme appears to provide us with a digital equivalent of cash. However, there is at least one notable difference. Because everything is digital, the bank can record the serial number when it provides the note to the user. Now, after the customer gives the note to a merchant, and the merchant remits it, the bank can then use the serial number to trace the transaction back to the original user. Thus, by doing some extra bookkeeping, the bank can learn its customers' spending habits. Their privacy may be violated. With regular paper cash, it is very difficult to do such tracing since the cash may go through several hands before finally being deposited. (There is the potential problem of double spending, which we do not discuss in detail here.)

Thus, the question arises whether we can restore privacy in such a system. David Chaum (Chaum 1982) proposed a very elegant solution to this problem, known as a *blind digital signature*. The core idea is to create a protocol between the bank and its customer whereby, at the end of the protocol, the customer receives a digitally signed monetary note of the appropriate amount, but the bank obtains no information that would enable it to link the note back to the customer. In the physical world, this might be achieved by taking the document to be signed, placing a piece of carbon paper over it, and then placing an opaque sheet of paper over the carbon paper. If the bank signs the opaque sheet, the carbon paper will transfer its signature to the document to be signed. However, because the bank did not see the underlying document, it will not be able to trace it back to the person requesting the signature, even though it can recognize the signature as its own.

To achieve this same concept mathematically, Chaum proposed a scheme with the following core idea. Let $((n, e), (p, q, d))$ be an RSA public – private key pair. The customer will actually create a monetary note, call it m. The customer then picks a random number r and computes $\mu = m \cdot r^e \bmod n$. The customer gives μ to the bank. The bank "signs" μ: $\sigma' = \mu^d \bmod n$ and gives the signature back to the customer. Now, observe that $\mu^d = m^d r^{ed} = m^d \cdot r \bmod n$. The customer computes $\sigma = \sigma' \cdot r^{-1} \bmod n$. Then $\sigma = m^d \bmod m$, which is a valid signature on the message m. However, the bank only sees $m \cdot r^e$, which gives it no information about m since r was chosen randomly by the user and was effectively used to blind m.

We remark that the protocol, as we have just presented it, should *not* be directly used in an actual system. In particular, there is nothing to prevent the user from constructing a message that says "this note is worth \$100,000,000" and getting the bank to sign it. There are a number of mechanisms to deal with this type of issue. One is to use a "cut-and-choose" protocol, in which the user essentially provides a number of blinded messages (with, perhaps, different serial numbers) to the bank, the bank picks which one it will sign and requires the user to reveal the blinding factors for the unpicked messages. If those revealed messages are legitimately formed, the bank has confidence that the user is behaving properly, and it will be more amenable to signing the remaining unrevealed message. Yet another technique is for the bank to associate a specific verification key to a specific type of transaction. For example, it may specify a verification key that can only be used for withdrawals of \$100.

Beyond their uses in electronic cash, blind digital signatures may also be used for electronic voting. Here, providing anonymity is easily seen to be a big concern.

Another construct that can be achieved using techniques from public-key cryptography is that of a *ring signature*. Such a signature allows a person to effectively sign a document, but only have the signature be traceable back to a group of public keys, one of which belongs to the original signer. The construct is motivated by the following situation. Suppose a corporate official wishes to divulge a story to the press regarding some wrongdoing within the company, but he does not want the press to find out who he is. Yet, he may want to provide evidence to the press that he is indeed an official within the company. He can use a ring signature where the "ring" consists of all corporate officials; the press would know that the story came from a corporate official, but it would not be able to figure out from which. A simple ring signature scheme can be constructed using RSA. We illustrate this scheme here, but remark that there are better schemes. Suppose that there are r ring members and that their keys pairs are:

$$((n_1, e_1), (p_1, q_1, d_1)), ((n_2, e_2), (p_2, q_2, d_2)), \ldots, ((n_r, e_r), (p_r, q_r, d_r)).$$

Suppose that the jth person wants to construct a ring signature on the message M. In this case, he knows all the public keys: $(n_1, e_1), (n_2, e_2), \ldots, (n_r, e_r)$, but he only knows his own private key: (p_j, q_j, d_j). The signing process works as follows. He first picks values s_i at random for $1 \leq i \leq r, i \neq j$. For each such s_i, he sets $m_i = s_i^{e_i} \bmod n_i$. Next, he computes $T = m_1 \oplus \ldots \oplus m_{j-1} \oplus m_{j+1} \oplus \ldots \oplus m_r$, and he sets $m_j = T \oplus M$. He next uses his signing exponent d_j to sign m_j by computing $s_j = m_j^{d_j} \bmod n$. The ring signature on M consists of $s_1, \ldots, s_r$. To check the validity of the signature, the verifier checks that $M = m_i \oplus \ldots \oplus m_r$, where $m_i = s_i^{e_i} \bmod n$ for $1 \leq i \leq r$. The verifier cannot determine which signing key the signer used, and so his identity is hidden. However, one can show that only someone with knowledge of one of signing exponents d_i could have signed (assuming that the RSA signature scheme is secure). Such a proof is beyond our scope.

Ring signatures have two noteworthy properties:

1. The verifier must know the public verification keys of each ring member.

2. Once the signature is issued, it is impossible for anyone, no matter how powerful, to determine the original signer; that is there is no anonymity escrow capability.

Another related property is that a ring signature requires the signer to specify the ring members, and hence the number of bits he transmits may be linear in the ring size. One can imagine that in certain settings these properties may not always be desirable.

Group signatures, which predate ring signatures, are a related cryptographic construct that address these issues. Naturally, we stress that there are situations in which a ring signature is preferable to a group signature.

Group signature schemes allow members of a given group to digitally sign a document as a member of – or on behalf of – the entire collective. Signature verification can be done with respect to a single group public key. Furthermore, given a message together with its signature, only a designated group manager can determine which group member signed it.

Because group signatures protect the signer's identity, they have numerous uses in situations where user privacy is a concern. Applications may include voting or bidding. In addition, companies wishing to conceal their internal corporate structure may use group signatures when validating any documents they issue such as price lists, press releases, contracts, financial statements, and the like.

Moreover, Lysyanskaya and Ramzan (Lysyanskaya and Ramzan 1998) showed that by blinding the actual signing process, group signatures could be used to build digital cash systems in which multiple banks can securely issue anonymous and untraceable electronic currency.

See Figure 10.4 for a high-level overview of a group signature scheme in which an individual Bob requests a signature from a group and receives it anonymously from a group member Alice. During a dispute, the group manager can open the signature and prove to Bob that Alice did indeed sign the message.

Group signatures involve the following six procedures:

INITIALIZE: A probabilistic algorithm that takes a security parameter as input and generates global system parameters $\mathcal{P}$.

SETUP: A probabilistic algorithm that takes $\mathcal{P}$ as input and generates the group's public key $\mathcal{Y}$ as well as a secret administration key $\mathcal{S}$ for the group manager.

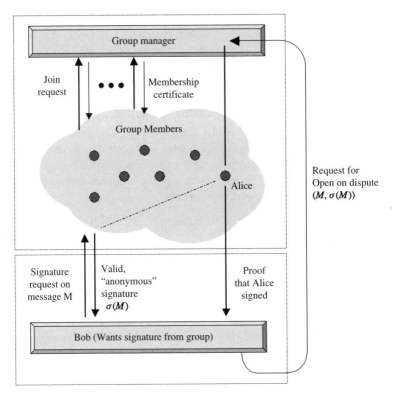

Figure 10.4 A high-level overview of a group signature scheme. Bob requests a signature from a group and receives it anonymously from group member Alice. If a dispute arises, the group manager can open the signature and prove to Bob that Alice did indeed sign the message

JOIN: An interactive protocol between the group manager and a prospective group member Alice by the end of which Alice possesses a secret key s_A and her membership certificate v_A.

SIGN: A probabilistic algorithm that takes a message m, as well as Alice's secret key s_A and her membership certificate v_A, and produces a group signature σ on m.

VERIFY: An algorithm that takes $(m, \sigma, \mathcal{Y})$ as input and determines whether σ is a valid signature for the message m with respect to the group public key $\mathcal{Y}$.

OPEN: An algorithm that on input $(\sigma, \mathcal{S})$ returns the identity of the group member who issued the signature σ together with a publicly verifiable proof of this fact.

In addition, group signatures should satisfy the following security properties:

Correctness: Any signature produced by a group member using the SIGN procedure should be accepted as valid by the VERIFY procedure.

Unforgeability: Only group members can issue valid signatures on the group's behalf.

Anonymity: Given a valid message-signature pair, it is computationally infeasible for anyone except the group manager to determine which group member issued the signature.

Unlinkability: Given two valid message-signature pairs, it is computationally infeasible for anyone except the group manager to determine whether both signatures were produced by the same group member.

Exculpability: No coalition of group members (including, possibly, the group manager) can produce valid-looking message-signature pairs that do not identify any of the coalition members when the OPEN procedure is applied.

Traceability: Given a valid message-signature pair, the group manager can always determine the identity of the group member who produced the signature.

While we have listed the above properties separately, one will notice that some imply others. For example, unlinkability implies anonymity. Traceability implies exculpability and Unforgeability.

PERFORMANCE PARAMETERS. The following parameters are used to evaluate the efficiency of group signature schemes:

- The size of the group public key $\mathcal{Y}$

- The length of signatures

- The efficiency of the protocols SETUP, JOIN, and SIGN, VERIFY

- The efficiency of the protocol OPEN.

Group Digital Signatures were first introduced and implemented by Chaum and van Heyst (Chaum and van Heyst 1991). They were subsequently improved upon in a number of papers (Camenisch 1997; Chen and Pederson 1995). All these schemes have the drawback that the size of group public key is linear in the size of the group. Clearly, these approaches do not scale well for large group sizes.

This issue was resolved by Camenisch and Stadler (Camenisch and Stadler 1997), who presented the first group signature scheme for which the size of the group public key remains independent of the group size, as do the time, space, and, communication complexities of the necessary operations. The construction of Camenisch and Stadler (1997) is still fairly inefficient and quite messy. Also, the construction was found to have certain potential security weaknesses, as pointed out by Ateniese and Tsudik (Ateniese and Tsudik 1999). These weaknesses are theoretical and are thwarted by minor modifications. At the same time, the general approach of Camenisch and Stadler is very powerful. In fact, all subsequent well-known group signature schemes in the literature follow this approach.

By blinding the signing process of the scheme in Camenisch and Stadler (1997), Lysyanskaya and Ramzan (Lysyanskaya and Ramzan 1998) showed how to build electronic cash systems in which several banks can securely distribute digital currency; the conceptual novelty in their schemes is that the anonymity of both the bank and the spender is maintained.

Their techniques also apply to voting. Ramzan (Ramzan 1999) further extended the ideas by applying the techniques of Ateniese and Tsudik (1999) to enhance security.

Subsequently, Camenisch and Michels developed a new scheme whose security could be reduced to a set of well-defined cryptographic assumptions: the strong RSA assumption, the Discrete Logarithm assumption, and the Decisional Diffie – Hellman assumption.

Thereafter, Ateniese, Camenisch, Joye, and Tsudik (Ateniese et al. 2000) came up with a more efficient scheme that relied on the same assumptions as Camenisch and Michels (1998). This scheme is the current state of the art in group signatures.

Ring signatures, group signatures, and privacy-enhancing cryptographic techniques in general, have substantially broadened the purview of cryptography, permitting the reconciliation of security with privacy concerns, with a rich variety of financial applications. In the next subsection, we focus on the effort, which came to full fruition in the 1990s, to place the security of these cryptographic constructs on a firm foundation.

10.6.2   Coping with Heterogeneity

One of the significant challenges of XG, particularly in the area of network value-added services, is achieving "mass customization" – personalization of content for a huge clientele. Currently, it is unclear what this will mean in practice. However, we can attempt to extrapolate from current trends.

One of these trends is multifaceted heterogeneity. The Internet is becoming accessible to an increasingly wide variety of devices. As these devices, ranging from mobile handheld devices to desktop PCs, differ substantially in their display, power, communication and computational capabilities, a single version of a multimedia object may not be suitable for all users. This heterogeneity presents a challenge to content providers, particularly when they want to multicast their content to users with different capabilities. At one extreme, they could store a different version of the content for each device, and transmit the appropriate version on request. At the other extreme, they could store a single version of the content, and adapt it to a particular device on the fly. Neither option is compatible with multicast, which achieves scalability by using a "one-size-fits-all" approach to content distribution.

Instead, what we need are approaches that not only have the scalability of multicast for content providers but also efficiently handle heterogeneity at the user's end. Of course, we also need security technologies that are compatible with these approaches.

BENDING END-TO-END SECURITY. One way to deal with this problem is through the use of *proxies*, intermediaries between the content provider and individual users that adapt content dynamically on the basis of the user needs and preferences. For example, let us consider multimedia streams, which may be transmitted to users having devices with different display capabilities as well as different and time-varying connection characteristics. Since one size does not always fit all, media streams are often modified by one or more intermediaries from the time they are transmitted by a source to the time they arrive at the ultimate recipient. The purpose of such modifications is to reduce the amount of data transmitted at the cost of quality in order to meet various resource constraints such as network congestion and the like. One mechanism for modifying a media stream is known as *multiple file switching or simulcast*. Here, several versions are prepared: for example, low, medium, or high quality. The intermediary decides on the fly which version to send and may decide to switch dynamically on the fly. Another mechanism is to use a scalable video coding scheme. Such schemes have the property that a subset of the stream can be decoded and the quality is commensurate with the amount decoded. These schemes typically encode video into a base layer and then to zero or more "enhancement" layers. Just the base layer alone would be sufficient to view the stream; the enhancement layers are utilized to improve the overall quality. An intermediary may decide to drop one or more enhancement layers to meet constraints.

Naturally, these modifications make it rather difficult to provide end-to-end security from the source to the recipient. For example, if the source digitally signs the original media and the intermediary modifies it, then any digital-signature verification by the receiver will fail. This poses a major impediment to the source authentication of media streams.

What is needed here is a scheme that allows proxies to "bend" end-to-end security without breaking it. For example, the content source may sign its content in such a way that source authentication remains possible after proxies perform any of a variety of transformations to the content – dropping some content, adding other content, modifying content in certain ways – as long as these transformations fall within a *policy* set by the content source.

The obvious ways of achieving such flexible signing tend to be insecure or highly inefficient. For example, the source can provide the intermediary with any necessary signing keys. The intermediary can then re-sign the data after any modifications to it. There are three major disadvantages to this approach. First, the source must expose its secret signing key to another party, which it does not have any reason to trust. If the intermediary gets hacked and the signing key is stolen, this could cause major problems for the source. Second, it is computationally expensive to sign an entire stream over again. The intermediary may be sending multiple variants of the same stream to different receivers and may not have the computational resources to perform such cryptographic operations. Finally, this approach does not really address the streaming nature of the media. For example, if a modification is made and the stream needs to be signed again, when is that signature computed and when is it transmitted? Moreover, it is not at all clear how to address the situation of multiple file switching with such an approach.

An alternative approach is to sign every single packet separately. Now, if a particular portion of the stream is removed by the intermediary, then the receiver can still verify the other portions of the stream. However, this solution also has major drawbacks. First of all, it is computationally expensive to perform a digital-signature operation. Signing each packet would be rather expensive. Not to mention that it might not be possible for a low-powered receiving device to constantly verify each signature, imagine how unpleasant it would be to try watching a movie with a pause between each frame because a signature check is taking place. Second, signatures have to be transmitted and tend to eat up bandwidth. Imagine if a 2048-bit RSA signature is appended to each packet. Given that the point of modifying a media stream is to meet resource constraints, such as network congestion, it hardly seems like a good idea to add 256 bytes of communication overhead *to each packet.*

What is needed here is an exceptionally flexible signature scheme that is also secure and efficient. In particular, since transcoding is performed dynamically in real time, transcoding must involve very low computational overhead for the proxy, even though it cannot know the secret keys. The scheme should also involve minimal computational overhead for the sender and receiver, even though the recipients may be heterogeneous. Wee and Apostopoulos (Wee,

S.J and Apostolopoulos, J.G. 2001) have made some first steps in considering an analogous problem in which proxies transcode encrypted content without decrypting it.

MULTICAST. Multicast encryption schemes (typically called *broadcast encryption* (BE) schemes in the literature) allow a center to transmit encrypted data over a broadcast channel to a large number of users such that only a select subset P of privileged users can decrypt it. Traditional applications include Pay TV, content protection on CD/DVD/Flash memory, secure Internet multicast of privileged content, such as video, music, stock quotes, and news stories. BE schemes can, however, be used in any setting that might require selective disclosure of potentially lucrative content. BE schemes typically involve a series of prebroadcast transmissions at the end of which the users in P can compute a broadcast session key bk. The remainder of the broadcast is then encrypted using bk. There are a number of variations on this general problem.

Let us examine two simple, but inefficient, approaches to the problem. The first is to provide each user with its own unique cryptographic key. The advantage of this approach is that we can transmit bk to any arbitrary subset of the users by encrypting it separately with each user's key. However, the major disadvantage is that we need to perform a number of encryptions proportional to the number of nonrevoked users. This approach does not scale well. The second simple approach is to create a key for every distinct subset of users and provide users keys corresponding to the subsets to which they belong. The advantage now is that bk can be encrypted just once with the key corresponding to the subset of nonrevoked users. However, there are $2n - 1$ possible nonempty subsets of an n-element. So, the complexity of the second approach is exponential in the subscriber set size, and also does not scale well.

For the "stateless receiver" variant of the BE problem, in which each user receives a set of keys that never need to be updated, Asano (Asano 2002) presented a BE scheme using RSA accumulators that only requires each user to store a single master key. Though interesting, the computational requirements for the user and the transmission requirements for the broadcast center are undesirably high; thus, one research direction is to improve this aspect of his result. Another research direction is to explore what efficiencies could be achieved in applying his approach to the dynamic BE problem. In general, there are many open issues in BE relating to group management – how to join and revoke group members efficiently, how to assign keys to group members on the basis of correlations in their preferences, and so on.

SUPER-FUNCTIONAL CRYPTOSYSTEMS. Currently, cryptography consists of a collection of disparate schemes. Separately, these schemes can provide a variety of "features" – confidentiality, authentication, nonrepudiability, traceability, anonymity, unlinkability, and so forth. Also, some schemes allow a number of these features to be combined – for example, group signatures allow a user to sign a message as an anonymous member of a well-defined group, and certificate-based encryption allows a message sender make a ciphertext recipient's ability to decrypt contingent on its acquisition of a digital signature from a third party.

In general, we would like security technologies to be maximally flexible and expressive, perhaps transforming information (data information, identity information, etc.) in any manner that can be expressed in formal logic (without, of course, an exponential blowup in computational complexity). Ideally, a user or application developer could calibrate the desired features and set their desired interrelationships in an essentially *a lá carte* fashion, and an appropriate cryptosystem or security protocol could be designed dynamically, perhaps as a projection of a single super-functional cryptosystem. Currently, cryptosystems are not nearly this flexible.

10.6.3   Efficient Cryptographic Primitives

With the seemingly inexorable advance of Moore's Law, PCs and cell phones have better processing speed than ever; memory capacity and transmission speed have also advanced substantially. However, at least for cell phones, public-key operations can be computationally expensive, delaying the completion of transactions and draining battery power. Moreover, the trend toward putting increased functionality on smaller and smaller devices – wrist watches, sensor networks, nano-devices – suggests that the demand for more efficient public-key primitives will continue for some time.

Currently, RSA encryption and signing are the most widely used cryptographic primitives, but ECC, invented independently by Victor Miller and Neil Koblitz in 1985, is gaining wider acceptance because of its lower overall computational complexity and its lower bandwidth requirements. Although initially there was less confidence in the hardness of the elliptic curve variants of the discrete logarithm and Diffie – Hellman problems than in such mainstays as factoring, the cryptographic community has studied these problems vigorously over the past two decades, and our best current algorithms for solving these problems have even higher computational complexity than our algorithms for factoring. Interestingly, the US military announced that it will secure its communications with ECC.

NTRU (Hoffstein et al. 1996), invented in 1996 by Hoffstein, Pipher, and Silverman, is a comparatively new encryption scheme that is orders of magnitude faster than RSA and ECC, but which has been slow to gain acceptance because of security concerns. Rather than relying on exponentiation (or an analog of it) like RSA and ECC, the security of NTRU relies on the assumed hardness of finding short vectors in a specific type of high-dimensional lattice.[2] Although the arbitrariness of this assumed hard problem does not help instill confidence, no polynomial-time algorithms (indeed, no subexponential algorithms) have been found to solve it, and the encryption scheme remains relatively unscathed by serious attacks. The inventors of the NTRU encryption scheme have also proposed signature schemes based on the "NTRU hard problem," but these have been broken repeatedly (Gentry and Szydlo 2002; Gentry et al. 2001; Mironov 2001); however, the attack on the most recent version of "NTRUSign" presented at the rump session of Asiacrypt 2001 requires a very long transcript of signatures.

ESIGN (Okamoto et al. 1998) is a very fast signature scheme, whose security is based on the "approximate eth root" problem – that is, the problem of finding a signature s such that $|s^e - m(\mathrm{mod}\, n)| < n^\beta$, where n is an integer of the form $p^2 q$ that is hard to factor, m is an integer representing the message to be signed, and where typically e is set to be 32, and β to 2/3. While computing *exact* eth roots, as in RSA, is computationally expensive ($O((\log n)^3)$), the signer can use its knowledge of n's factorization to compute approximate eth roots quickly ($O((\log n)^2)$) when e is small. Like NTRU, ESIGN has been slow to gain acceptance because of security concerns. Clearly, the approximate eth root problem is no

---

[2] NTRU's security in not *provably based* on this assumption, however.

harder than the RSA problem (extracting exact eth roots), which, in turn, is no harder than factoring. Moreover, the approximate eth root problem has turned out to be easy for $e = 2$ and $e = 3$. The security of ESIGN for higher values of e remains an open problem.

Aggregate signatures, invented in 2002 by Boneh, Gentry, Lynn and Shacham (Boneh et al. 2003), are a way of compressing multiple digital signatures by multiple different signers S_i on multiple different messages M_i into a signed short signature; from this short aggregate signature, anyone can use the signer's public keys PK_i to verify that S_i signed M_i for each i. The first aggregate signature scheme (Boneh et al. 2003), which uses "pairings" on elliptic curves, allows *anyone* to combine multiple individual pairing-based signature into a pairing-based aggregate signature. The security of this aggregate signature scheme is based on the computational hardness of the Diffie – Hellman problem over supersingular elliptic curves (or, more generally, over elliptic curves or abelian varieties for which there is an "admissible" pairing), which is a fairly well-studied problem, but not as widely accepted as factoring. In 2003, Shacham et al. (Lysyanskaya et al. 2003) developed an aggregate signature scheme based on RSA. Since computing pairings is somewhat computationally expensive, their scheme is faster than the pairing-based version, but the aggregate signatures are longer (more bits), and the scheme is also *sequential* – that is, the signers embed their signatures into the aggregate in sequence; it is impossible for a nonsigner to combine individual signatures post hoc. Since aggregate signatures offer a huge bandwidth advantage – namely, if there are k signers, it reduces the effective bit length of their k signatures by a factor of k – they are useful in a variety of situations. For example, they are useful for compressing certificate chains in a hierarchical PKI.

10.6.4 Cryptography and Terminal Security

There are some security problems that cryptography alone cannot solve. An example is DRM (digital rights management). Once a user decrypts digital content for its personal use (e.g., listening to an MP3 music file), how can that user be prevented from illegally copying and redistributing that content? For this situation, pure cryptography has no answer.

However, cryptography can be used in combination with *compliant hardware* – for example, *trusted platforms* or *tamper-resistant devices* – to provide a solution. Roughly speaking, a trusted platform uses cryptography to ensure compliance with a given *policy*, such as a policy governing DRM. Aside from enforcing these policy-based restrictions, however, a trusted platform is designed to be flexible; subject to the restrictions, a user can run various applications from various sources.

Although we omit low-level details, a trusted platform uses a process called *attestation* to prove to a remote third party that it conforms to a given policy. In this process, when an application is initiated, it generates a public key/private key pair (PK_A, SK_A); obtains a certificate on (PK_A, A_{hash}) from the trusted platform, which uses its embedded signing key to produce the certificate, and where A_{hash} is the hash of application's executable; and then it authenticates itself by relaying the certificate to the remote third party, which verifies the certificate and checks that A_{hash} corresponds to an approved application. The application and the remote third party then establish a session key.

Trusted platforms are most often cited as a potential solution to the DRM problem, since "compliant" devices can be prevented from copying content illegally. Other notable applications of trusted platforms are described in Garfinkel et al. (2003), including a distributed firewall architecture in which the security policy is defined centrally but enforced at well-regulated endpoints, the use of rate limiting to prevent spam and DDoS attacks (e.g., by limiting the rate at which terminals can open network connections), and a robust reputation system that prevents identity switching through trusted platforms.

If trusted platforms become truly feasible, they may change how we view cryptography. For example, "formal methods" for security protocol evaluation, such as BAN logic (Burrows et al. 1989) and the Dolev – Yao model (Dolev and Yao 1983), assume that the adversary is prohibited from performing arbitrary computations; instead, it is limited to a small number of permitted operations. For example, the adversary may be prohibited from doing anything with a ciphertext other than decrypting it with the correct key. Since a real-world adversary may not obey such restrictions, a proof using formal methods does not exclude the possibility that the adversary may be successful with an unanticipated attack. This is why cryptography uses the notion of "provable security," which does not directly constrain the adversary from performing certain actions, but instead places general limits on the adversary's capabilities. Recent work has begun to bridge the gap between these two approaches to "provable security" by *enforcing* the restrictions using the cryptographic notion of *plaintext awareness* (Herzog et al. 2003), but the prospect of trusted platforms may cause a much more dramatic shift toward the formal methods approach, since trusted platforms could enforce the restrictions directly.

Another issue at the interface of cryptography and terminal security concerns "side-channel attacks." Suppose we assume that a device is tamper resistant; does this imply that the adversary cannot recover a secret key from the hardware? Not necessarily. An adversary may be able to learn significant information – even an entire secret key – simply by measuring the amount of time the device takes to perform a cryptographic operation, or by measuring the amount of power that the device consumes. Amazingly, such "side-channel" attacks were overlooked until recently (Kocher 1996) (Kocher et al. 1999), when they were applied to implementations of Diffie – Hellman and other protocols. (See Ishai et al. (2003) and Micali and Reyzin (2004) for a description of how such attacks may be included in the adversarial model.) We need general ways of obviating such attacks, while minimally sacrificing efficiency.

10.6.5 Other Research Directions

There are many other exciting research directions in cryptography; it is virtually impossible to give a thorough treatment of all of them. Many of the fundamental questions of cryptography are still open. Is factoring a hard problem? Are discrete logarithm and Diffie – Hellman (in fields or on elliptic curves) hard problems? Is RSA as hard to break as factoring? Is Diffie – Hellman as hard as discrete logarithm? Are there any hard problems at all; does P = NP? Can the average-case hardness of breaking a public-key cryptosystem be based on an NP-complete problem? With these important questions still unanswered, it is remarkable that cryptography has been as successful as it has been.

Interestingly, the progress of quantum mechanics is relevant to the future of cryptography. In particular, quantum computation (which does not fall within the framework of Turing computation) enables polynomial-time algorithms for factoring and discrete logarithm. Many current cryptosystems – RSA, Diffie – Hellman, ECC, and so forth. – could be easily broken if quantum computation on a sufficiently large scale becomes possible. Oddly, other public-key cryptosystems – for example, lattice-based and knapsack-based cryptosystems – do not yet appear vulnerable to quantum computation. In general, an important research question for the future of cryptography is how quantum complexity classes relate to traditional complexity classes and to individual "hard" problems.

A more mundane research direction is to expand the list of hard problems on which cryptosystems can be based. This serves two purposes. By basing cryptosystems on assumptions that are weaker than or orthogonal to current assumptions, we hedge against the possibility than many of our current cryptosystems could be broken (e.g., with an efficient factoring algorithm). On the other hand, as in ESIGN, we may accept stronger assumptions to get better efficiency.

Autonomous mobile agents have been proposed to facilitate secure transactions. However, Goldreich et al. (Barak et al. 2001) proved the impossibility of complete program obfuscation, suggesting that cryptographic operations performed by mobile agents may be fundamentally insecure, at least in theory. Because mobile agents may nonetheless be desirable, it is important to assess the practical impact of the impossibility result.

Spam and the prospect of distributed denial of service (DDoS) attacks continue to plague the Internet. There are a variety of approaches that one may use to address these problems – ratelimiting using trusted platforms, Turing-test-type approaches such as "CAPTCHAs," using accounting measures to discourage massive distributions, proof-of-work protocols, and so forth. – and each of these approaches has advantages and disadvantages. The importance of these problems demands better solutions.

10.7 Conclusion

We considered the prospect of designing cryptographic solutions in a XG world. We began by identifying some existing techniques such as anonymity-providing signatures and provable security. Next, we described the challenges of securing XG and identified some fundamental problems in cryptography, such as certificate revocation and designing lightweight primitives, that currently need to be addressed. Finally, we considered current research directions, such as coping with a heterogeneous environment and achieving security at the terminal level.

It is clear that securing the XG world is a daunting task that will remain a perpetual work in progress. While we have a number of excellent tools at our disposal, the ubiquity and heterogeneity of XG has introduced far more problems. However, these problems represent opportunities for future research directions. Furthermore, as we continue to advance the state of the art in cryptography, we will not only address existing problems but will likely create tools to *enable* even greater possibilities.

11

Authentication, Authorization, and Accounting

Alper E. Yegin and Fujio Watanabe

Providing a secure and manageable service requires the ability to authenticate and authorize legitimate users and collect associated accounting information. The architectural component that is responsible for these functionalities is called *Authentication, Authorization, and Accounting* (AAA or "triple-A") module.

Authentication is the verification of a claimed attribute.

Authorization is the process of determining whether a particular right should be granted to an entity.

Accounting is the act of collecting usage information for billing and resource-management purposes.

These three elements are the essential components of data network security. Whether it is an enterprise network used for employees' access to the Internet or an ISP network used for public access, clients must be authenticated before they are authorized to access the data (IP) services.

Generally, authentication and authorization are integrated. Authorization of a requested service by a user must be accompanied by verification of the claimed identity. Authentication is a necessary, but not sufficient, step for the overall AAA process. Many factors, such as access control and resource usage, play a role in determining whether an authenticated user should be granted access to the service. For example, an authenticated user might not be allowed to access a network just because she is not allowed to use it during business hours.

Next Generation Mobile Systems. Edited by Dr. M. Etoh
© 2005 John Wiley & Sons, Ltd

A successful user authorization enables the requested service, and also initiates accounting mechanisms. Accounting allows the network operator to keep track of network usage for various reasons, such as usage-based billing, trend analysis, auditing, and resource allocation.

Overall, AAA is responsible for protecting services from unwanted users, collecting service charges, and obtaining insight into the network usage. A secure IP service cannot be achieved without using a solid AAA system. Today, some form of AAA is built into any given data service, such as WLAN hotspots and enterprise networks, cellular IP services, and dial-up ISP services. For example, when a user dials up her ISP, she is engaged in a user login process. Simple exchange of user ID and password accomplishes the authentication and authorization steps. Subsequently, the usage information is collected during the session. In today's mostly flat-rate dial-up services, the accounting information does not impact the billing. On the other hand, it produces necessary data for the ISP to efficiently run its network.

11.1   Evolution of AAA

AAA technologies are rapidly evolving as the overall Internet scenery changes. AAA is getting significant attention within the industry as the backbone of the service-providing business and a requirement of any secure network. This interest is leading to significant industry, government, and academic research and development.

One high-impact factor in the evolution of AAA has been the development of wireless access technologies and mobility. Unlike their wired predecessors, wireless networks cannot depend on the presence of physical security. Preventing eavesdropping and spoofing on radio traffic requires airtight security features from the AAA technologies. Some of the research and development activities have been directed at identifying vulnerabilities of the newly deployed systems, which, in many cases, are quickly integrating the existing technologies that are not suitable to wireless environments. Mobility enables users to access the Internet at any one of the many service Access Points (AP), such as WLAN hotspots. This gives rise to performance issues with the AAA processing. A typical authentication and authorization process involves the access network consulting a centralized server for verification. The need to access the centralized AAA centers each time a user moves is a bottleneck for seamless mobility. Therefore, optimizing AAA has been a fertile and essential research subject in recent years.

Despite achieving similar functionalities, AAA technologies used in today's networks vary significantly among themselves. This is due to varying architectural bases (for example, 3GPP, 3GPP2, and WLAN hotspots), deployment considerations (for example, uniformity in standards-based 3G terminals versus variability in WLAN access devices), and the availability of several standards-based and ad hoc solutions. As the network operators find themselves running multiple types of access networks, such as cellular and WLAN, they realize that managing and integrating these incompatible AAA systems poses a challenge. Converging AAA under a unified umbrella is a key goal for network operators; associated research and development has been actively pursued in the industry. It is important to harmonize AAA for data services, and also to integrate AAA for other types of services (such as application and content delivery) in addition to network access. Assuming that XG networks will entertain more heterogeneity in terms of access technologies and terminals, and aim for enhanced user experience, an integrated AAA system emerges as one of the most important research topics in this area.

The authentication and authorization aspect of AAA in mobile networks is directly related to cryptography. Identity verification involves possession and proof of secret keys. The cryptographic method used during authentication often has a direct impact on the performance of this process. For example, a shared-secret-based authentication would incur a long-haul communication with a centralized AAA server, whereas a public-key-based authentication can be processed without consulting such a third party. Cryptography research as outlined in Chapter 10 is expected to have a direct impact on the AAA systems for XG.

While AAA is a must-have technology for any network operator, industry is also starting to see it is as a service of its own. Acting as a trusted entity that can broker a data service between a client and a service provider is a revenue-generating business today. The high cost of building and maintaining the infrastructure needed to provide AAA, combined with the ability to separate it from the actual service itself, gave birth to this new business area. In some deployment scenarios, relying on third-party AAA service providers turned out to be the only feasible way to provide data services. For example, with the introduction of unlicensed WLAN hotspot services, several service providers emerged in overlapping locations. Normally, a user should obtain an account from every one of the possible service providers that she might use, but this is not a practical solution. Instead, the user can have an account with a so-called Virtual Network Operator (VNO), such as Boingo[1] or iPass[2], which does not own any data service infrastructure but instead maintains business relations with those who do. The VNO helps the user get authorized for accessing any of the affiliated operators' networks. Effectively, what a VNO provides is a AAA brokerage service. It is expected that this paradigm will evolve as we progress to the next generation of mobile networks.

Overall, AAA is an area that will shape the XG services in significant ways. Aside from being an essential component of the overall architecture, it will directly contribute to service differentiation and new service generation. Research activities in this field are expected to increase as we move toward XG networks.

11.2 Common AAA Framework

Any AAA system can be analyzed under a common framework despite the differences among such systems (see Figure 11.1).

In this framework, one of the entities is the client. The client is a host that connects to an access network for sending and receiving IP packets. This host can be an employee laptop connected to the enterprise WLAN, or a pedestrian's phone connected to a cellular IP network. The client is configured with a set of credentials, such as a username and password. These credentials are used in authentication and authorization phases during network connection. Additionally, the client should also be configured with a service-selection criteria. There may be more than one service available in a given location, and the client must know how to pick one among these. The associated services may differ in capabilities and cost. Furthermore, some of these networks might also be malicious. Having a service-selection criteria that enables early elimination of potentially malicious networks is a useful

---

[1] See http://www.boingo.com for more information on Boingo Wireless.

[2] See http://www.ipass.com for more information on iPass.

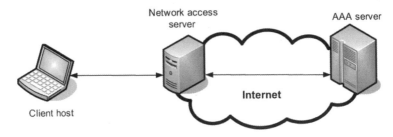

Figure 11.1 AAA framework

feature. Careful design in this area becomes more important with wireless networks, where physical security does not exist.

The other endpoint of a typical AAA exchange is a AAA server. This entity verifies the authentication and authorization of clients for network service access, and collects accounting information. A AAA server maintains the credentials and the associated authorization information of its clients. There exists a preestablished trust relation between a AAA server and its clients stemming from business relations, such as a service subscription. From the perspective of network service providers, these servers are leveraged as a trusted third party. When a client attempts to gain access to a network, the AAA server is consulted for the verification process. These servers are generally located in data centers behind several levels of security (including physical security, such as guards and dogs).

The third entity in this framework is the Network Access Server (NAS, pronounced "nas") (Mitton and Beadles 2000). The NAS' responsibility is to act as an intermediary between the client and the AAA server as the representative of the visited access network. A NAS is located on the access network, for example on a WLAN AP or a 3GPP2 Access Router (AR). It acts as a local point of contact for the client during the AAA process. It obtains a subset of credentials from the client and consults with an appropriate AAA server to authenticate and authorize the client for the network access service. The NAS should have a direct or indirect trust relation with the AAA server in order to engage in a secure AAA communication. Upon successful authorization, the NAS is responsible for notifying appropriate policy Enforcement Points (EP) on the visited network that allow the client's traffic, and also for collecting usage information.

The client, the NAS, and the AAA server are located on different nodes. This separation requires a set of communication protocols for carrying the AAA traffic among the entities. A client directly interacts only with the NAS. This leg of communication is considered the front end of AAA and is handled by protocols like PPP (Simpson 1994), IEEE 802.1X (IEEE 2001a), and Protocol for Carrying Authentication for Network Access (PANA) (PANA n.d.). On the back end, the NAS interacts with the AAA server using another protocol, such as Remote Authentication Dial-In User Service (RADIUS) (Aboba and Calhoun 2003) or Diameter (Calhoun et al. 2003). Both the front-end and back-end protocols are needed to establish a AAA session between the client and the AAA server that goes through NAS.

The initial phase of a AAA session carries out the authentication of the client by means of an authentication method. CHAP (Simpson 1996) and TLS (Aboba and Simon 1999) are two popular authentication methods that are used in wired and wireless networks respectively. These methods are in charge of authenticating endpoints to each other. They achieve this by carrying various credentials among them. The authentication methods are encapsulated within the front-end and back-end AAA protocols using a "shim" layer called the *Extensible Authentication Protocol* (EAP) (Blunk et al. 2003). EAP is a generic authentication method encapsulation used for carrying arbitrary methods inside any of the communication protocols. Authorization is engaged as soon as the AAA server verifies the credentials of the client. Authorization data, such as allowed bandwidth and traffic type, is transferred from the AAA server to the NAS by the help of back-end protocols. The same back-end protocols later carry accounting data from the NAS to the AAA server.

Internet access service is generally provided by the combination of a Network Access Provider (NAP) and an Internet Service Provider (ISP). An NAP is the owner of the access network that allows clients to physically attach to the network and enables IP packet forwarding between the ISP and the client. The clients only subscribe to an ISP service, and they can connect to the Internet via any NAP that has a roaming agreement with that particular ISP. In this configuration, an NAP hosts a NAS at each access network. This NAS consults appropriate AAA servers on the ISP networks during clients' AAA process. The NAS may or may not have a direct trust relationship with the particular AAA server. In cases where this relationship is not preestablished, a AAA broker server can be used as a meeting point between these servers. The process of identifying the right AAA broker or server, and directing the AAA traffic accordingly, is called *AAA routing*. The collection of AAA servers, NAS, and AAA brokers form an Internet-wide AAA *web of trust*. Only through the existence of this web is it possible for a user to hop from one coffee shop to another in a city and be able to reach the Internet using an account with a single ISP.

An essential aspect of the network access AAA process is the binding between the authorized client identity and the subsequent data traffic. In wireless networks, unless an authenticated client is cryptographically bound to its data traffic, service theft cannot be prevented. The shared medium allows any client to assume the role of an authorized client and send data packets on its behalf unless some secret is used as part of data transmission. For this reason, the AAA process must generate a local trust relationship between the NAS and the client, in the form of a Security Associations (SA) with shared secrets. Master secrets are delivered as part of the AAA process. These secrets are used in conjunction with another protocol exchange between the client and the NAS (for example, IEEE 802.11i (IEEE 2003b) 4-way handshake or IKE (Harkins and Carrel 1998a)) for producing keys for data traffic ciphers. Cryptographically protected data traffic can prove its origin authenticity and additionally provide confidentiality. Any wireless access network that lacks the technology or deployment of this cryptographic binding effect cannot achieve true security.

11.3 Technologies

Mobile data service providers and vendors have already developed and deployed a number of technologies that form today's AAA systems. These systems are undergoing constant evolution. The ongoing research and standardization efforts are changing the AAA landscape.

Widely deployed RADIUS and emerging Diameter are the IETF-defined AAA back-end protocols. Any large-scale AAA architecture relies on the presence of one of these protocols. EAP has been taking the center stage as the generic authentication method encapsulation. It is carried end to end between a client host and the authentication server by front-end and back-end AAA protocols. PANA is an ongoing IETF development that aims to provide a link-layer agnostic AAA front-end protocol. The combination of authentication methods encapsulated in EAP and carried over PANA and RADIUS/Diameter forms a complete AAA system. Although a unified AAA architecture can be defined by these components, the currently deployed wireless access networks vary significantly. WLAN-based networks not only differ from the cellular networks but also come with an array of solutions within themselves. This fact can be attributed to the standards not being in place when the deployment needed them. Lack of standards usually leads to development of multiple ad hoc solutions by the leading industry players. On the other hand, although AAA design of 3GPP and 3GPP2 are not the same, at least they are uniform and well defined within the respective cellular architectures.

11.3.1   RADIUS and Diameter

Since the number of roaming and mobile subscribers has increased dramatically, ISPs need to handle thousands of individual dial-up connections. A network administrator in a company has to deal with more remote users accessing the company's LAN through the Internet. To handle this situation, an ISP can deploy many Remote Access Servers (RAS, or NAS) over the Internet. It can then use the RADIUS (Rigney 1997; Rigney et al. 1997) for centralized authentication, authorization, and accounting for network access through a RAS.

RADIUS remote access has three components: users, RAS, and the RADIUS (AAA) server. Each user is a client of an RAS; each RAS is both a server to the user and a client of the RADIUS server. Figure 11.2 illustrates a typical configuration using the RADIUS server (Davies 2002; Metz n.d.). Although the RADIUS server can support accounting services, the main function of RADIUS is authentication. An example of a RADIUS procedure is:

1. A user uses a dial-up to connect to one of the ISP RASs. PPP negotiation begins.

2. The RAS (client of the RADIUS) sends user credential and connection parameter information in the form of the RADIUS message (Access-request) to the RADIUS server. It also sends the RADIUS accounting messages to the RADIUS servers. Security for RADIUS messages is provided based on a common shared secret configured between the RAS and the RADIUS server.

3. If the RADIUS server can authenticate the user, it issues an accept response (Access-accept) to the RAS, along with profile information required by the RAS to set up the connection.

4. If the RADIUS server cannot authenticate the user, it issues a reject response (Access-reject) to the RAS.

5. The RAS completes PPP negotiation with the user. It can allow the user to begin communication. Otherwise, it terminates the connection.

RADIUS was originally designed for small networks supporting just a few end users requiring simple server-based authentication. Roaming and large numbers of concurrent users accessing network service required a new AAA protocol, so Diameter was developed by IETF as a next-generation AAA protocol (Calhoun et al. 2003). Diameter was designed to support roaming and mobile IP networks from the beginning. The primary improvements in Diameter are:

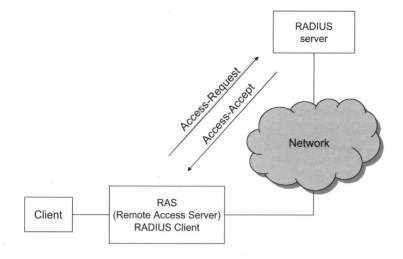

Figure 11.2 Overview of RADIUS environment

- Flexibility of the attribute data

- Better transport

- Better proxying

- Better session control

- Better security.

An attribute carried in a RADIUS message has variable length, but the size is one octet, with a maximum value of 255. A Diameter attribute has a variable length with three octets, for a maximum value of over 16 million. In addition, Diameter uses a more-reliable transport protocol than RADIUS. RADIUS operates over User Datagram Protocol (UDP), a simple connectionless datagram delivery transport protocol, while Diameter operates over Transmission Control Protocol (TCP) or Stream Control Transmission Protocol (SCTP), connection-oriented transport protocols.

11.3.2 Extensible Authentication Protocol

EAP is an authentication framework that can support multiple authentication methods. Having been developed originally for PPP, EAP provides an abstraction that allows any authentication method and access technology to work together without requiring a tight integration between the two.

The basic idea behind EAP design is that, as long as the network is capable of carrying EAP packets, it can use any authentication method that is implemented as an EAP method. Networks become more heterogeneous as we move toward XG. Access technologies, user profiles and access devices, and network policies are all diversifying. IEEE 802.1 architecture has already adopted EAP as part of its IEEE 802.1X protocol. By carrying EAP over the

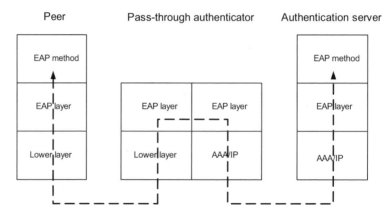

Figure 11.3 Pass-through authenticator scenario

link layer, currently any one of the more than 50 authentication methods can be used over an Ethernet network. The availability of new methods is expected to grow without requiring any change in the underlying access technology (that is, the link layer and physical layer). The EAP framework defines three entities (see Figure 11.3).

- The peer is the client that desires to engage in authentication for gaining access to a network. The peer engages in an EAP conversation with an EAP server.

- The EAP server authenticates and authorizes the peer for network access service.

- The authenticator acts as an EAP relay and resides on the access network. The authenticator forwards EAP packets between the peer and the EAP server.

This framework is built along the same lines as the generic AAA framework. When considered within this framework, a peer resides on a client, the authenticator on a NAS, and the EAP server on a AAA server. This is the most common layout, to allow for roaming scenarios, but a peer can also communicate directly with the EAP server. Currently, EAP is defined as part of several AAA systems. It is carried over PPP and IEEE 802.1X at the front end, and over RADIUS and Diameter at the back end.

Each one of the protocol entities implements an EAP stack, where EAP methods are carried over the EAP layer, and in turn over a lower layer. The lower layer is responsible for carrying EAP packets between the peer and the authenticator. PPP and IEEE 802.1X are two relatively well-established and standardized EAP lower layers. These protocols are isolated from any authentication method details. The EAP layer is responsible for passing EAP messages between the EAP methods and the lower layer. Finally, EAP methods implement authentication algorithms. For example, EAP-MD5 implements MD5-based challenge-response authentication method, EAP-TLS (Aboba and Simon 1999) implements public-key-based TLS authentication.

The strength of EAP comes from the fact that once EAP is built into an architecture, adding new authentication methods only requires adding new EAP methods on the peer and the EAP server. Not having to make modifications on the authenticators that reside on the access networks makes Internet-wide changes more manageable. Authentication methods are expected to change and multiply as part of the wireless evolution.

There are various research and development activities related to EAP. Having been developed for wired PPP networks originally, simple application of EAP to the wireless world has raised significant issues. Many details that were not relevant to PPP networks or simply not well thought out originally started to impact the EAP implementations over the IEEE 802.11 networks. As a result, the EAP Working Group was formed under the IETF to update the EAP specification. Even though many loose ends have been fixed under this effort, the requirement that the new specification be backward compatible has limited the extent of problems solved in this effort. It is anticipated that a new version of EAP, EAPv2, would be designed to solve the lingering problems of EAP.

Although it is not really a road block for the current deployments, EAP problems can easily be solved in a fresh design. For example, the lock-step request-response style of EAP causes high latency for certificate-based authentication methods, such as TLS. Up to 20 round-trip exchanges might be required between a peer and its EAP server, and they might be located several hops away. Such latency can easily constitute a bottleneck for seamless mobility services. Lack of a large identifier field and more importantly a Message Authentication Code (MAC) field reduce the protection of EAP conversation against various active attacks on the wireless networks. Another issue is the lack of ability to separate the authentication result from the authorization result. Such a separation would enable a more informative interaction between the client and the access network. Current EAP frameworks also lack a service advertisement and selection facility. EAP assumes this process takes place out-of-band prior to the EAP conversation. Building this into EAP frameworks would have the added benefit of providing a bundled solution for the overall AAA process. The EAPv2 effort has not officially started in the industry; however, when it starts, it is expected to be tackled in IRTF prior to IETF standardization.

Another area of research activity is the design of new EAP methods. The portfolio of authentication methods is increasing as new deployment scenarios are considered for wireless networks. For example, the desire to authenticate a client the same way, whether it is accessing a GPRS network or a WLAN network, led to the development of the EAP-SIM (Haverinen and Salowey 2003) method, the SIM authentication method defined in terms of the EAP framework. The strength gained from this approach is the unification of AAA under various access technologies. Another good example is the development of the EAP-Archie (Walker and Housley 2003) method. Similar to its predecessor EAP-MD5, this method relies on static pre-shared secrets. But the strength of EAP-Archie is its capability to derive session keys. These keys are used for cryptographic binding of data traffic to client authentication. Lack of this capability prohibits the use of EAP-MD5 on WLAN networks.

Finally, another type of activity in this area is the development of new lower layers for EAP. Designing an EAP lower layer that runs above IP is a recipe for allowing EAP on any link layer. This approach is taken by the IETF PANA Working Group.

11.3.3 PANA

One of the most fundamental aspects of XG networks is expected to be their heterogeneity. The new generation access networks will incorporate a wide array of radio access technologies, user types, and network policies. This variety is likely to create complexity both for the users and the service providers unless some actions are taken to harmonize various components under a common umbrella.

If we look at today's systems, we realize that the current AAA picture is not so pretty. AAA mechanisms and credentials used in various networks differ considerably. For example, the username-password pair provided through a login web page for accessing a WLAN hotspot service is not the same pair that is provided to a 3GPP2 network via PPP. Having to maintain multiple sets of credentials and deal with different user interfaces is a hassle for the users. Similarly, supporting a multiplicity of protocols and disjoint AAA systems on the network is a costly operation for the service providers. For these reasons, a unified AAA system is one of the must-haves for the XG networks.

The back-end protocols, such as RADIUS and Diameter, already contribute to the unification of AAA. They provide a common framework that can support varying access technologies and user types. Another useful protocol is EAP, which enables encapsulation of any authentication method in a generic way. Generally, the choice of authentication method is determined by the user type, network policy, and access technology. The only missing component to achieve a unified AAA for network access was a generic front-end protocol that can carry EAP on any link layer. This need gave birth to the on-going development effort of the PANA protocol (PANA n.d.).

PANA is currently being designed as a link-layer-agnostic network access authentication protocol. It aims at enabling any authentication method on any type of link layer. It achieves this goal by carrying EAP over IP. Along with this basic principle, it also introduces various powerful features, such as enabling separate NAP and ISP authentication, bootstrapping local trust relations, fast reauthentication, secure EAP exchanges, extensibility via additional protocol payloads, flexible placement of AAA and access control entities, and so on. It is expected that PANA will become a necessary component for the AAA architecture of IP-based XG networks.

PANA and IEEE 802.1X are similar to each other since they both carry EAP between the clients and the network. The most important difference between the two is that the former can be used on any link layer whereas the latter is only applicable to IEEE 802 links. IEEE 802.1X also lacks the additional PANA features mentioned above.

The IETF PANA Working Group has been in charge of developing and standardizing the PANA protocol. At the time of writing, the working group has completed the identification of usage scenarios (Ohba et al. 2003), requirements (Patil et al. 2003), and relevant security threats (Parthasarathy 2003b) of PANA in respective Internet drafts and is concentrating on designing the protocol (Forsberg 2003) based on these documents.

PANA Framework

The PANA protocol runs on the last IP hop between a PANA Client (PaC) and an authentication agent (PANA Authentication Agent – PAA) (see Figure 11.4). The PAA would reside on a NAS in the access network. This NAS would bridge the AAA session between the client and the AAA servers by using PANA and the RADIUS/Diameter protocols.

The PANA framework also defines an entity called *an enforcement point* (EP). The EP controls access by disallowing network access for unauthorized clients. It achieves this with packet filtering. Filtering can be based on simple selectors, such as source and destination addresses, but in general, this type of filtering is not adequate in multiaccess wireless networks; cryptography-based methods, such as IPsec-based access control, are

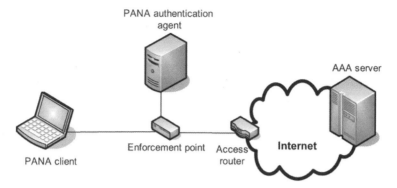

Figure 11.4 PANA framework

typically required. An EP must be located at a choke point in the network so that it has full control over the traffic in both directions. An example EP is the AP in a WLAN network. Alternatively, when there is no network entity between the client and the AR, access control can be implemented on the AR.

In many deployments, the authentication agent and EP are colocated (for example, in IEEE 802.1X-based networks). PANA allows separation of the EP from the PAA, and allows multiple EPs per PAA. This separation requires another protocol to run between the PAA and the EP. This protocol would be engaged as soon as a new client is authorized to access the network by the PAA, which must inform the EP about the client's identity. The PANA Working Group has decided that SNMP can be used for this purpose. Work is currently in progress to define necessary extensions to SNMP for supporting PAA-to-EP communication.

The separation of the EP from the PAA, and the ability to place a PAA on any node in the last hop (not just on the immediate link-layer access device), provides flexible PANA deployment. In one scenario, an EP and PAA can be colocated with the link-layer access device (for example, switch) that sits between the client and the AR. In another, a PAA can be located with the AR while the EP is on the switch. Access router, EP, and PAA can also be colocated on a single node in the access network. And finally, they can all be on separate nodes. In that case, the PAA can be a dedicated server connected to the link between the AR and the client. In summary, PANA enables creative ways to organize the network depending on the specific needs of the deployment.

PANA is an "EAP lower layer" that encapsulates EAP packets (see Figure 11.5). It is defined as a UDP-based protocol that runs between two IP-enabled nodes on the same IP link. It also provides an ordering guarantee to deliver EAP messages, as required by the base EAP specification (Blunk et al. 2003).

Protocol Flow

The protocol execution consists of a series of request-response type message exchanges. Some of the messages simply carry EAP traffic between the network and the client, while others manage a PANA authentication session (see Figure 11.6).

The discovery stage involves either the PaC soliciting for PAAs on the link, or a PAA detecting the attachment of the client and sending an unsolicited message. Either way, a

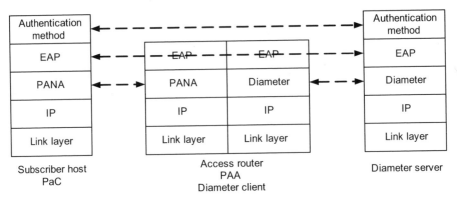

Figure 11.5 PANA stack

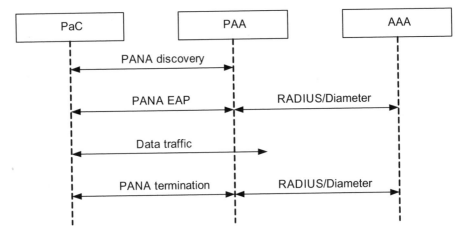

Figure 11.6 PANA flow

PANA-Start exchange marks the beginning of a new PANA session. PANA-Start is followed by a series of PANA-Auth message exchanges. These messages simply carry EAP payloads between the peer and the authenticator. Since PANA is just an EAP lower layer, it does not concern itself with the detailed content of these payloads. Only the final message, EAP Success and EAP Failure, has significance for PANA. This marks the end of the authentication phase, and the EAP packet must be carried in a PANA-Bind message. If the authentication is a success, this message also establishes a common agreement between the PaC and PAA on the device identifiers (such as IP or MAC addresses) to be used and the associated per-packet protection. The device identifier of the PaC will later be provided to the EP for access control. Meanwhile, the peers can also decide if link-layer ciphers or IPsec will be enabled for additional cryptography-based per-packet protection. This type of mechanism can only be enabled when the EAP method generates cryptographic keys. The available keys are also used to generate a PANA Security Association (PANA SA) that is used to protect subsequent PANA exchanges.

Once the PaC is authorized, it can start sending and receiving any IP packets on the access network. During this phase, PaC and PAA can verify the liveness of the other end by sending asynchronous PANA-Ping messages. This message is useful for detecting disconnections.

A PANA session is associated with a session lifetime. At the end of the session, the PAA must engage in another round of PANA authentication with the PaC. If the PAA decides to disconnect the PaC prior to that, or the PaC decides to leave the network, a PANA-Terminate message can be used to signal this event. Transmission of this message marks the end of a PANA session.

Supported Environments

PANA can be deployed in a variety of environments of these three types:

Physically secured: For example, the DSL networks that run over physically secured telephone lines. It is assumed that eavesdropping and spoofing threats are negligible in these networks.

Cryptographically secured before PANA: For example, the cdma2000 networks that enable link-layer ciphering prior to IP connectivity. Although these are wireless links, by the time PANA is run, eavesdropping and spoofing are no longer a threat.

Cryptographically secured after PANA: A wireless network that relies on PANA to bootstrap a SA for enabling link-layer ciphering or IPsec would fall into this category. An example would be bootstrapping WEP-based security on a WLAN network. Eavesdropping and spoofing are concerns during PANA authentication. An EAP method that can generate cryptographic keys must be used.

The PANA protocol executes the same way regardless of the environment it operates in. Less secured environments, such as the third type, require carefully chosen EAP methods. Methods that can provide mutual authentication and cryptographic key generation are needed in these networks. Furthermore, generated keys must be bound to the data traffic, which is accomplished by additional protocol exchanges following a successful PANA authentication.

PANA and IPsec

IPsec-based access control is deemed necessary in networks where eavesdropping and spoofing are threats but link-layer ciphering is not available. Using IPsec between a client and the network requires an IPsec Security Association (IPsec SA), which may not normally exist between two arbitrary nodes. A dynamically generated SA is needed before IPsec can be engaged.

PANA enables IPsec-based access control by helping an IPsec SA created after successful PANA authentication (Parthasarathy 2003a). The cryptographic keys generated by the EAP method are used to create a PANA SA. PANA SA cannot be readily used as an IPsec SA. The latter requires traffic selectors and other parameters that are not available in the former. But nevertheless, PANA SA represents a local trust relation between the PaC and PAA, which can be used as the basis of a "preshared secret" for generating a dynamic IPsec SA.

This approach leads to the use of IKE for turning a PANA SA into an IPsec SA for IPsec-based access control. Preshared keys are driven from PANA SA and simply fed into the IKE protocol. The resultant IPsec SA is used to create a tunnel between the PaC and EP for providing an authenticated (and optionally encrypted) data channel.

Use of IPsec-based access control along with PANA also appears as another deliberate choice for an all-IP architecture. Such a design can be applied to any IP network regardless of the link-layer technologies used.

Building Local Trust

One of the pressing issues with securing network-layer protocols on the access network is the lack of trust relation between the clients and the network. Well-known mechanisms, such as using messages authentication codes, rely on availability of a shared secret between two entities. PANA protocol in conjunction with EAP and EAP methods is one way to facilitate creation of dynamic SA.

The current thinking is that PANA SA can be transformed into purpose-specific SAs as needed. An example is how PANA SA between PaC and PAA is used to generate a DHCP SA between a DHCP client running on the same host as a PaC and a DHCP server in the access network (Tschofenig 2003). It is assumed that the PAA and DHCP server already have a preestablished trust relation. The dynamically created trust relation between the PaC and PAA can be used by the PAA to introduce the PaC to the DHCP server.

Currently, this model is being analyzed for its security aspects. It relies heavily on EAP keying framework, which is going through a formal redesign in the IETF EAP Working Group. If this model proves to be valid, it will be used for solving similar problems, such as securing fast Mobile IP handovers (Koodli 2003b) and hierarchical mobility protocols (Soliman et al. 2003).

11.3.4   WLAN

This section addresses current WLAN AAA schemes, which include conventional IEEE 802.11 standard authentication and IEEE 802.1X authentication for the next IEEE 802.11i security, the Wireless Internet Service Provider (WISP) proprietary solution, and other proprietary solutions based on the SIM authentication.

Wired Equivalent Privacy (WEP)

The IEEE 802.11 standard (ISO 1999) defines Wired Equivalent Privacy (WEP) as an authentication method and a cryptographic confidentiality algorithm. Its first purpose is to prevent an unauthorized user from accessing the wireless LAN network. A secondary purpose is to protect authorized users of a wireless LAN from malicious eavesdropping.

In IEEE 802.11, two types of authentications are implemented: one is open and the other uses a shared key. Open authentication does not provide access control; anyone can access the wireless network. The shared-key authentication mechanism exchanges a challenge (a pseudorandom number sequence) and a response message (the encrypted challenge) between Stations (STAs). If the response message is correctly decrypted at the AP, access to the wireless network is granted to the STA. The response message (128 octets) is encrypted by the WEP. These message exchanges are illustrated in Figure 11.7.

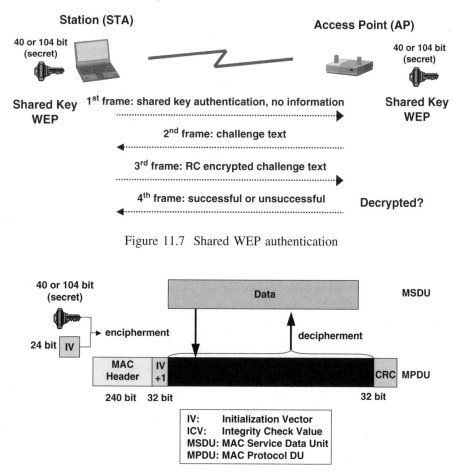

Figure 11.7 Shared WEP authentication

Figure 11.8 WEP encapsulation and decapsulation

The same confidential WEP key used for authentication is also used to encipher and decipher the message, as shown in Figure 11.8.

The WEP key length is either 64 bits or 128 bits. Forty out of 64 bits or 104 out of 128 bits are a shared secret between the AP and STA. The WEP uses the RC4 encryption algorithm, which is known as a *stream cipher*. The encrypted data is generated using the SWEP Pseudorandom Number Generator (PRNG) with the shared secret and a random Initialization Vector (IV). An integrity check value is attached to the MPDU, as well as the 32-bit Cyclic Redundancy Check (CRC), to ensure that packets are not modified. Unfortunately, it has recently been proven that breaking WEP is easily within the capabilities of any laptop (Arbaugh et al. 2002a; Borisov et al. 2001a,b; Walker 2000). A security vulnerability allows hackers to intercept and alter transmissions passing through wireless networks. Therefore, researchers at UC Berkeley recommend that anyone using an 802.11 wireless network not rely on WEP, but employ other security (such as a Virtual Private Network (VPN) or additional encryption software) to protect their wireless network from snooping.

J. Walker (Walker 2000) pointed out that increasing the length of shared keys does not help to achieve privacy with the WEP encapsulation. His conclusion is to replace RC4 with different cryptographic primitives. Therefore, a new security standard 802.11i addresses a counter mode CBC-MAC protocol (CCMP) algorithm based on an Advanced Encryption Standard (AES).

Even though WEP was proven to be vulnerable, the market for IEEE 802.11b devices has exploded. Currently, many laptop computers have WLAN with either embedded 802.11b (operating at 2.4 GHz) or 802.11a (operating at 5 GHz), because it is so easy to deploy a simple, plug-and-play wireless network.

WISP Captive Portal

The low cost of equipment has accelerated entry into the hotspot business. In the past few years, many companies have set up wi-fi access points (also known as "hotspots") in hotels, airports, cafes and many other public spaces. Well-recognized WISPs are Wayport[3], Boingo[4], and Cometa.[5] In addition, mobile phone operators such as T-mobile, Sprint, Verizon, and AT&T wireless have started to operate hotspot services in Starbucks, book stores, Kinko's stores, and airports. However, these hotspot operators have a problem controlling access. In terms of wireless security, access control may be more important than privacy (encryption). What the hotspot WISP needs is a simple way to block access from unauthorized individuals. The most common access-control mechanisms for WLAN besides WEP authentication are client MAC address authentication and browser-based authentication. When the user tries to go to any website, the user will automatically be taken to the WISP's sign up or login page, referred to as a captive portal.

In the client MAC address authentication, each AP has a list of MAC addresses of clients. If a client's MAC address is in the list, it is permitted access to the network. If the address is not listed, access to the network is prevented. However, WLAN hotspot operators generally prefer to have flexible access control based on who a user is, rather than on which particular device is used. Therefore, browser-based authentication is widely used.

Browser-based authentication requires a user to enter authentication information on a login web page before the user can access other resources of the network. This login and logout web page is returned to the installed web browser via an HTTP redirect. To prevent sniffing, SSL is used to protect the user's authentication credentials. In addition, the certificate at the access gateway (a hotspot's entrance point to the Internet) can be used to protect users from rogue APs. A typical login request procedure is described in Figure 11.9 (iPass 2004).

IPsec VPN

When the link-layer security mechanisms are insufficient or completely absent, a higher layer mechanism must be used. One of the commonly used solutions is to rely on VPN technologies (a Layer 3 solution). There are basic trade offs between Layer 2 and Layer 3 solutions, and currently many enterprise users choose IPsec VPN for flexibility in access

---

[3] See http://www.wayport.net/ for the Wayport homepage.

[4] See http://www.boingo.com/index.htm for the Boingo Wireless home page.

[5] See http://www.cometanetworks.com/ for the Cometa Networks home page.

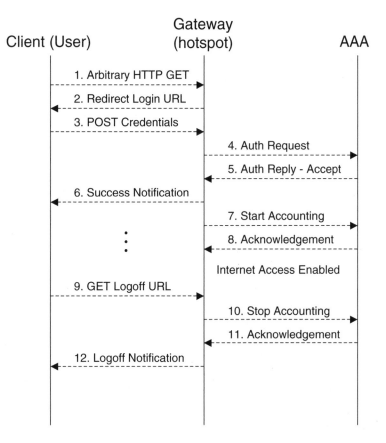

Figure 11.9 Login request: successful case – no proxy or polling

control (such as filtering at Layer 3). VPN is currently considered best for secure wireless access because it provides additional protection from malicious attacks, and it is easy to deploy.

The IPsec, as defined in IETF, provides a number of security features such as data confidentiality, device's authentication and credentials, data integrity, address hiding, and security association and key management. In this scheme, a VPN gateway located on the edge of the WLAN network acts as a security gateway. Clients cannot access beyond the last-hop network until they are successfully authenticated to the VPN gateway. Successful authentication and authorization allows the client to set up a tunnel to the gateway, which securely transports data traffic from and to the client. In addition to providing access control, an IPsec VPN can also provide confidentiality by the Triple Data-Encryption Standard (3DES) 168-bit encryption or the new AES.

Other Proprietary Solutions

Two other proprietary WLAN access control solutions (Ala-Laurila et al. 2001b; Bostršm et al. 2002b) are worthy of attention. Some of the GSM/GPRS operators provide WLAN as a complementary service to their cellular customers. The operator WLAN solutions are based on a single Subscriber Identity Module (SIM). Although two major proprietary schemes are listed here, there are many similar solutions based on the GSM SIM authentication.

These solutions provide for easy roaming. Mobile operators have the infrastructure and the culture to support roaming between different access networks as well as between operator's networks. International roaming based on GSM is widely used in over 200 countries. This roaming feature makes it possible to combine the WLAN and GSM accounts into a single account (GSM authentication over WLAN), so that the user needs only one subscription while using both services.

The GSM/SIM solution gives the user a single bill and gives the operator a reliable authentication scheme based on the GSM system. In order to authenticate the user (WLAN embedded SIM), the main design challenge is to transport standard GSM subscriber authentication signaling from the terminal to the cellular site using the IP protocol framework. See Nokia's architecture in Figure 11.10.

Ericsson (Bostršm et al. 2002b) has developed an operator WLAN system that employs an authentication system using a web login combined with SIM. This scheme uses SIM-based authentication with a One-Time Password (OTP) delivered via SMS, or static password-based authentication. The SIM-based authentication model uses a secure and authenticated GSM channel to distribute OTPs for WLAN access service. This secure password delivery step is followed by the static password-based authentication mechanism for accessing the WLAN, which is exactly the same as the captive-portal solution. The operator solutions can also provide localized content access and service differentiation. For example, the user can access localized content, such as departure information at the airport, the menu of the day at the hotel, and local advertisements. Service differentiation can offer, for example, Gold, Silver, or Bronze type of services to different users.

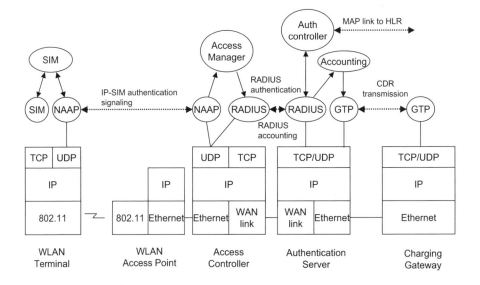

Figure 11.10 The Nokia operator WLAN solution (Ala-Laurila et al. 2001b)

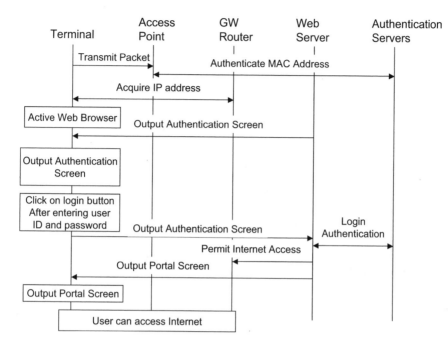

Figure 11.11   Mzone authentication sequence (Osugi et al. 2002)

For example, NTT DoCoMo Mzone WLAN service employs a combination of a MAC address with a web-based static password authentication scheme. The sequence is shown in Figure 11.11.

DoCoMo's MZone WLAN uses two schemes; it authenticates a WLAN terminal with a preregistered MAC address, and also authenticates the user ID and password for further access control in case the WLAN terminal is stolen.

IEEE 802.1X

The IEEE 802.1X standard (IEEE 2001a) has been developed to prevent unauthorized access to networks for users connecting inside the administrative domain, as opposed to connecting from an untrusted network via a firewall checkpoint. The IEEE 802.1X standard originally focused on IEEE 802.3 (Ethernet) and IEEE 802.5 (Token-ring) LANS, but it is now being used in wireless security with the IEEE 802.11 standard.

The IEEE 802.1X uses the MAC layer to carry Extensible Authentication Protocol (EAP) messages. These messages enable and disable the port to which a host is attempting to connect on a wired or wireless LAN. The authentication can be unidirectional or bidirectional; that is, the protocol requires that the peer seeking access to the network identify itself, but the peer may also demand that the network port to which it is connecting also identify itself. The network port has two virtual ports, a controlled port and an uncontrolled port. The controlled port normally requires authorization to operate, while the uncontrolled port is allowed access to permit the initial authentication messages and possibly DCHP and other configuration initialization traffic to pass through prior to authentication. Once authentication is done, the controlled port is changed from closed to open. IEEE 802.1X (Aboba et al. 2000; IEEE 2001a) provides its services using these methods:

1. Creates two logical ports for each point-to-point link: a controlled port and an uncontrolled port. By default, unauthenticated traffic is carried over the uncontrolled port and authenticated traffic is carried over the controlled port.

2. The entity that processes authentication requests from the supplicant is called the *authenticator Port Access Entity* (PAE).

3. Either port can be forced open or closed, or authorized or unauthorized, by management action via SNMP. Direction can be IN or BOTH.

4. The uncontrolled port's primary duty is to open the controlled port by processing the authentication traffic between the Supplicant and the Authentication Server.

5. Authentication is conducted by means of EAP frames that are carried directly via the MAC service. EAP over LAN is called *EAPOL*.

6. The uncontrolled port can be used to carry any traffic desired, such as DHCP or redirection to a registration server.

7. The two ports' traffic filters are analogous to the mechanisms of a packet filter or firewall.

8. By conducting EAP, sufficient information is generated to create a session key. This session key is sent to the authenticator from the authentication server to permit link encryption. The IEEE 802.1X sequence is illustrated in Figure 11.12.

11.3.5 IP-based Cellular Networks

All-IP cellular networks are currently being considered by the third generation standards groups, 3GPP[6] and 3GPP2 (*cdma2000 Wireless IP Network Standards-Draft* 2001). This section briefly addresses current 3G AAA schemes for all-IP networks.

3GPP

The 3rd Generation Partnership Project (3GPP) was formed in December 1998, bringing together a number of telecommunication standards bodies including ARIB, CCSA, ETSI, T1, TTA, and TTC. The 3GPP amends its draft standard four times a year since the freezing of Release 99. The 3GPP specification is based on the evolved GSM core networks, and the issue of security is now a much larger topic than it was when considered for the basic GSM system. In GSM, security is mainly concerned with the wireless part, but in 3G networks, especially WCDMA-based networks, many other aspects related to the total security are

---

[6]See http://www.3gpp.org/ for the 3GPP Initiative web page.

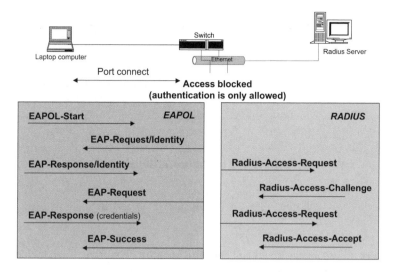

Figure 11.12 IEEE 802.1X sequence diagram (Aboba et al. 2000)

important. Secure user access to 3G radio networks uses the basic GSM mechanism, with some enhancements. The security features of the basic GSM (Kaaranen et al. 2001b; Mouly and Pautet 1992) system are:

- Authentication of the user based on the SIM card

- Encryption of the radio interface

In 3G networks, additional new features are being considered and improved:

- Mutual authentication of the user and network

- Radio access network encryption

- Use of temporary identities

- Protection of signaling integrity inside UMTS Terrestrial Radio Access Network (UTRAN).

Preventing unauthorized user access to networks is important for both users and operators. The authentication scheme uses the same challenge-and-response mechanism that GSM and most 2G systems employ. For a 3G network, the mutual secret, called the "*Master Key*," is a 128-bit value shared between the Authentication Center (AuC) and a Universal Subscriber Identity Module (USIM). 3G offers mutual authentication, so that the user side (USIM) can authenticate the validity of the current access network. A unique, increasing Sequence Number (SQN) is used to prove that a generated authentication vector has not been used before.

In DoCoMo's 3G FOMA services, the subscriber module, called the *User Identity Module* (UIM), has more features than the USIM (as defined by the ETSI TS 102.221) and SIM. The main additional features of UIM are:

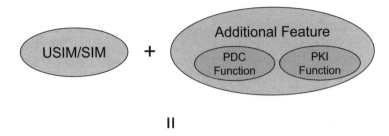

II

UIM (user identity module)

Figure 11.13 UIM versus USIM and SIM

- Network access management and authentication for networks, which includes not only voice/data communication but also recognition of different access networks such as 3G networks, GSM, and the 2G system called *PDC* in Japan.

- Function of providing additional services, which includes DoCoMo's own services such as the First Pass (Secure Sockets Layer (SSL) client authentication) digital certificate service.

Figure 11.13 shows the relationship between DoCoMo's UIM and USIM.

As mobile Internet access is more widely used, user authentication becomes more important. Users will access a broad range of services (such as Internet shopping, the stock exchange and accessing protected intranets), through the mobile phone. In June 2003, DoCoMo launched a digital certificate service called *First Pass* (Nakamura et al. 2003). This service makes use of a Public-Key Infrastructure (PKI) mechanism, which is widely used over the Internet. The DoCoMo First Pass center (DoCoMo CA) issues the user's certificate, which the user can then download to the FOMA's UIM card, which is highly protected.

When FOMA users need to authenticate, they send their certificates to the Content Provider (CP). This is simpler than exchanging user name and password between the user and the CP. The CP can use a Secure Sockets Layer (SSL) client authentication to verify the certificate and prevent disguise. Figure 11.14 shows this procedure.

1. The FOMA terminal downloads the user certificate from DoCoMo's CA.

2. When the user wishes to use the CP content, the FOMA First Pass compliant terminal accesses the CP web server.

3. The CP sends back the CP server certificate in order to authenticate itself.

4. After the CP server certificate is verified, the FOMA terminal sends the user certificate to the CP server for mutual authentication.

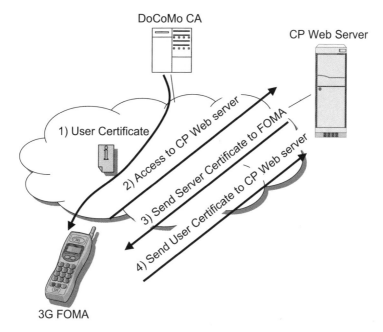

Figure 11.14 Service procedure

3GPP2

The 3GPP2 (*cdma2000 Wireless IP Network Standards-Draft* 2001) architecture is designed to provide high-speed IP data service to mobile users in a wide coverage area. The architecture features radio technologies of varying capabilities (such as 1xRTT, 1xEV-DO, 1xEV-DV), circuit-switched voice capability, and packet-switched data capability for simple IP and mobile IP services (see Figure 11.15).

Network access AAA in 3GPP2 is a two-tier process. First, the mobile terminal is authenticated to the network by a set of 3GPP2-specific servers called *MSC* and *VLR*. This authentication provides access to the circuit-switched voice service only. If the terminal wants to access data services, it has to follow up with another layer of AAA exchange. PPP and Mobile IPv4 authentication mechanisms are used for this purpose. Although a PPP connection is established both in simple IP and mobile IP services, PPP authentication takes place only for the simple IP service. Either the CHAP or PAP (Perkins and Hobby 1990) authentication method is used during this phase. The front-end AAA protocol is coupled with the RADIUS back end in order to authenticate and authorize data services for the mobile terminal. The PPP end-point (Packet Data Serving Node – PDSN) acts as the default IP gateway and NAS at the same time. Both IPv4 and IPv6 services are enabled in this fashion. In the case of Mobile IPv4 service, PPP authentication is skipped in favor of Mobile IPv4 authentication. The AAA functionality built into the mobility protocol is used to authenticate and authorize the terminal for accessing data services. In this scenario, the PDSN acts as a default IP gateway, NAS, and Mobile IPv4 foreign agent.

Neither PPP authentication nor mobile IP authentication generates keys for cryptographic binding of data traffic to client authentication. However, this does not present a

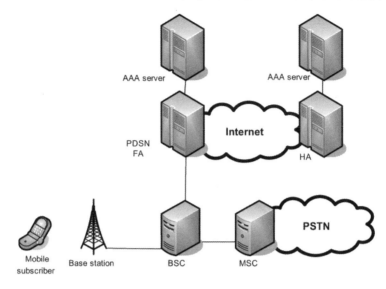

Figure 11.15 3GPP2 architecture

security threat to 3GPP2 networks, because these protocols and data traffic are run over an encrypted channel.

3GPP2 is an evolving architecture. Introduction of Mobile IPv6 service is planned for the upcoming release of the architecture. One significant problem this integration imposes on the architecture is in the area of AAA. While a foreign agent in Mobile IPv4 service performs NAS functionality on the access network, Mobile IPv6 does not define such an agent and therefore it cannot be used as the front-end AAA protocol. The architects are currently considering two alternative solutions to this problem. One is based on using PPP authentication even for Mobile IPv6 service. The other relies on introduction of PANA into the 3GPP2 architecture for this functionality. The latter appears to be an architecturally cleaner and forward-looking solution. PANA can be used as the unified AAA solution that can handle Simple IP and Mobile IP services, for both versions of IP on any access technology (cellular and WLAN). It also helps remove PPP from the 3GPP2 architecture by assuming the AAA functionality. Removal of PPP is perceived necessary to simplify the architecture and move in the direction of an all-IP network.

11.4   Emerging Research

This section briefly introduces further research in the area of AAA. These discussions are meant to give an idea of additional areas of research that will shape XG networks.

11.4.1   AAA for Access Networks with Ad hoc Extensions

Traditional access networks are built in a framework where the client hosts are directly connected to the service provider's access infrastructure. For example, a user laptop establishes the radio link with the hotspot provider's AP using IEEE 802.11b technology. The AP acts as the NAS throughout the AAA exchanges, and as the link-layer gateway to the Internet afterward. No other client or unknown node is allowed to insert itself between the clients and the access network in this model.

It has been envisioned that adding an ad hoc extension to the access network will benefit both the service providers and the users. Imagine an ad hoc network cloud that extends the access network to multiple hops away. While some clients access the Internet by directly connecting to the AP, others might be sending and receiving data packets through intermediate clients relaying them. One of the benefits of this approach is the increased coverage area for the service. The limited coverage area of an IEEE 802.11b AP can be extended by the help of WLAN-enabled clients relaying packets for each other (see Figure 11.16).

Furthermore, A WLAN-and-cellular-enabled client can form a WLAN-coverage area when there is only a cellular data service in a location. This kind of access technology bridging will be in demand as most of the devices are produced WLAN-enabled and the coverage area of compatible services are very limited. Increasing the coverage area benefits users because now they can gain Internet access where they would not otherwise. Operators benefit as well by increasing the utilization of their fixed infrastructure. And finally, the relay clients can be convinced to collaborate by receiving credits for the service they provide. Even though the benefits are clear to anyone, design and implementation of a working system presents several challenges. In particular, the presence of nontrusted intermediaries in both the AAA routing and data traffic routing makes an

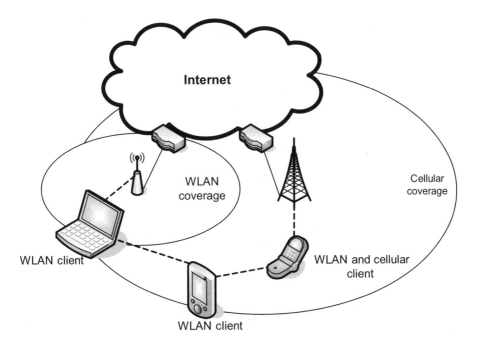

Figure 11.16 Wireless access network with ad hoc extension

important difference. These nontrusted parties are both selfish and resource constrained, in general. They will have the motive to cheat the system to maximize their gains while minimizing their cost.

There are a number of fundamental requirements that must be satisfied by a proposed solution. First of all, the solution must be relatively simple and efficient. Otherwise the cost of gaining the additional benefits might not be well justified. Implementation of this scheme is expected on a wide range of mobile devices. Secondly, it should yield low latency in the face of high mobility. A lengthy AAA process each time a client moves might severely impact ongoing data traffic. Finally, it must impose minimal impact on the existing AAA systems to be readily deployable on the Internet.

This continues to be an unsolved problem for researchers. Only a limited number of publications exist that attempt to solve some part of the problem (Salem et al. 2003; Zhang et al. 2002). It is expected that a complete solution that satisfies the requirements will see immediate deployment on the Internet.

11.4.2 802.11i Handover Optimizations

Within the IEEE 802.11 Working Group, the Fast Roaming Study Group was formed in the November 2003 Plenary Meeting, to minimize any disconnection while the STA is in handoff process.[7]

Since WLAN applications are mostly data based or web-access based, WLAN does not currently support real-time communication with QoS support. If a user tries to use Voice over IP (VoIP) over a WLAN handset terminal, and moves from one AP to another, the user will need handoff support to keep the voice connection. However, the original IEEE 802.11 standard allows only a single association. This means that only one AP can notify the Distribution System (DS) of the mapping between itself and an STA. This makes a seamless handoff difficult, unless the association process is very fast.

To minimize the handoff processing time, a couple of proposals have been presented in the IEEE 802.11 standardization meetings. The research group from the University of Maryland (Mishra et al. 2002a, 2003b) proposes to use the neighboring AP graph map, which dynamically captures the mobility topology of an AP location structure through real-time handoff experience. A scheme can use the neighboring AP graph map to find the next set of potential APs for the handoff, then use proactive context transfer and key distribution to minimize the handoff processing time. The neighboring AP graph map (Mishra et al. 2002a, 2003b) is built on top of the IEEE Internet Access Point Protocol (IAPP) standardized by Task Group F. The proactive caching method can significantly reduce the latency of reassociation time. A similar idea was presented from a WLAN-only network to heterogeneous access networks. The geographical access network topology map is estimated on the basis of the real vertical handoff experience and then built on the basis of the IP address or MAC address (Watanabe et al. 2002). Figure 11.17 illustrates the process of map building when the mobile terminal hands off from one AR to another. The access network topology map coordinator tracks the IP address or MAC address changes from the mobile terminal, and can then track the mobility path from one access network to another network. Based on this topology map, context transfer and security key distribution can be arranged easily.

---

[7] See http://www.ieee802.org/11/ for the IEEE 802.11 working group web page.

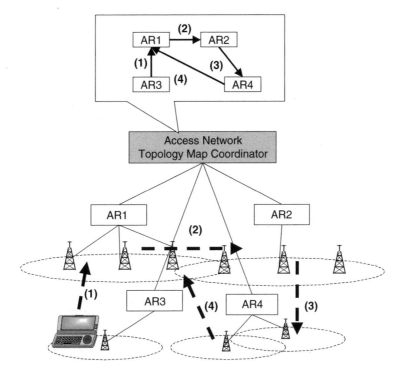

Figure 11.17 Geographical access network topology map building

Currently, one AP notifies the DS of one AP-STA mapping when the STA associates. Another new approach, called *"Make-Before-Break"* (Chaplin 2003), splits this into two association actions. In addition to the association notification, there are "Association Triggered Actions," notifications that are triggered by the STA sending a tentative Association Request to the new AP. During the tentative association request, the DS is notified of the STA to current AP mapping, but not the STA to new AP mapping.

The fast roaming study group is newly established, and is planning to define a new standard to complement the IEEE 802.11 standard some time in 2004.

11.4.3 Unified AAA

Heterogeneous wireless access technologies coexist today. For example, NTT DoCoMo has already started deploying wireless LAN services called *M-zone* alongside 3G wireless networks. A significant proportion of research activities has been devoted to create seamless wireless access network services on top of the intrinsic network heterogeneity. However, Mobile Internet application services are rapidly gaining popularity among consumers, for example DoCoMo's i-mode services, multimedia messaging services, generic web services, and so on. This creates a strong need for adaptive authentication, authorization, and payment services that integrate access services and application services. To provide a truly seamless experience to the user, there is a need for new frameworks and infrastructure in the application layer.

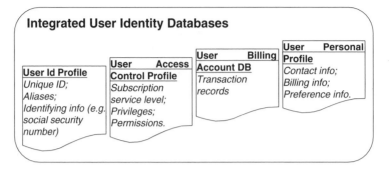

Figure 11.18 IUID internal data categories

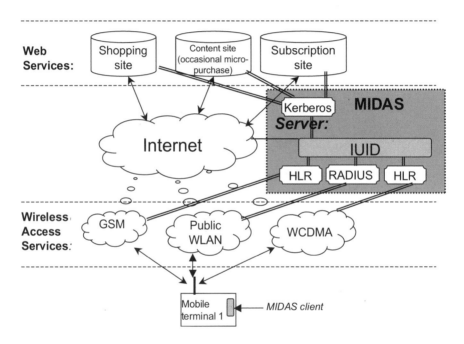

Figure 11.19 MIDAS in heterogeneous wireless networks

Currently, web services can provide a level of seamlessness by using the Microsoft Passport[8], AOL Quick Checkout[9], or Project Liberty[10]; however, for the truly seamless experience, future mobile subscribers should be able to use their mobile subscription accounts to go on-line through a public WLAN hotspot, sign in at a mobile website to perform various tasks such as shopping or music download, and direct all charges to the same account so

---

[8] See http://www.passport.net/Consumer/default.asp?lc=1033 for the Passport web page.

[9] See https://payment.aol.com for the AOL Quick Checkout Service web page.

[10] See http://www.projectliberty.org/ for the Liberty Alliance Project web page.

that the user receives only one bill. An application-level framework could be used to bridge the heterogeneous systems to present seamless services to consumers.

An application-level framework called *MIDAS* (Cao et al. 2002) (mobile Internet distributed authentication system) is being studied at DoCoMo USA Labs. MIDAS enables a consumer to use a single account to access any kind of service, as long as the service provider supports third-party authentication. The consumer can use the same account to handle all payment processing. In coordination with service providers, she can choose to bill various services to a single account, or use the MIDAS system as an electronic wallet to store credit-card information and transfer the information to service providers when conducting transactions.

At the core of MIDAS is the Integrated User Identity Database (IUID), illustrated in Figure 11.18.

This database is implemented at the application layer. It resides on the Internet, with a web interface for system administration tasks, and a separate interface with restricted access rights for consumer account self-management. AAA services to various service providers are handled by a group of gateways, each masquerading as an appropriate AAA server for corresponding service providers.

Figure 11.19 illustrates how MIDAS offers AAA services to various access and application services. Each gateway and/or proxy, corresponding to a particular type of application or access service, can require subscribers to have an ID of a particular format, which is registered inside an IUID as an alias ID to the same user. The gateways and IUID coordinate to always locate the appropriate subscriber accounts through any aliases.

12

Security Policy Enforcement for Downloaded Code

Nayeem Islam, Ph.D. and David Espinosa, Ph.D.

12.1 Introduction

Unlike their predecessors, the next generation of mobile phones will feature an *open and extensible platform*. By downloading many forms of executable content, their functionality will grow beyond the limits of the software originally included by the manufacturer. As usual, with this freedom comes responsibility. In this chapter, we examine techniques for controlling the behavior of downloaded programs. These techniques allow users to run downloaded applets and applications without corruption of data or loss of privacy.

Downloaded programs can take many forms, ranging from a significant application from a trusted source (e.g., a new word processor) to a small applet from an unknown source (e.g., an animated web page). In the latter case, the user may not even realize he is downloading a program. Other examples include codecs for new multimedia formats (Chapter 8), proxies for new protocols and services (Chapter 9), single and multiplayer games, and application and system software updates. In 2003, users are already downloading large applications; soon users will download *all* their software.

Modern cryptographic techniques (Chapter 10) can *help* to determine whether a code is safe. Indeed, code authors can sign their programs, and users can be sure that their downloaded programs have not been modified in transit. However, these techniques cannot *guarantee* that a code is safe. For example,

- the code author must spend time and money to obtain a certificate;

- the certificate authority (CA) only provides the name and address of the code author;

Next Generation Mobile Systems. Edited by Dr. M. Etoh
© 2005 John Wiley & Sons, Ltd

- the code author's name and address do not prove that the code is safe;

- the user may not have time to investigate the code author's reputation;

- even if the code author is well intentioned, he may have written unsafe code by accident;

- if the code is unsafe, many users must complain before the CA revokes its certificate, and these users' systems have already been damaged by the code;

- once the CA revokes the certificate, news of the revocation must reach the user.

For the these reasons, the user cannot rely solely on authority and reputation. He therefore requires a *security manager* to inspect or monitor the actual downloaded code. A security manager must satisfy several security requirements (see (Saltzer and Schroeder 1975) for a more complete list):

Time: The security manager must be as fast as possible. Because consumers judge phones on the basis of price, a phone's processor power and memory are critical resources. That is, any "overhead" use of processor or memory increases the cost of the phone.

Space: For similar reasons, the security manager must be as small as possible. Since most managers do not store significant amounts of data, the manager's size is determined mainly by the size of its code, which grows with complexity.

Flexibility: Users and administrators need to specify security policies in considerable detail, so the more control they have, the better. At the same time, users do not want to pay for unnecessary features.

Binary code: Because speed is critical, the security manager must safely execute downloaded machine code, not just bytecode.

TCB size: In order to be as reliable as possible, the security manager's trusted computing base (TCB) must be small and easy to verify.

We can divide the safety checks that a security manager performs into several broad categories:

Type safety: Operations must conform to published interfaces.

Memory safety: Downloaded code can only access certain memory regions. The safety policy may describe regions coarsely (e.g., a bound and a length) or finely (e.g., a data structure field).

Stack safety: Code must not overflow or underflow the stack.

Array bound safety: Array references must not exceed their bounds; otherwise, they can overwrite or expose important data.

System call safety: Downloaded code can only perform certain actions, such as reading, writing, and deleting files, and only under certain conditions.

Quota safety: Downloaded code can only use limited amounts of CPU, memory, network, and disk. These limits can apply to any aspect of resource use, such as current use, total use, current rate of use, and total rate of use.

In the following sections, we review a collection of security managers from the research literature. First, we discuss the Java 2 security manager, which is a standard dynamic monitor. Then, we discuss selective dynamic monitors, which are more flexible than standard dynamic monitors. Finally, we discuss static managers, which verify code safety *before* execution begins.

We evaluate each manager along various dimensions and compare it to other managers when the comparison makes sense. We do not evaluate each manager along all axes because many properties either do not apply or are not described in the literature. Instead, we mention what is substantially new or different about each approach.

12.2 Standard Dynamic Monitors: Java 2

Gong (1999) discusses the Java 2 security manager, the only standard dynamic monitor that we examine. A dynamic security monitor works as follows. Just before a program invokes each potentially unsafe system call, it calls the security monitor to check whether the program has permission to make the call. The security monitor examines the current context, including who wrote the code, who is running the code, and the particular call and its parameters. If the call is not allowed, the monitor raises a security exception or terminates the program.

For example, Figure 12.1 shows a dynamic security monitor running a safe program. The program tries to open the file /tmp/f, an action that the user's policy allows. The program calls the dynamic monitor, and the monitor in turn calls the runtime system to open the file.

In contrast, Figure 12.2 shows a dynamic security monitor running an unsafe program. The program tries to open the file /etc/passwd, an action that the user's policy prohibits. The program calls the dynamic monitor, but instead of opening the file, the monitor aborts the program. Note that in order for the dynamic monitor to work, it must intercept all potentially unsafe system calls.

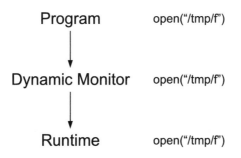

Figure 12.1 A dynamic monitor running a safe program

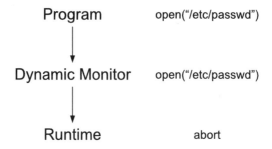

Figure 12.2 A dynamic monitor running an unsafe program

The designers of the Java 2 library have fixed the set of system calls that the security manager can intercept; the security policy author cannot extend it. Furthermore, if the security policy allows all invocations of a monitored call, the monitoring still requires some performance overhead. This extra cost encourages the security policy author to monitor as few methods as possible, leading to possible security holes. These shortcomings are the primary motivation for the selective dynamic monitors described in Section 12.3. Java 2 bases its security decisions on:

- The code source (a URL)

- The code signer (a principal)

- The action requested (class, method, and arguments)

- The local security policy

- The stack of currently executing methods.

When a method tries to perform a potentially unsafe action, such as opening a file, the `openFile` method checks to see whether the user has installed a security manager. If no manager is installed, it opens the file. If there is a manager, it asks the manager whether it can open the file.

The standard Java security manager is quite complex. Each potentially unsafe system method calls `checkPermission` with an appropriate permission object. The `checkPermission` method collects the set of methods on the current call stack. If each method has the required permission, then the security manager allows the call, otherwise it denies it.

In Java 2, a method has a permission if its class has the permission. A class has a permission if the code source from which it was downloaded has the permission. A code source has permission if its URL and digital signature match those specified in the user's security policy, and the policy grants it the permission. In essence, the security manager allows a system call if and only if the user's security policy allows all the methods on the call stack to perform the call.

In Java, users can install their own security managers, and these managers can make decisions based on many different criteria. This aspect of the security mechanism is very flexible and is a significant improvement on the previous version. Furthermore, when a library developer writes a new method, he can create a permission for it, which the security policy author can reference. However, the set of methods from the standard Java class library has already been fixed. The security policy author cannot add new permission checks to existing methods or remove undesired checks for performance.

The size of the J2SE security manager depends strongly on whether we include the cryptographic infrastructure required for authentication. The main security directories in the Java 2SE 1.4.1 source contain 125,000 lines of code in 447 files. However, the security manager by itself consists of 22,000 lines of code in 99 files. According to Sun's documentation, there are approximately 200 methods with security checks.

The standard security manager is quite complex, involving numerous classes and levels of indirection. For example, while performing benchmarks, we discovered that by default, nonsystem classes can read local files but cannot write them. We tried for several hours to determine why, but finally gave up!

12.2.1   Stack-inspecting Dynamic Monitors

The standard Java 2 security manager's monitoring decisions depend not only on the caller and the callee but also on the caller's caller, and so on up the call chain. It allows a method call if and only if the security policy permits *all* classes on the call chain to call it. This approach is called *stack inspection* because the stack represents the call chain. Stack inspection tries to prevent a malicious class from invoking a sensitive method indirectly, through other benign classes.

Some researchers contend that the Java 2 stack inspection mechanism is too slow. Slow security checks not only waste CPU cycles but, more significantly, encourage library authors to omit them to increase performance. We present some benchmarks that we collected and some collected by Erlingsson and Schneider (2000). We also describe an alternative approach to stack inspection developed by Wallach et al. (2000).

Benchmarks: Islam and Espinosa

Table 12.1 shows the speed of three security managers: the null manager, a stub manager that counts the number of security checks, and the standard manager. The benchmark performs recursive calls to a predetermined stack depth (either 0 or 8000), then opens two files 10,000 times each. Times are shown for Sun's J2SE version 1.4.1_03 running a bytecode interpreter on a 2.0 GHz Intel Pentium 4.

Why the null and stub managers slow down at high stack depths is unclear, since recursing to depth 8000 costs less than one millisecond. But the null and stub managers are acceptably fast in all cases. The standard manager is considerably slower, even at stack depth zero, because of its overall complexity, and its performance is particularly bad at depth 8000.

Table 12.1   Java security manager timings (Islam and Espinosa)

Security Manager	Time at Depth 0 (s)	Time at Depth 8000 (s)
Null	21.6	23.2
Stub	22.2	23.8
Standard	29.6	57.0

Table 12.2   Java security manager timings (Erlingsson and Schneider)

Benchmark	Description	Overhead (%)
jigsaw	Web server	6.2
javac	Java compiler	2.9
tar	File archive utility	10.1
mpeg	MPEG2 decoder	0.9

Benchmarks: Erlingsson and Schneider

Erlingsson and Schneider (2000) compare the standard Java security manager, which performs stack inspection, to a null security manager. Under the null security manager, each potentially dangerous method still performs a null pointer check to see whether a security manager is installed. They obtain the timings shown in Table 12.2, which seem consistent with the measurements for the file-open benchmark in Table 12.1.

The Java 2 security manager is fairly inefficient. For example, although its implementation is not included in the Sun source distribution, the primitive

```
getStackAccessControlContext
```

appears to return the entire list of protection domains currently on the stack, without removing duplicates. This inefficiency probably causes the factor of two slowdown observed for large stack depths.

Unfortunately, the benchmarks described above refer to interpreted code. Benchmarking interpreted code makes little sense, because users who are serious about speed will run a JIT compiler, or perhaps even a whole-program compiler. A typical JIT compiler should yield at least a factor of ten speed-up.

Wallach et al.: Security-passing Style

Wallach et al. (2000) describe a more flexible version of stack inspection called *security passing style*. Instead of inspecting the stack, Wallach passes a security argument to each method. This approach makes security more amenable to program analysis, since most analyzers can handle function arguments, but few analyzers maintain a representation of the call stack. Wallach also tries to determine when the security argument is unnecessary. However, since security passing requires additional argument to each method call, it is slower than standard stack inspection.

12.3   Selective Dynamic Monitors

A dynamic monitor is *selective* if the security policy determines the set of monitored calls rather than the library. In a selective monitor, the policy can monitor *any* of the library's public methods, and nonmonitored calls incur *no* run time overhead. Java 2's security manager is *not* selective, since its library fixes the set of monitored calls, and each monitored call incurs run time overhead, even when the policy allows it unconditionally.

The idea of a selective monitor is apparently both good and obvious, because we found eight different systems that implement it, the earliest of which is Wallach et al. (1997). Several of these systems transform bytecode programs using the JOIE bytecode rewriting toolkit described in Cohen et al. (1998).

12.3.1 Wallach et al.: Capabilities and Namespaces

Wallach et al. (1997) discuss the merits of three approaches to Java security: capabilities, stack inspection, and namespace management.

Capabilities perform two main functions. First, they cache permissions. If a program calls the same potentially unsafe method many times, the security manager performs a complete security check on the first call and then issues a capability that it verifies quickly on the remaining calls. Second, capabilities allow a method to grant a permission to another method by passing a capability to it. This feature is dangerous because the capability can easily fall into the wrong hands. It is doubly dangerous if the system allows methods to copy and store capabilities.

The goal of namespace management is to control the classes that a downloaded program can reference. For example, instead of the real System class, the program sees a wrapped System class whose potentially unsafe methods check their arguments before executing. Evans and Twyman (1999) and Chander et al. (2001) also wrap classes in this way. Indeed, namespace management is generally useful for building programs from abstract classes and interfaces. See, for example, the ML module system (Milner et al. 1997). Bauer et al. (2003) also describe a module system for Java that can construct security wrappers.

12.3.2 Erlingsson and Schneider: Security Automata

Erlingsson and Schneider (1999) implement Schneider's notion of a *security automaton*. The alphabet of a security automaton is the set of actions of the monitored system. The automaton rejects a word over the alphabet if that sequence of actions leads to an unsafe state, at which point it stops the monitored program. It accepts all (possibly infinite) sequences that it does not reject. That is, before each action, it decides whether to stop or continue.

Erlingsson allows the automaton to make a transition before each instruction of the monitored system. This design is flexible in theory, because it can monitor anything, but is difficult in practice, because operations such as method calls are difficult to recognize at the instruction level. However, the system can implement memory bounds safety by monitoring each memory reference. Competitive systems, whose events are higher-level system calls, cannot perform such fine-grain checks.

Erlingsson also enumerates the automaton's states explicitly. Thus, if the monitor stops the program after one million memory references, it needs one million explicit states. Erlingsson and Schneider (2000) implement a more realistic system that computes the automaton's state in Java code and defines its events using a Java-like language. The TCB of this system includes 17,500 lines of Java code.

12.3.3 Evans and Twyman: Abstract Operating Systems

Like the other authors, Evans and Twyman (1999) add security checks to Java programs using bytecode rewriting. With their system, the security policy author specifies events and checks in terms of a single "abstract operating system" that maps to multiple concrete OSs. Indeed, they can run the same security policies on both Java and Windows. They implement this idea by transforming each concrete call into an abstract call, but only for purposes of security checking.

12.3.4   Pandey and Hashii: Benchmarks

Pandey and Hashii (2000) describe another tool for instrumenting Java programs via byte-code editing. Their monitors can raise a security exception whenever one of the following events happen:

- Any method creates instance of class C.

- A specific method $C_1.M_1$ creates instance of class C.

- Any method calls a method $C.M$.

- A specific method $C_1.M_1$ calls a method $C_2.M_2$.

These events are also conditional on the current state, and the policy can attach new state variables to classes so that conditions can depend on per-instance state. For example, Pandey and Hashii show a rule that allows clients to call the f method of each instance of class C at most ten times. These conditions can invoke arbitrary Java code. In another example, they show how to inspect the call stack to determine the method call chain.

Using a simple microbenchmark, Pandey and Hashii compare their system to Sun's JDK 1.1.3 security manager. This benchmark limits the number of calls to an empty function to be less than one million. They run the benchmark with the constraint in place and also with no constraint. In their approach, no constraint means that the code is unaltered (a plain Java function call). In the JVM, no constraint means that the code still contains a null-pointer check to see whether a security manager has been installed. Table 12.3 shows the times relative to a plain Java function call.

12.3.5   Kim et al.: Languages of Events

Kim et al. (Kim et al. 2001) present another implementation of run time monitoring. Their system allows the security policy author to specify the abstract set of events he wants to monitor. These events serve as the interface between the program and the security policy. This additional level of indirection allows the policy author to specify several policies for the same set of events and to extract several sets of events from the same program. Thus, the relation between programs and policies is many-to-many, but it is mediated by sets of

Table 12.3   Bytecode   editing   versus   JDK (Pandey and Hashii)

System	Constrained	Unconstrained
Binary editing	2.0	1.0
JDK 1.1.3	3.0	2.0

abstract events. For instance, several real-time programs can generate the same language of time-stamped events, and the policy author can impose several sets of timing requirements on these events.

12.3.6   Chander et al.: Renaming Classes and Methods

Chander et al. (Chander et al. 2001) demonstrate another system of run time monitoring by bytecode instrumentation. Following the namespace management approach of Wallach et al. (1997), they redirect class and method references from potentially unsafe versions to known safe versions. They use class renaming as much as possible, since it is simple. However, for final classes and interfaces, where class renaming is impossible, they rename individual method invocations.

For standard browsers, Chander et al. perform class renaming in an HTTP proxy. For the JINI framework, they perform class renaming in the client's class loader, since JINI does not use a specific transport protocol for downloaded code.

12.3.7   Ligatti et al.: Edit Automata

Ligatti et al. (2003) extend Erlingsson and Schneider (2000) by allowing security automata not only to stop program execution but also to suppress actions and insert new ones. In this respect, edit automata resemble Common Lisp's before, after, and around methods, which also form the inspiration for aspect-oriented programming.

From a theoretical point of view, Bauer and Walker study which policies edit automata can enforce. However, Bauer is currently implementing a tool to apply edit automata to Java bytecode programs.

In an example, Bauer and Walker show how to add "transaction processing" to a pair of `take` and `pay-for` calls. This automaton prevents a program from taking an object without paying for it.

12.3.8   Colcombet and Fradet: Minimizing Security Automata

Colcombet and Fradet (2000) present a general method for assuring that a program respects a safety property. First, they define a map from program entities, including function calls, to an abstract set of events. Next, they express the desired property as a finite state automaton over the alphabet of abstract events. Finally, they express the program as an abstract graph, whose nodes are program points and whose edges are instructions that generate events.

Instead of executing the original program, they execute the product of the graph with the automaton. The resulting instrumented graph (I-graph) has the same behavior as the original program but only allows execution traces that satisfy the property specified by the automaton. By statically analyzing the I-graph, they minimize the number of inserted safety checks. They express the algorithms for minimization in terms of NP-complete problems and suggest heuristics for solving them.

Unfortunately, they do not present any performance measurements, so it is not clear whether their approach is useful in practice. They also consider only properties captured by finite state automata, which cannot easily account for resource use.

12.4   Static Security Managers

Unlike dynamic monitors, both standard and selective, static security managers operate purely by analyzing program text. Static managers detect unsafe code earlier than dynamic monitors, so that a program cannot cause a security violation in the middle of a critical operation, such as saving a file. Also, once a static manager verifies a program, it executes it with no checks whatsoever, so the program runs faster than under a dynamic monitor. Since a static manager predicts the behavior of a program before running it, it performs a complex analysis that is difficult to implement and is therefore likely to have errors. Also, since it is impossible for a static manager to make perfect predictions, it always rejects some safe programs.

For example, Figure 12.3 shows a static security manager examining a safe program. The program tries to open the file /tmp/f, an action that the user's policy allows. If the policy allows all the program's actions, the static manager passes the program unchanged. The resulting program then runs without any further intervention and calls the runtime system directly to open the file.

In contrast, Figure 12.4 shows a static security manager running an unsafe program. The program tries to open the file /etc/passwd, an action that the user's policy prohibits. Since the policy prohibits this action, the static manager rejects the program and never executes it. Note that the boundary between safe and unsafe programs may be arbitrarily complex, so the static manager must err on one side or the other. Thus, if it rejects all unsafe programs, it necessarily rejects some safe programs as well.

Static managers are much more complex than dynamic monitors and use a wide variety of sophisticated implementation techniques drawn from type theory, program analysis, model checking, and theorem proving. We do not describe each technique in complete detail, but we try to present the essence of each approach. Also, although many such systems appear in the research literature, we have chosen a representative cross section.

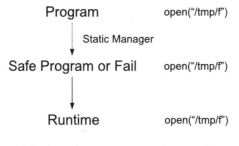

Figure 12.3 A static manager running a safe program

Figure 12.4 A static manager running an unsafe program

12.4.1 Gosling et al.: Java Bytecode Verifier

One of Java's most important contributions is a type system for bytecode programs. This system statically verifies memory safety and stack underflow safety, once and for all. However, it does not guarantee array bounds safety, quota safety, stack overflow safety, or system call safety, so these forms of safety are usually checked at run time in a Java system.

12.4.2 Morrisett et al.: Typed Assembly Language

Morrisett et al. (1998) check a large class of machine code programs for memory safety by providing a type annotation at each label. This annotation describes the memory and stack layout that holds when control transfers to that label. In their system, programs can access data via the registers, the stack, or a global data area. They describe memory layout via tuple types, arrays, and tagged unions.

This system does for real assembly language what the Java bytecode verifier did for Java bytecodes. The authors also describe a simple type-preserving compiler for a typed functional language that targets this architecture.

12.4.3 Xi: Dependent Types

Just as Morrisett et al. (1998) show how to handle memory safety with a type system, Xi and Pfenning (1998) show how to handle array bounds elimination using dependent types, that is, types that are parameterized by values. Dependent types occur fairly often in mathematics and are used in several theorem provers intended for mathematical applications.

Using dependent types, we can refine the type $array[t]$ into the indexed family of types $array[n, t]$ of arrays of length n. Similarly, we can refine the integers into $int[a, b]$, the integers between a and b (inclusive). The array reference operation then has type

$$\texttt{ref} : array[n, t] \times int[0, n - 1] \to t$$

The difficulty is that this approach requires a theorem prover to show that the bounds are correct, and the prover might need human assistance. Also, complete array bounds checking requires a means of reasoning about the entire language, since the index and array may have come from anywhere. Thus, a dependent type system is more complex than most simple or polymorphic type systems.

12.4.4 Crary and Weirich: Dependent Types

Crary and Weirich (Crary and Weirich 2000) describe a system that uses dependent types to account for CPU usage. For example, if $\texttt{sort}$ is a function that sorts an array of size n of integers in time $3n^2$, then it has type

$$\texttt{sort} : \forall t, n.(array[n], t) \to (void, t + 3n^2)$$

This type shows that $\texttt{sort}$ starts at time t and finishes at time $t + 3n^2$.

To specify a dependently typed language, the designer must decide on which expressions types can depend. Also, he needs to connect program execution with the expressions appearing in the types. Crary and Weirich encode dependent types using a system of "sum and inductive kinds." This system serves to connect recursion on treelike data types with the time it takes to compute these recursions. Like the programs, the times are expressed using recurrences, but on natural numbers rather than trees. This idea seems reasonable, although limited to recursions on treelike data.

By allowing programs to "waste cycles," Crary and Weirich obtain a form of subtyping, but since they do not compare recurrences or render them in closed form, this facility seems of little use. For example, the developer cannot use a function that satisfies the Fibonacci recurrence in place of a function that runs in time n^2, even though it is always faster. Also, all cost functions are exact rather than asymptotic.

Their system seems more complex than Xi's more direct version of dependent types. They call both ordinary types and natural numbers "constructors" and distinguish them by their *kind* (essentially a type system for types). Overall, the complexity of their formalism make their ideas fairly difficult to understand.

12.4.5   Necula and Lee: Proof-carrying Code

Necula (1998) proposed that if a program property is hard to verify, then the program should come with a proof that this property holds. This approach succeeds because it is usually easier to check a proof than to rediscover it. Necula focuses mainly on memory safety and array bounds checking, but his system can also handle system call safety in a straightforward way.

Necula encodes these properties in first-order logic (FOL), which is a fairly general reasoning system that is implemented by many theorem provers. As Morrisett et al. (1998) argue, FOL is overly expressive for memory safety. Indeed, both the Java bytecode verifier and their TAL system provide memory safety without using a full theorem prover.

Array bounds checking requires more serious reasoning about indices, so FOL seems appropriate for this purpose. FOL is also useful for system-call safety, since a security policy may impose complex preconditions on sensitive methods.

12.4.6   Sekar et al.: Model-carrying Code

Sekar et al. (2001, 2003) present a variant of PCC in which each binary includes an abstract model of its security behavior. A security policy inspects the model to verify that the program is safe. Naturally, the manager also checks that the model correctly represents the behavior of the code. This approach allows the code author to provide a single model that suffices for all security policies, instead of providing a separate safety proof for each policy.

Models capture sequences of system calls using finite-state and push-down automata. The authors build FSAs by learning them from execution traces, but they build PDAs by extracting them from the program source.

12.4.7   Xia and Hook : Abstraction-carrying Code

Xia and Hook (2003a,b,c) describe abstraction-carrying code, which is similar to model-carrying code. In this framework, the producer sends a predicate abstraction of the program along with the program. A predicate abstraction is a simpler program whose execution traces *include* those of the original program, and possibly more. Thus, if a behavior cannot happen in the abstract program, it cannot happen in the original. The consumer verifies that the abstraction has all the behaviors of the concrete program and that none of these behaviors is unsafe. Since the abstract program is simpler, it is hopefully easier to analyze, but if it is too simple, it will have unsafe behaviors that the original program avoids.

Instead of tracking precise variable values, the abstract program tracks only certain predicates of the variables. For example, instead of x, the program may only maintain the predicate $x \geq 0$. Thus, if the original program branches on the test $x \geq 100$, the abstract program cannot tell which way to go and must therefore explore both alternatives. For this reason, the abstract program is nondeterministic (i.e., has many possible executions), even though the original program is deterministic (i.e., has only one possible execution).

Xia and Hook refer to other work for actually computing the abstract program. Their main contribution is a technique for encoding the abstraction using dependent types. This method appears to work well, since the annotations are small, and the abstraction is easy to extract and verify.

12.4.8 Fitzgerald et al.: The Marmot Java Compiler

Fitzgerald et al. (2000) describe Marmot, a native-code optimizing compiler for Java. Their work provides useful data on eliminating array bounds checks and method synchronizations. We discuss their compiler in this section because it uses static analysis to reduce the number of dynamic safety checks.

Marmot provides several sophisticated optimizations to remove array bounds checks. After optimization, bounds checks cost an average of 4 % on a large collection of Java benchmarks, compared to no bounds checks at all. Unfortunately, they do not say what bounds checks cost with no optimization at all, or with only simpler optimizations enabled. Earlier work on array bounds elimination demonstrated a factor of two increase in performance over code with all bounds checks enabled, but only for array-intensive code on older architectures.

Marmot also provides synchronization elimination, but only when the entire program can be proven to be single-threaded. Synchronization elimination speeds up some small single-threaded benchmarks by a factor of three. But on a more representative collection of larger benchmarks, it yields roughly a 40% increase in speed. It is also hard to measure the impact of synchronization, because it cannot simply be turned off if a program requires it for correctness. But it is clear that reducing synchronization costs is extremely important for multithreaded Java code.

12.5   Conclusion

We have examined several different approaches to security for downloaded code: standard dynamic monitoring, dynamic monitoring with stack inspection, selective dynamic monitoring, and static verification. No single approach is uniformly better than the others, but we can draw some general conclusions.

First, the flexibility afforded by selective monitoring makes it clearly superior to the standard approach of monitoring a fixed set of system calls, with no drawbacks. The eight systems surveyed provide a variety of different realizations of this important idea.

Second, although stack inspection sounds like a good idea, we have not seen a clear presentation of the numerous assumptions behind it and of how it should best be used. Until then, it seems premature to place it at the center of a widely used system.

Third, although complex reasoning is not required to verify basic memory safety, it is essential to verify system call safety. Abstraction-based approaches are useful when combined with other techniques, but they do not seem sufficient by themselves.

Like the Marmot Java compiler, future research will undoubtedly combine both static and dynamic approaches to build hybrid security managers. As next- generation cell phone users download new classes of applications to a flexible, open platform, they will need robust security managers to keep their data safe and their phones operating securely.

Bibliography

n.d.b http://www.trustedcomputinggroup.org.

3GPP 1999a Mobile execution environment (MExE); service description, stage 1.

3GPP 1999b Network architecture (release 5). Technical Report TS 23.002, Technical Specification Group Services and Systems Aspect.

3GPP 1999c Technical Realization of Short Message Service (SMS).

3GPP 1999d *TS 26.071: AMR Speech Codec; General Description*, 3rd Generation Partnership Project; Technical Specification Group Services and Systems Aspects.

3GPP 1999e Unstructured Supplementary Service Data (USSD); stage 3.

3GPP 1999f User-to-user signaling (UUS); stage 1.

3GPP 1999g User-to-user signaling (UUS); stage 3.

3GPP 1999h User-to-user signaling (uus) supplementary service; stage 2.

3GPP 2000a Customized applications for mobile network enhanced logic (CAMEL) phase 3 – stage 2.

3GPP 2000b Customized applications for mobile network enhanced logic (CAMEL); service description, stage 1.

3GPP 2000c Mobile execution environment (mexe); functional description, stage 2.

3GPP 2000d Support of optimal routing (SOR); technical realization.

3GPP 2000e Unstructured Supplementary Service Data (USSD); stage 1.

3GPP 2000f Unstructured Supplementary Service Data (USSD); stage 2.

3GPP 2000g USIM/SIM application toolkit (USAT/SAT); service description; stage 1.

3GPP 2000h Virtual Home Environment (VHE) / Open Service Access (OSA); stage 2.

3GPP 2001a Ip multimedia subsystem (ims), stage 2.

3GPP 2001b *QoS for Speech and Multimedia Codec; Quantitative Performance Evaluation of H.324 Annex C over 3G*, 3rd Generation Partnership Project; Technical Specification Group Services and System Aspects. TR 26.912.

3GPP 2002a Service aspects; the virtual home environment; stage 1.

3GPP 2002b *TS 22.140: Multimedia Messaging Service (MMS); stage 1*, 3rd Generation Partnership Project; Technical Specification Group Services and Systems Aspects.

3GPP 2002c *TS 22.223: Transparent End-to-End Packet-switched Streaming Service Stage 1*, 3rd Generation Partnership Project; Technical Specification Group Services and Systems Aspects.

3GPP 2002d *TS 26.110: Codec for Circuit Switched Multimedia Telephony Service; General Description*, 3rd Generation Partnership Project; Technical Specification Group Services and Systems Aspects.

Next Generation Mobile Systems. Edited by Dr. M. Etoh
© 2005 John Wiley & Sons, Ltd

3GPP 2002e *TS 26.140: Multimedia Messaging Service (MMS); Media Formats and Codecs (Release 5)*, 3rd Generation Partnership Project; Technical Specification Group Services and Systems Aspects.

3GPP 2002f *TS 26.233: Transparent End-to-End Packet-switched Streaming Service (PSS); General Description*, 3rd Generation Partnership Project; Technical Specification Group Services and Systems Aspects.

3GPP 2003a *Packet switched Conversational Multimedia Applications; Transport protocols*, 3rd Generation Partnership Project; Technical Specification Group Services and System Aspects. TS 26.236.

3GPP 2003b *Quality of Service (QoS) Concept and Architecture*, 3rd Generation Partnership Project; Technical Specification Group Services and System Aspects. TS 23.107.

3GPP 2003c *Transparent End-to-End Packet Switched Streaming Service (PSS); RTP Usage Model*, 3rd Generation Partnership Project; Technical Specification Group Services and System Aspects. TR 26.937.

3GPP 2003d *TS 23.140: Multimedia Messaging Service (MMS); Media Formats and Codecs (Release 5)*, 3rd Generation Partnership Project; Technical Specification Group Services and Systems Aspects.

3GPP 2003e *TS 26.111: Codec for Circuit Switched Multimedia Telephony Service; Modifications to H.324*, 3rd Generation Partnership Project; Technical Specification Group Services and Systems Aspects.

3GPP 2003f *TS 26.234: Transparent End-to-End Packet-switched Streaming Service (PSS); Protocols and Codecs*, 3rd Generation Partnership Project; Technical Specification Group Services and Systems Aspects.

3GPP 2003g *TS 26.911: Codec(s) for Circuit Switch Multimedia Telephony Service; Terminal implementor's Guide*, 3rd Generation Partnership Project; Technical Specification Group Services and Systems Aspects.

3GPP 2003h *TS 26.912: QoS for Speech and Multimedia Codec; Quantitative Performance Evaluation of H.324 Annex C over 3G*, 3rd Generation Partnership Project; Technical Specification Group Services and Systems Aspects.

3GPP2 2001 *Selectable Mode Vocoder Service Option for Wideband Spread Spectrum Communications Systems*, version 2.0 3GPP2.

Aboba B et al 2000 IEEE 802.1X for wireless LANs. Technical Report, IEEE 802.11-00/035, IEEE.

Aboba B and Calhoun P 2003 RADIUS (Remote authentication dial in user service) support for extensible authentication protocol (EAP). Technical Report RFC 3579, IETF.

Aboba B and Simon D 1999 PPP EAP TLS authentication protocol. Technical Report RFC 2716, IETF.

Adachi F, Sawahashi M and Suda H 1998 Wideband DS-CDMA for next-generation mobile communication systems. *IEEE Commun. Mag.* **36**(9), 56–69.

Adya A, Bolosky WJ, Castro M, Cermak G, Chaiken R, Douceur JR, Howell J, Lorch JR, Theimer M and Wattenhofer RP 2002 FARSITE: Federated, available, and reliable storage for an incompletely trusted environment. *USENIX, 5th Symposium on Operating Systems Design and Implementation*, Boston, MA, December 9-11, 2002.

Akamai White Paper 2002 *Applications for Akamai EdgeScape*. http://www.akamai.com.

Al-Riyami S and Paterson K 2003 Certificateless Public-Key cryptography. *Proceedings of Asiacrypt 2003*.

Ala-Laurila J, Mikkonen J and Rinnemaa J 2001a Wireless LAN access network architecture for mobile operators. *IEEE Commun. Mag.* **39**(11), 82–89.

Ala-Laurila J, Mikkonen J and Rinnemaa J 2001b Wireless LAN access network architecture for mobile operators. *IEEE Commun. Mag.* **39**(11), 82–89.

Andrews T, Curbera F, Dholakia H, Goland Y, Klein J, Leymann F, Liu K, Roller D, Smith D, Thatte S, Trickovic I and Weerawarana S 2003 Business process execution language for web services version 1.1. http://www-106.ibm.com/developerworks/library/ws-bpel/.

Angin O, Campbell AT, Kounavis ME and Liao RRF 1998 The mobiware toolkit: Programmable support for adaptive mobile netoworking. *IEEE Personal Commun.* **4**(5), 32–43.

Anjum F, Caruso F, Jain R, Missier P and Zordan A 2001 Cititime: A system for rapid creation of telephony services using third-party software components. *Computer Networks* **35**(5), 579–595.

Arbaugh W n.d. Improving the latency of probe phase during 802.11 handoff. http://www.umiacs.umd.edu/partnerships/ltsdocs/Arbaug_talk2.pdf.

Arbaugh W, Shankar N, Wan Y and Zhang K 2002a Your 802.11b network has no clothes. *IEEE Wireless Commun.* **9**(6), 44–51.

Arbaugh WA, Shankar N, Wan Y and Zhang K 2002b Your 802.11b network has no clothes. *IEEE Wireless Commun.* **9**(6), 44–51.

Arkin A, Askary S, Fordin S, Jekeli W, Kawaguchi K, Orchard D, Pogliani S, Riemer K, Struble S, Takacsi-Nagy P, Trickovic I and Zimek S n.d. Web service choreography interface (WSCI) 1.0. http://www.w3.org/TR/wsci/.

Arkko J, Kempf J, Sommerfeld B, Zill B and Nikander P 2004 SEcure Neighbor Discovery (SEND). Internet draft, work in progress.

Asano T 2002 A revocation scheme with minimal storage at receivers. *Proceedings of Asiacrypt 2002.*

Atarashi H, Abeta S and Sawahashi M 2001 Broadband packet wireless access appropriate for high-speed and high-capacity throughput. *IEEE VTC 2001-Spring*, 566–570.

Atarashi H, Abeta S and Sawahashi M 2003a Variable spreading factor orthogonal frequency and code division multiplexing (VSF-OFCDM) for broadband packet wireless access. *IEICE Trans. Commun.* **E86-B**(1), 291–299.

Atarashi H, Maeda N, Kishiyama Y, Higuchi K and Sawahashi M 2003b (3) Broadband wireless access technology using VSF-OFCDM and VSCRF-CDMA. *NTT DoCoMo Tech. J.* **5**(2), 24–32.

Ateniese G and Tsudik G 1999 Some open issues and new directions in group signatures. *Proceedings of Financial Cryptography 1999.*

Ateniese G, Camenisch J, Joye M and Tsudik G 2000 A practical and provably secure Coalition-Resistant group signature scheme. *Proceedings of Crypto 2000*, 255-270, LNCS 1880. Springer-Verlag.

Aurea T 2004 Cryptographically Generated Addresses (CGA). Internet draft, work in progress.

Backgrounder

Backgrounder 2003 https://www.trustedcomputinggroup.org/downloads/TCG_Backgrounder.pdf.

Bahl P and Padmannabhan V 2000 RADAR: An in-building RF-based user location and tracking system. IEEE INFORCOM, Tel Aviv, Israel.

Bahl P, Padmanabhan V and Balachandran A 2000 Enhancements to the RADAR user location and tracking system. Technical Report MSR-TR-00-12, Microsoft Research.

Balachandran A, Campbell AT and Kounavis ME 1997 Active filters: Delivering scaled media to mobile devices. *Proceedings of International Workshop on Network and Operating System Support for Digital Audio and Video (NOSSDAV)*, 133–142.

Banavar G, Beck J, Gluzberg E, Munson J, Sussman JB and Zukowski D 2000 An application model for pervasive computing. *6th ACM MOBICOM.*

Barak B, Goldreich O, Impagliazzo R, Rudich S, Sahai A, Vadhan S and Yang K 2001 On the (Im)possibility of obfuscating programs. *Proceedings of Crypto 2001.*

Barkan E, Biham E and Keller N 2003 Instant Ciphertext-only cryptanalysis of GSM encrypted communication. *Proceedings of Crypto 2003*, 600–616. Springer-Verlag.

Barnes M 2000 Layered functional architecture. MWIF 2000.138.9 Mobile Wireless Internet Forum (MWIF).

Bauer L, Appel A and Felten E 2003 Mechanisms for secure modular programming in Java. *Software Pract. Experience* **33**(5), 461–480.

Bequet H 2001 *Professional Java SOAP*. Wrox Press.

Bharghavan V, Lee KW, Lu S, Ha S, Li JR and Dwyer D 1998 The TIMELY adaptive resource management architecture. *IEEE Pers. Commun.* **4**(5), 20–31.

Bing B 2002 *Wireless Local Area Networks*. Wiley-Interscience.

Bitfone n.d. Bitfone. http://www.bitfone.com.

Blunk L, Vollbrecht L, Aboba J, Carlson B, Levkowetz J and Levkowetz H 2003 Extensible Authentication Protocol (EAP). draft-ietf-eap-rfc2284bis-06 (work in progress).

Bollella et al. G 2002 Real-time specification for java. http://www.rtj.org.

Boneh D and Franklin M 2001 Identity-based encryption from the weil pairing. *Proceedings of Crypto 2001*, 213–229. Springer-Verlag.

Boneh D, Gentry C, Lynn B and Shacham H 2003 Aggregate and verifiably encrypted signatures from bilinear maps *Proceedings of Eurocrypt 2003*.

Borisov N, Goldberg I and Wagner D 2001a Intercepting mobile communications: the insecurity of 802.11. *Proceedings of ACM/SIGMOBILE 7th Annual International Conference on Mobile Computing and Networking*.

Borisov N, Goldberg I and Wagner D 2001b Security of the WEP algorithm. http://www.isaac.cs.berkeley.edu/isaac/wep-faq.html.

Borisov N, Goldberg I and Wagner D 2002 Security of the WEP algorithm. http://www.isaac.cs.berkeley.edu/isaac/wep-faq.html.

Borman C, Burmeister C, Degermark M, Fukushima H, Hannu H, Jonsson L-E, Hakenberg R, Koren T, Le K, Liu Z, Martensson A, Miyazaki A, Svanbro K, Wiebke T, Yoshimura T, Zheng H 2001 Robust Header Compression (ROHC). *RFC 3095*.

Bostršm T, Goldbeck-Lšwe T and Keller R 2002a Ericsson mobile operator WLAN solution. *Ericsson Rev.* **1**, 36–43.

Bostršm T, Goldbeck-Lšwe T and Keller R 2002b Ericsson mobile operator WLAN solution. *Ericsson Rev.* **1**, 36–43.

Brooks F 1995 *The Mythical Man-Month: Essays on Software Engineering*. Addison-Wesley. Anniversary Edition (2nd ed.).

Burrows M, Abadi M and Needham R 1989 A logic of authentication *Proc. R. Soc.* **A426**, 233–271.

Calhoun P, Loughney J, Guttman E, Zorn G and Arkko J 2003 Diameter Base Protocol. Technical Report RFC 3588, IETF.

Camenisch J 1997 Efficient and generalized group signatures. *Proceedings of Eurocrypt 1997*, 465–479. Springer-Verlag.

Camenisch J and Michels M 1998 A group signature scheme with improved efficiency. *Proceedings of Asiacrypt 1998*, 160–174. Springer-Verlag.

Camenisch J and Stadler M 1997 Efficient group signature schemes for large groups. *Proceedings of Crypto 1997*, 410–424. Springer-Verlag.

Cao J, Watanabe F and Kurakake S 2002 MIDAS: An integrated user identity management system for future wireless operators. *Proceedings of 2002 3G Wireless Conference*, 265–269.

Capra L, Mascolo C and Emmerich W 2002 Exploiting reflection in mobile computing middleware. *Mobile Comput. Commun. Rev.* **6**, 34–44.

Carpenter B 1996 The Internet Architecture. *RFC 1958*.

Castro P 2001 A probabilistic location service for wireless network environment (Nibble). *Ubiquitous Computing 2001*, Atlanta, Georgia.

cdma2000 Wireless IP Network Standards-Draft

cdma2000 Wireless IP Network Standards-Draft 2001 http://pulse.tiaonline.org/ . TIA/EIA/IS-835-1.

Chander A, Mitchell J and Shin I 2001 Mobile code security by Java bytecode instrumentation. *DARPA Information Survivability Conference and Exposition*.

Chaplin C 2003 Make before break. Technical Report, IEEE 802.11-03/770r1, IEEE.

Chase D 1985 Code combining – a maximum-likelihood decoding approach for combining an arbitrary number of noisy packets. *IEEE Trans. Commun.* **33**(5), 385–393.

Chaum D 1982 Blind signatures for untraceable payments. *Proceedings of Crypto 1982*.

Chaum D and van Heyst E 1991 Group signatures. *Proceedings of Eurocrypt 1991*, 257–265. Springer-Verlag.

Chen L and Pederson T 1995 New group signature schemes. *Proceedings of Eurocrypt 1995*, 171–181. Springer-Verlag.

Chu WC 2003 *Speech Coding Algorithms: Foundation and Evolution of Standardized Coders*. John Wiley & Sons.

Clark D 1988 The design philosophy of the DARPA Internet Protocols. *Proceedings of SIGCOMM 88*, ACM CCR 18:4, 106–114.

Cocks C 2002 An identity-based encryption scheme based on quadratic residues. *Proceedings of Cryptography and Coding*.

Cohen G, Chase J and Kaminsky D 1998 Automatic program transformation with JOIE. *USENIX Conference*, New Orleans, Louisiana.

Colcombet T and Fradet P 2000 Enforcing trace properties by program transformation. *Principles of Programming Languages*, Boston, Massachusetts.

Counter with CBC-MAC (CCM)

Counter with CBC-MAC (CCM) 2003 IETF RFC3610.

Cox LP and Noble BD 2001 Fast reconciliations in fluid replication. *21st International Conference on Distributed Computing Systems*.

Crary K and Weirich S 2000 Resource bound certification, *Principles of Programming Languages*, Boston, Massachusetts.

Dahlman E, Gudmundson B, Nilsson M and Skold J 1998 UMTS/IMT-2000 based on wideband CDMA. *IEEE Commun. Mag.* **36**, 70–80.

Davies J 2002 RADIUS protocol security and best practices. Technical Report, Microsoft Corp.

Decasper D, Dittia Z, Parulkar G and Plattner B 2000 Router plugins: A software architecture for next generation routers. *ACM Trans. on Networking* **8**(1), 2–15.

Deering S 2001 Watching the waist of the protocol hourglass. *Proceedings of IAB Meeting 51st IETF*, London, UK.

Demers A, Greene D, Hauser C, Irish W, and Larson J 1987 Epidemic algorithms for replicated database maintenance. *Proceedings of the Sixth Annual ACM Symposium on Principles of Distributed Computing*, 1–12.

Dolev D and Yao A 1983 On the security of Public-Key protocols. *IEEE Trans. Inf. Theory* **29**, 198–208.

Douglis F and Ousterhout JK 1991 Transparent process migration: Desing alternatives and the sprite implementation. *Software Pract. Experience* **21**, 757–785.

Droms R, Bound J, Volz B, Lemon T, Perkins C and Carney M 2003 Dynamic host configuration protocol for IPv6 (DHCPv6), RFC 3315.

Ebling MR and Satyanarayanan M 1998 On the importance of translucence for mobile computing. Technical Report.

Ecma 2002 Common language infrastructure (CLI). http://www.ecma-international.org/publications/standards/Ecma-335.htm.

Edwards WK 1999 Core Jini. The Sun Microsystems Press.

El Malki K 2004 Low Latency Handoffs in IPv4. Work in progress.

Elwailly F, Gentry C and Ramzan Z 2004 QuasiModo: Efficient certificate validation and revocation. *Proceedings of Public-Key Cryptography 2004.*

Erlingsson U and Schneider F 1999 SASI enforcement of security policies: a retrospective. *New Security Paradigms Workshop*, Caledon, Canada.

Erlingsson U and Schneider F 2000 IRM enforcement of Java stack inspection. *Security and Privacy*, Oakland, California.

Etoh M and Takeshi Y 2004 Advances in wireless video delivery. *Proc. IEEE*, to appear.

ETSI 2002a Open Service Access (OSA); Application Programming Interface (API); Part 1: Overview V1.1.1 . ES 202 915-1.

ETSI 2002b Open Service Access (OSA); Application Programming Interface (API); Part 3: Framework V1.1.1. ES 202 915-3.

Evans D and Twyman A 1999 Flexible policy-directed code safety. *Security and Privacy*, Oakland, California.

Fell S 2004 PocketHTTP. http://www.pocketsoap.com/pocketHTTP/.

Finlayson R, Mann T, Mogul JJ and Theimer M 1984 A Reverse Address Resolution Protocol. RFC 903 (STD 23), IETF.

FIPS 2001 *Advanced Encryption Standard (AES)*. FIPS PUB 197 edn.

Fitzek F, Angelini D, Mazzini G and Zorzi M 2003 Design and performance of an enhanced IEEE 802.11 MAC protocol for multihop coverage extension. *IEEE Wireless Commun.* **10**(6), 30–39.

Fitzgerald R, Knoblock T, Ruf E, Steensgaard B and Tarditi D 2000 Marmot: an optimizing compiler for Java. *Software Pract. Experience* **30**(3), 199–232.

Ford B and Lepreau J 1994 Evolving Mach 3.0 to a migration thread model. *Usenix.*

Forsberg D 2003 Protocol for Carrying Authentication for Network Access. draft-ietf-pana-01.txt (work in progress).

Forsberg D 2004 Protocol for Carrying Network Access (PANA). Internet draft, work in progress, 2004.

Forum WWR 2001 The book of visions 2001. http://www.wireless-world-research.org/.

Frank W, Reger R and Appel U 1992 Loudspeaker nonlinearities-analysis and compensation. *26th Asilomar Conference on Signals, Systems and Computers*, 756–760.

Fuller V, Li T, Yu J and Varadhan K 1993 Classless Interdomain Routing (CIDR). RFC 1338.

Furusawa H, Hamabe K and Ushirokawa A 2000 SSDT – Site selection diversity transmission power control for CDMA forward link. *IEEE J. Selected Areas Commun.* **18**(8), 1546–1554.

Garfinkel T, Rosenblum M and Boneh D 2003 Flexible OS support and applications for trusted computing. *Proceedings of the 9th Hot Topics in Operating Systems (HOTOS-IX).*

Garg V 2000 *IS-95 CDMA and cdma 2000: Cellular/PCS Systems Implementation*. Prentice Hall.

Gartner I n.d. The service station: A P2P web services usage model. http://www4.gartner.com/DisplayDocument?doc_cd=103925.

Gentry C 2003 Certificate-Based encryption and the certificate revocation problem. *Proceedings of Eurocrypt 2003.* Springer-Verlag.

Gentry C and Silverberg A 2002 Hierarchical ID-based cryptography. *Proceedings of Asiacrypt 2002*, 548–566. Springer-Verlag.

Gentry C and Szydlo M 2002 Cryptanalysis of the revised NTRU signature scheme. *Proceedings of Eurocrypt 2002*.

Gentry C, Jonsson J, Stern J and Szydlo M 2001 Cryptanalysis of the NTRU signature scheme. *Proceedings of Asiacrypt 2001*.

Goldreich O 1999 *Modern Cryptography, Probabilistic Proofs and Pseudorandomness*. Springer-Verlag.

Golmie N 2003 Bluetooth adaptive frequency hopping and scheduling. *Proceedings of MILCOM '03*, Boston, MA.

Golmie N, Chevrollier N and Rebala O 2003 Bluetooth and WLAN coexistence: challenges and solutions. *IEEE Wireless Commun.* **10**(6), 22–29.

Gong L 1999 *Inside Java 2 Platform Security*. Addison-Wesley.

Goto Y, Kawamura T, Atarashi H and Sawahashi M 2003 Variable spreading and chip repetition factors (VSCRF) – CDMA in Reverse Link for Broad Band Wireless access. IEICE Technical Report.

Gottlieb Y and Peterson L 2002 A comparative study of extensible routers. *OpenArch 2002*.

Grimm R, Davis J, Lemar E and Bershad B 2002 Migration for pervasive applications. *Technical Report*.

Groves C, Pantaleo M, Ericsson L, Anderson T and Taylor T 2003 Gateway control protocol, version 1. Technical Report RFC 3525, Internet Engineering Task Force (IETF).

Gruber R, Kaashoek F, Liskov B and Shrira L 1994 Disconnected operation in the thor object-oriented database system. *IEEE Workshop on Mobile Computing Systems and Applications*.

Guttman E, Perkins C, Veizades J and Day M 1999 Service Location Protocol, version 2. IETF, RFC 2608, http://www.rfc-editor.org/rfc/rfc2608.

Guy RG, Heidemann JS, Mak W, Page TW, Popek JGJ and Rothmeir D 1990 Implementation of the ficus replicated file system *Proceedings of the Summer 1990 USENIX Conference*, 63–72.

Gwon Y, Jain R and Kawahara T 2004 Robust indoor location estimation of stationary and mobile users. IEEE INFOCOM, Hong Kong.

Hara S and Prasad R 1997 Overview of multicarrier CDMA. *IEEE Commun. Mag.* **35**(12), 126–133.

Harkins D and Carrel D 1998a The Internet Key Exchange (IKE). Technical Report RFC 2409, IETF.

Harkins D and Carrel D 1998b The Internet Key Exchange (IKE). *RFC 2409*.

Haverinen H and Salowey J 2003 EAP SIM Authentication. draft-haverinen-pppext-eap-sim-12.txt (work in progress).

Henning M and Vinosky S 1999 Advanced CORBA Programming with C++. Addison-Wesley.

Herre J and Purnhagen H 2002 General audio coding. In *The MPEG-4 Book* (eds. Pereira F and Ebrahimi T), Prentice Hall.

Herzog J, Liskov M and Micali S 2003 Plaintext awareness via key registration. *Proceedings of Crypto 2003*.

Higuchi K, Andoh H, Okawa K, Sawahashi M and Adachi F 2000 Experimental evaluation of combined effect of coherent rake combing and SIR-based fast transmit power control for reverse link of DS-CDMA mobile radio. *IEEE J. Selected Areas Commun.* **18**(8), 1526–1535.

Hinden R and Deering S 1995 IP Version 6 Addressing Architecture. RFC 1884, IETF.

Hirata S, Nakajima A and Uesaka H 1995 Pdc mobile packet data communication network. *Proceedings on 1995 Fourth IEEE International Conference on Universal Personal Communications*, 644–648.

Hodges J 1997 Introduction to Directories and LDAP. http://www.stanford.edu/%7Ehodges/talks/mactivity.ldap.97/index2.html.

Hoffstein J, Pipher J and Silverman J 1996 A new high speed (Ring-Based) public key cryptosystem. Preprint presented at the *Crypto 1996 Rump Ssession*.

Hohler E, Morris R, Chen B, Jannotti J and Kaashoek M 2000 The click modular router. *ACM Trans. Comput. Syst.*

IBM n.d. Web Services Security (WS-Security). http://www-106.ibm.com/developerworks/webservices/library/ws-secure/.

ICC 1998 *Interleaved FDMA – A New Spread-spectrum Multiple-access Scheme.*

IEEE 2003 *Part11: Wireless LAN Medium Access Control (MAC) and Physical Layer (PHY) Specifications: Medium Access Control (MAC) Security Enhancements.* IEEE Std 802.11i/D7.0.

IEEE 1999a *Part11: Wireless LAN Medium Access Control (MAC) and Physical Layer (PHY) Specifications.* Std 802.11-1999.

IEEE 1999b *Part11: Wireless LAN Medium Access Control (MAC) and Physical Layer (PHY) Specifications: Higher-Speed Physical Layer Extension in the 2.4 GHz Band.* IEEE Std 802.11b-1999.

IEEE 1999c *Part11: Wireless LAN Medium Access Control (MAC) and Physical Layer (PHY) Specifications: Higher-Speed Physical Layer Extension in the 5 GHz Band.* IEEE Std 802.11a-1999.

IEEE 1999d *Recommended Practice for Multi-Vendor Access Point Interoperability via an Inter-Access Point Protocol Across Distribution Systems Supporting IEEE 802.11 Operation.*

IEEE 1999e *Wireless LAN Medium Access Control (MAC) and Physical Layer (PHY) specifications.*

IEEE 2001a *Standard for Local and Metropolitan Area Networks: Port Based Network Access Control.* IEEE Std. 802.1X.

IEEE 2001b *Port-based Network Access Control.* IEEE Std 802.1X.

IEEE 2001c *Standards for Port-based Network Access Control.*

IEEE 2002 *Part11: Wireless LAN Medium Access Control (MAC) and Physical Layer (PHY) Specifications: Specification for Enhanced Security.* IEEE Std 802.11i/D3.0.

IEEE 2003a *Part11: Wireless LAN Medium Access Control (MAC) and Physical Layer (PHY) Specifications: Amendment 4: Further Higher Data Rate Extension in the 2.4 GHz Band.* IEEE Std 802.11g(-2003.

IEEE 2003b *Part11: Wireless LAN Medium Access Control (MAC) and Physical Layer (PHY) Specifications: Medium Access Control (MAC) Security Enhacements.* Technical Report, IEEE Std 802.11i/D7.0, IEEE.

IEEE 2003c *Part11: Wireless LAN Medium Access Control (MAC) and Physical Layer (PHY) specifications, Medium Access Control (MAC) Security Enhancement.* IEEE Std 802.11i/D7.0.

IEEE 2003d *Recommended Practice for Multivendor Access Point Interoperability via an Inter-access Point Protocol Across Distribution Systems Supporting IEEE 802.11 Operation.* IEEE Std 802.11f-2003.

IEEE 2003e *Specification for Radio Resource Measurement* (Draft Supplement to IEEE Std 802.11, 1999 Edition). Std 802.11k/D0.6.

IEEE 2004 *802.11 TGn Functional Requirements.* IEEE 802.11-03/0813-12-000n.

IETF 1998 PPP Extensible Authentication Protocol (EAP). IETF RFC2284.

IETF 1999 PPP EAP TLS Authentication Protocol. IETF RFC2716.

IETF n.d. Simple mail transfer protocol. RFC 821, http://www.ietf.org/rfc0821.txt.

Intelligent wireless software manager

Intelligent Wireless Software Manager n.d. http://www.doongo.com.

iPass 2004 Generic interface specification. http://www.ipass.com/platform/platform_whitepapers.html.

Isenberg D 1997 The rise of the stupid network. *Comput. Telephony* 16–26.

Ishai Y, Sahai A and Wagner D 2003 Private Circuits: Securing Hardware Against Probing Attacks. *Proceedings of Crypto 2003.*

Islam N, Zhou D, Shoaib S, Ismael A and Kizhakkiniyil S 2004 AOE: A mobile operating environment for web-based application. *Proceedings of SAINT 2004,* to appear.

ISO 1999 *Part 11: Wireless LAN Medium Access Control (MAC) and Physical Layer (PHY) Specifications*. ISO/IEC 8802-11.

ISO/IEC 1993a *ISO/IEC 11172-2:1993 Information Technology – Coding of Moving Pictures and Associated Audio for Digital Storage Media at up to About 1,5 Mbit/s – Part 2: Video*.

ISO/IEC 1993b *ISO/IEC 11172-3: Coding of Moving Pictures and Associated Information – Part 3: Audio*. ISO/IEC.

ISO/IEC 1997 *ISO/IEC 13818-7: MPEG-2 Advanced Audio Coding*. ISO/IEC.

ISO/IEC 1999 *ISO/IEC 14496-3: Information Technology – Coding of Audio-Visual Objects – Part 3: Audio*. ISO/IEC.

ISO/IEC 2000 *ISO/IEC 13818-2:2000 Information Technology – Generic Coding of Moving Pictures and Associated Audio Information: Video*.

ISO/IEC 2001 *ISO/IEC 14496-2:2001 Information Technology – Coding of Audio-visual Objects – Part 2: Visual*.

ISO/IEC 2003 *ISO/IEC 14496-10:2003 Information Technology – Coding of Audio-visual Objects – Part 10: Advanced Video Coding*.

ITU-R Working Party 8F 2003 *Framework and Overall Objectives of the Future Development of IMT-2000 and Systems Beyond IMT-2000*. Recommendation ITU-R M.1645.

ITU-T 1986 *ITU-T Recommendation G.722 – 7 kHz Audio Coding Within 64 kbit/s*.

ITU-T 1988 *ITU-T Recommendation G.711 – Pulse Code Modulation (PCM) of Voice Frequencies*.

ITU-T 1990 *ITU-T Recommendation G.726 – 40, 32, 24, 16 kbit/s Adaptive Differential Pulse Code Modulation (ADPCM)*.

ITU-T 1993 *ITU-T Recommendation H.261 – Video Codec for Audiovisual Services at p x 64 kbit/s*.

ITU-T 1995 *ITU-T Recommendation G.729 – Coding of Speech at 8 kbit/s using CS-ACELP*.

ITU-T 1998 *ITU-T Recommendation H.263 – Video Coding for Low Bit Rate Communication*.

ITU-T 2000a Activities on imt-2000. http://www.itu.int/home/imt.html.

ITU-T 2000b *Recommendation H. 248 – Media Gateway Control Protocol*.

ITU-T 2002 *ITU-T Recommendation G.722.2 – Wideband Coding of Speech at Around 16 kbit/s using Adaptive Multi-Rate Wideband (AMR-WB)*.

Jain R 2003 4G services, architectures and networks; speculation and challenges. Keynote address. International Conference on Mobile Data Management (MDM), London, UK.

Jain, R, Anjum F and Bakker JL 2004a *Programming Converged Networks: Call Control APIs in JTAPI, JAIN and Parlay/OSA*. John Wiley & Sons .

R. Jain and J.-L. Bakker and F. Anjum 2005 *Programming Converged Networks: Call Control in Java, XML, and Parlay/OSA*. John Wiley & Sons.

Jain R, bin Tariq M, Kempf J and Kawahara T 2004b The All-IP 4G architecture. *DoCoMo Tech. J.*

JCP 2003 JSR-00172 J2ME web servics specification (final release). http://jcp.org/aboutJava/communityprocess/final/jsr172/index.html.

JCP n.d.a Jsr 118 – mobile information device profile 2.0. http://www.jcp.org/en/jsr/detail?id=118.

JCP n.d.b Uddi4j. http://www-124.ibm.com/developerworks/oss/uddi4j/.

Jeong M, Watanabe F and Kawahara T 2003a *Fast Active Scan for Measurement and Handoff*. IEEE 802.11-03/416.

Jeong M, Watanabe F, Kawahara T and Zhong Z 2003b *Fast Active Scan Proposals*. IEEE 802.11-03/623.

Jepsen T (ed.) 2001a *Java Telecommunications: Solutions for Next Generation Networks*. John Wiley & Sons.

Jepsen T 2001b *Java in Telecommunications*. John Wiley & Sons.

Jepsen T, Bhat R and Tait D 2001 Java APIs for integrated networks. *Java Telecommunications: Solutions for Next Generation Networks*. John Wiley & Sons.

Johansson A, Grbic N and Nordholm S 2003 Direction-of-arrival estimation using the far-field SRP-Phat in conference telephony. *ICASSP 2003*.

Johnson D, Perkins C and Arkko J 2004 Mobility Support in IPv6. Internet Proposed Standard, RFC 3775.

Joseph AD, de Lespinasse AF, Tauber JA, Gifford DK and Kaashoek MF 1995 Rover: a toolkit for mobile information access *Proceedings of the Fifteenth ACM Symposium on Operating Systems Principles*, 156–171.

Joseph AD, Tauber JA, and Kaashoek MF 1997 Mobile computing with the Rover toolkit. *IEEE Trans. Comput.* **3**(46), 337–352.

Jun J and Sichitiu M 2003 The nominal capacity of wireless mesh networks. *IEEE Wireless Commun.* **10**(5), 8–14.

Kaaranen H, Ahitainen A, Laitinen L, Naghian S and Niemi V 2001a *UMTS Networks: Architecture, Mobility and Services*. John Wiley & Sons.

Kaaranen H, Ahtiainen A, Laitinen L, Naghian S and Niemi V 2001b *UMTS Networks*. John Wiley & Sons.

Kempf J and Yegani P 2002 OpenRAN: A New Architecture for Mobile Wireless Internet Radio Access Networks. *IEEE Commun. Mag.*

Kempf S and Alstein R 2004 The Rise of the Middle and the Future of End to End: Reflections on the Evolution of the Internet Architecture. RFC 3724.

Kent S and Atkinson R 1998 Security Architecture for the Internet Protocol. RFC 2041, IETF.

kHTTP n.d. kHTTP CLDC compliant HTTP server. http://khttp.enhydra.org/.

Kim M, Kannan S, Lee I and Sokolsky O 2001 Java-MaC: a run-time assurance tool for Java programs. *Electronic Notes in Theoretical Computer Science*.

Kim P and Bohm W 2003 Support of real-time applications in future mobile networks: the IMS approach. *Proceedings of Wireless Personal Mobile Communications*.

Kinoshita K 2001 Easy IMT-2000, the third generation mobile communication system. TTA. In Japanese.

Kistler JJ and Satyanarayanan M 1991 Disconnected operation in the coda file system. *Thirteenth ACM Symposium on Operating Systems Principles*, Vol. 25, 213–225.

kObjects n.d.a kSOAP 2 Project. http://kobjects.dyndns.org/kobjects/index.html.

kObjects n.d.b kXML 2 Project. http://kobjects.dyndns.org/kobjects/index.html.

Kocher P 1996 Timing Attacks on Implementations of Diffie-Hellman, RSA, DSS and Other Systems. *Proceedings of Crypto 1996*.

Kocher P 1998 On certificate revocation and validation. *Proceedings of Financial Cryptography 1998*.

Kocher P, Jaffe J and Jun B 1999 Differential power analysis. *Proceedings of Crypto 1999*.

Koin G and Haslestad T 2003 Security aspects of 3G-WLAN interworking. *IEEE Commun. Mag.* **41**(11), 82–88.

Koodli R 2003a Fast Handovers for Mobile IPv6. Internet draft, work in progress, 2004.

Koodli R (ed.) 2003b Fast Handovers for Mobile IPv6. draft-ietf-mobileip-fast-mipv6-05.txt (work in progress).

Kozono S 1994 Received signal level characteristics in a wideband mobile radio channel. *IEEE Trans. Veh. Technol.* **43**(3), 480–486.

KUDDI n.d. kUDDI Project. http://kuddi.enhydra.org/.

Kuenning GH and Popek GJ 1997 Automated hoarding for mobile computers. *Symposium on Operating Systems Principles*, 264–275.

Lampsal P n.d. J2ME architecture and related embedded technologies. http://www.cs.helsinki. fi/u/campa/teaching/j2me/papers/J2ME.pdf.

LAP 2003 Liberty architecture overview.

Lazar AA 1997 Programming telecommunication networks. *IEEE Network* Sept. 8–18.

Lazar A, Lim K and Marconcini F 1996 Realizing a foundation for programmability of ATM networks with the binding architecture. *IEEE J. Selected Areas Commun.*

Leibsch M and Singh A 2004 Candidate Access Router Discovery. Internet draft, work in progress.

Ligatti J, Bauer L and Walker D 2003 Edit automata: enforcement mechanisms for run-time security policies. *Int. J. Inf. Security.*

Loughney, J (ed) 2004 Context Transfer Protocol. Internet draft, work in progress, 2004.

Lysyanskaya A and Ramzan Z 1998 Group blind digital signatures: a scalable solution to electronic cash. *Proceedings of Financial Cryptography 1998*, 184–197 LNCS 1465. Springer-Verlag.

Lysyanskaya A, Micali S, Reyzin L and Shacham H 2003 Sequential aggregate signatures from trapdoor permutations manuscript.

Briceno M, Goldberg I and Wagner D. n.d. GSM Cloning. http://www.isaac.cs.berkeley.edu/isaac/gsm-faq.html.

Maeda N, Atarashi H, Abeta S and Sawahashi M 2002 VSF-OFCDM using two-dimensional spreading and its performance. *IEICE Technical Report.*

Mangold S, Choi S, Hiertz GR, Klein O and Walke B 2003 Analysis of IEEE 802.11e for QoS support in wireless LANs. *IEEE Commun. Mag.* **10**(6), 40–50.

Martin B and Jano B n.d. Wap binary Xml content format. W3C NOTE 24, June 1999. http://www.w3.org/1999/06/NOTE-wbxml-19990624/.

Matei R, Iamnitchi A and Foster I 2002 Mapping the gnutella network. *Internet Comput.* **6**, 50–57.

Matthews VJ 1991 Adaptive polynomial filters. *IEEE Signal Process. Mag.* **8**(3), 10–26.

McCree AV, Supplee LM, Cohn RP and Collura JS 1997 MELP: The new federal standard at 2400 bps. *ICASSP*, 1591–1594. IEEE.

McGrath R and Mickunas D 2000 Discovery and its discontents: Discovery protocols for ubiquitous computing. Technical Report UIUCDCS-R-99-2132, University of Illinois at Urbana-Champaign.

Metz C n.d. AAA protocol. http://www.computer.org/internet/v3n6/w6onwire2.htm?SMSESSION= NO.

Micali S 1996 Efficient certificate revocation. Technical Report LCS/TM 542b, Massachusetts Institute of Technology.

Micali S 1997 Efficient certificate revocation. *Proceedings of RSA Data Security Conference 1997.*

Micali S 2002 NOVOMODO: Scalable Certificate Validation and Simplified PKI Management. *Proceedings of PKI Research Workshop 2002.*

Micali S and Reyzin L 2004 A model for physically observable cryptography *Proceedings of Theory of Cryptography Conference 2004.*

Microsoft 2003a Understanding universal plug and play. http://www.upnp.org/download/UPNP_Understanding UPNP.doc.

Microsoft 2003b *WMV9 – An Advanced Video Codec for 3GPP*, 3GPP document S4 (03) 0613.

Microsystems S n.d. Java 2 Platform, Micro Edition (J2ME). http://java.sun.com/j2me/.

Miki N, Atarashi H, Abeta S and Sawahashi M 2001 Comparison of hybrid ARQ schemes and optimization of key parameters for high-speed packet transmission in W-CDMA forward link. *IEICE Trans. Commun.* **E84-A**(7), 1681–1690.

Milner R, Tofte M, Harper R and MacQueen D 1997 *The Definition of Standard ML (Revised)*. MIT Press.

Milojicic D, Zint W and Dangel A 1992 Task migration on top of the mach microkernel – design and implementation. Technical Report.

Mironov I 2001 A note on cryptanalysis of the preliminary version of the NTRU signature scheme. Cryptology ePrint archive, Report 2001/005.

Mishra A, Shin M and Arbaugh W 2002a Content caching using neighbor graphs for fast handoffs in a wireless network. Technical Report, University of Maryland.

Mishra A, Shin M and Arbaugh W 2002b An Empirical Analysis of the IEEE 802.11 MAC Layer Handoff Process. UMIACS-TR-2002-75.

Mishra A, Shin M and Arbaugh W 2003a Context caching using neighbor graphs for fast handoffs in a wireless network. http://www.cs.umd.edu/Library/TRs/CS-TR-4477/CS-TR-4477.pdf. UMIACS-TR-2003-46.

Mishra A, Shin M and Arbaugh W 2003b Proactive key distribution to support fast and secure roaming. Technical Report, IEEE 802.11-03/084r1, IEEE.

Mishra A, Shin M, Arbaugh W, Lee I and Jang K 2002c *Proactive Caching Strategies for IAPP Latency Improvement During 802.11 Handoff.* IEEE 802.11-02/758r1.

Mishra A, Shin M, Arbaugh W, Lee I and Jang K 2003c *Proactive Key Distribution to Support Fast and Secure Roaming.* IEEE 802.11-03/084r1.

Mitton D and Beadles M 2000 Network access server requirements next generation (NASREQNG) NAS Model. Technical Report RFC 2881, IETF.

Mockapetris P 1987 Domain Names Implementation and Specification. RFC 1034 (STD 13).

Moerdijk A and Klostermann L 2003 Opening the networks with Parlay/OSA: standards and aspects behind the APIs. *IEEE Network.*

Mohr W 2002 WWRF – the wireless world research forum. *Electron. Commun. Eng. J.* 283–291.

Morikura M and Matsue H 2001 Trends of IEEE 802.11 based wireless LAN. *IEICE Trans. Commun.* **J84-B**(11), 1918–1927.

Morrisett G, Walker D, Crary K and Glew N 1998 From system F to typed assembly language. *Principles of Programming Languages*, San Diego, California.

Mosberger D and Peterson L 1996 Making paths explicit in the scout operating system. *OSDI 1996*, 153-167. Operating Systems Design and Implementation (OSDI).

Mouly M and Pautet M 1992 The GSM system for mobile communication. *Cell & SYS.*

MPEG 2003 *N5701, Report on Call for Evidence on Scalable Video Coding (SVC) Technology.*

Myers M, Ankney R, Malpani A, Galperin S and Adams C 1999 X.509 Internet Public Key Infrastructure Online Certificate Status Protocol – OCSP. Internet RFC 2560.

Nakamura N et al 2003 Summary of first pass service. *NTT DoCoMo Tech. J.* **11**(3), 6–11.

Narten T and Draves R 2001 Privacy Extensions for Stateless Address Autoconfiguration in IPv6. RFC 3041, IETF.

Narten T, Nordmark E and Simpson W 1998 Neighbor Discovery for IP Version 6 (IPv6). RFC 2461, IETF.

Natsuno T 2003 *i-mode Strategy.* John Wiley & Sons.

Necula G 1998 *Compiling with Proofs.* PhD thesis, Carnegie Mellon.

Necula GC 1997 Proof-carrying code. *Proceedings of the 24th ACM SIGPLAN-SIGACT Symposium on Principles of Programming Languages*, 106–119.

Nikander P, Kempf J and Nordmark E 2004 IPv6 Neighbor Discovery Trust Models and Threats. RFC 3756, IETF, 2004.

Nishiguchi M and Edler B 2002 Speech coding. In *The MPEG-4 Book* (ed. Pereira F and Ebrahimi T), Prentice Hall.

Noble MB and Fleis B 1999 A case for fluid replication. *Netstore '99, The Network Storage Symposium.*

Oberg R 2001 Mastering RMI. *Wiley.*

Ohba Y, Das S, Patil B, Soliman H and Yegin A 2003 Problem Statement and Usage Scenarios for PANA. Internet draft, work in progress, 2004.

Ohm JR 1994 Three-dimensional subband coding with motion compensation. *IEEE Trans. Image Process.*

Ohrtman FD 2003 *Softswitch: Architecture for VoIP*. McGraw-Hill.

Okamoto T, Fujisaki E and Morita H 1998 TSH-ESIGN: Efficient Digital Signature Scheme Using Trisection Size Hash. IEEE P1363.

OMA 2003a Mapping of the OMA Architecture Requirements to Parlay. OMA-ARC-2003-0314R1-Parlay-OMA.

OMA 2003b The OMA Aarchitecture of Today. OMA-ARC-2003-0274R02.

OMA n.d. OMA The Open Mobile Alliance. http://www.openmobilealliance.org.

Orava P, Haverinen H and Black S 2003 *Adaptive Beaconing*. IEEE 802.11-03/610.

Oreilly 2001 The o'reilly peer-to-peer and web services conference. http://conferences.oreillynet.com/p2p/.

Osugi T et al 2002 Public wireless LAN service: Mzone. *NTT DoCoMo Tech. J.* **4**(3), 12–16.

PANA n.d. Protocol for carrying authentication for network access (PANA). http://ietf.org/html.charters/pana-charter.html.

Pandey R and Hashii B 2000 Providing fine-grained access control for Java programs via binary editing. *Concurrency: Pract. Experience* **12**, 1405–1430.

Pandya D, Jain R and Lupu E 2003 Indoor location using multiple wireless technologies. IEEE PIMRC 2003, Beijin, China.

Park J 2002 Wireless internet access for mobile subscribers based on the gprs/umts network. *IEEE Commun. Mag.* 38–49.

Park J 2003 WLAN security: current and future. *IEEE Internet Comput.* **7**(5), 60–65.

Parlay 2002a *Parlay API 4.0, Parlay X Web Services*. The Parlay Group. White paper.

Parlay 2002b *Parlay Web Service Overview*. The Parlay Group. White paper.

Parlay 2002c *Parlay Web Services Business Models*. The Parlay Group. White paper.

Parthasarathy M 2003a PANA Enabling IPsec-based Access Control. draft-ietf-pana-ipsec-00 (work in progress).

Parthasarathy M 2003b PANA Threat Analysis and Security Requirements. draft-ietf-pana-threats-eval-04 (work in progress).

Patil B, Tschofenig H and Yegin A 2003 Protocol for Carrying Authentication for Network Access (PANA) Requirements. draft-ietf-pana-requirements-07 (work in progress).

Perkins C 1998 *Mobile IP: Design Principles and Practices*. Addison-Wesley, Reading, Massachusetts.

Perkins C 2002a Ip mobility support for IPv4. Technical Report RFC 3344, Internet Engineering Task Force (IETF).

Perkins C 2002b IP Mobility Support for IPv4. RFC 3344, IETF.

Perkins D and Hobby R 1990 The Point-to-Point Protocol (PPP) initial configuration options. Technical Report RFC 1172, IETF.

Petersen K, Spreitzer MJ, Terry DB, Theimer MM and Demers AJ 1997 Flexible update propagation for weakly consistent replication. *Proceedings of the 16th ACM Symposium on Operating SystemsPrinciples (SOSP-16)*.

Plumber D 1982 An Ethernet Address Resolution Protocol. RFC 826 (STD 36).

Princen J and Bradley A 1987 Subband/transform coding using filter bank designs based on time domain aliasing cancellation. *ICASSP*, 2161–2164.

Rammer I 2002 Advanced .NET Remoting, 2nd edition. AI Press.

Ramstat T 1991 Cosine modulated analysis-synthesis filter bank with critical sampling and perfect reconstruction. *ICASSP*, 1789–1792.

Ramzan Z 1999 *Group Blind Digital Signatures: Theory and Applications*. Master's thesis, Massachusetts Institute of Technology. Available at http://theory.lcs.mit.edu /tilda zulfikar/MyResearch/homepage.html.

Recommendation WC 2002 Extensible Markup Language (XML) 1.1. http://www.w3c.org/TR/xml11/.

Red Bend n.d. Red Bend Software. http://www.redbend.com.

Rigney C 1997 RADIUS accounting. Technical Report RFC 2139, IETF.

Rigney C, Rubens A, Simpson W and Willens S 1997 Remote authentication dial in user service (RADIUS). Technical Report RFC 2138, IETF.

Rigney C, Willens S, Rubens A and Simpson W 2000 Remote Authentication Dial in User Service (RADIUS). RFC 2865.

Roman M and Campbell RH 2003 Providing middleware support for active space applications. *ACM/IFIP/USENIX International Middleware Conference*.

Salem NB, Buttyan L, Hubaux J and Jakobsson M 2003 A charging and rewarding scheme for packet forwarding in multi-hop cellular networks. Technical Report, MobiHoc 2003.

Salkintzis A and Passas N 2003 The evolution of wireless LANs and PANs. *IEEE Wireless Commun.* **10**(6), 4–5.

Saltzer J and Schroeder M 1975 The protection of information in computer systems. *Proc. IEEE* **63**(9), 1278–1308.

Satyanarayanan M 2002 The evolution of coda. *ACM Trans. Comput. Syst. (TOCS)* **2**(20), 85–124.

Satyanarayanan M, Kistler J, Kumar P, Okasaki M, Siegel E, and Steere D 1990 Coda: A highly available file system for a distributed workstation environment. *IEEE Trans. Comput.* **4**(39), 447–459.

Sawahashi M, Abeta S, Atarashi H, Higuchi K, Tanno M, Asai T and Ihara T 2003 (2) Broadband packet wireless access. *NTT DoCoMo Tech. J.* **5**(2), 11–23.

Sawahashi M, Higuchi K, Atarashi H and Miki N 2001 High-speed packet wireless access in W-CDMA and its radio link performance. *IEICE Trans. on Commun.* **J84-B**(10), 1725–1745.

Sekar R, Ramakrishnan CR, Ramakrishnan IV and Smolka SA 2001 Model-carrying code: a new paradigm for mobile code security. *New Security Paradigms Workshop*, Cloudcroft, New Mexico.

Sekar R, Venkatakrishnan VN, Basu S, Bhatkar S and DuVarney D 2003 Model-carrying code: a practical approach for safe execution of untrusted applications. *Symposium on Operating Systems Principles*, Bolton Landing, New York.

Seo S, Dohi T and Adachi F 1998 SIR-based transmit power control of reverse link for coherent DS-CDMA mobile radio. *IEICE Trans. Commun.* **E81-B**(7), 1508–1516.

SG8 IR 2003 Vision, Framework and Overall Objectives of the Future Development of IMT-2000 and Systems Beyond IMT-2000. Document 8/1022-E RA'03, Geneva.

Shamir A 1985 Identity-based cryptosystems and signature schemes. *Proceedings of Crypto 1984*, 47–53. Springer-Verlag.

Silva RD, Landfeldt B, Ardon S, Seneviratne A and Diot C 1999 Managing application level quality of service through TOMTEN. *Computer Networks* **7**(31), 20–31.

Simoens S, Pellati P, Gosteau J and Gosse K 2003 The evolution of 5 GHz WLAN toward higher throughputs. *IEEE Wireless Commun.* **10**(6), 6–13.

Simpson W 1994 The Point-to-Point protocol (PPP). Technical Report RFC 1661, IETF. STD 51.

Simpson W 1996 PPP challenge handshake authentication protocol (CHAP). Technical Report RFC 1994, IETF.

Soliman H, Castelluccia C, El-Malki K and L. B 2003 Hierarchical mobile IPv6 mobility management (HMIPv6) draft-ietf-mipshop-hmipv6-00.txt (work in progress).

Soliman H, Catelluccia C, Malki KE and Bellier L 2004 Hierarchical Mobile IPv6 Mobility Management (HMIPv6). Internet draft, work in progress, 2004.

Song H, Hua H, Islam N, Kurakake S and Katagiri M 2002 Browser state repository service. *Pervasive 2002*.

Squid-Cache n.d. Squid web proxy cache. http://www.squid-cache.org/.

Srivastava M and Misha PP 1997 On quality of service in mobile wireless networks. *Proceedings of International Workshop on Network and Operating System Support for Digital Audio and Video (NOSSDAV)*, 155–166.

Stockhammer T, Hannuksela MM and Wiegand T 2003 H.264/AVC in wireless environments. *IEEE Trans. Circuits Syst. Video Technol.* **13**(7), 657–673.

Sun n.d.a Java 2 platform, Micro Edition. http://java.sun.com/j2me/.

Sun n.d.b Web services reliability (WS-Reliability), version 1.0. http://developers.sun.com/sw/platform/technologies/ws-reliability.v1.0.pdf.

Süsstrunk S and Winkler S 2004 Color image quality on the Internet. *Proc. SPIE/IS&T Internet Imag.* **5304**, 118–131.

Symbian 2003 Symbian OS technology. http://www.symbian.com/technology/technology.html.

Tachikawa K (ed.) 2002a *W-CDMA: Mobile Communications System*. John Wiley & Sons.

Tachikawa K 2002b Chapter 2, Radio transmission systems. *W-CDMA Mobile Communication Systems*. John Wiley & Sons, 21–80.

Tachikawa K 2003a A perspective on the evolution of mobile communications. *IEEE Commun. Mag.* **41**(10), 66–73.

Tachikawa K 2003b A perspective on the evolution of mobile communications. *IEEE Commun. Mag.* **41**(10), 66–72.

Tachikawa K 2003c A perspective on the evolution of mobile communications. *IEEE Commun. Mag.* **41**(10), 66–73.

Tait CD, Lei H, Acharya S and Chang H 1995 Intelligent file hoarding for mobile computers. *Mobile Comput. Network.* 119–125.

Taniwaki Y 2003 Emerging broadband market and the relevant policy agenda in Japan. *J. Interact. Advertising.* http://www.jiad.org/vol4/no1/taniwaki/#Abstract.

Terry DB, Demers AJ, Petersen K, Spreitzer MJ, Theimer MM and Welch BB 1994 Session guarantees for weakly consistent replicated data. *Proceedings Third International Conference on Parallel and Distributed Information Systems*.

Terry DB, Petersen K, Spreitzer MJ and Theimer MM 1998 The case for non-transparent replication: Examples from bayou. *IEEE Data Eng.*

Terry DB, Theimer MM, Petersen K, Demers AJ, Spreitzer MJ and Hauser CH 1995 Managing update conflicts in bayou, a weakly connected replicated storage system. *Proceedings of the Fifteenth ACM Symposium on Operating Systems Principles*, 172–182.

Thatte S n.d. XLANG web services for business process design. http://www.gotdotnet.com/team/xml_wsspecs/xlang-c/default.htm.

The JAIN APIs n.d. http://java.sun.com/products/jain.

Thomson S and Narten T 1998 IPv6 Stateless Address Autoconfiguration. *RFC 2462*.

TR-45.2 1997 Cellular radiotelecommunications intersystem operations. Technical Report TIA/EIA-41-D, Telecommunications Industry Association/Electronics Industries Alliance (TIA/EIA).

TR-45.2 2001 Wireless intelligent network (1999) and addendum 1 (2001). Technical Report TIA/EIA-IS-771, Telecommunications Industry Association/Electronics Industries Alliance (TIA/EIA).

Tremain TE 1982 The goverment standard linear predictive coding algorithm: LPC-10. *Speech Technol.* April, 40–49.

Tschofenig H 2003 Bootstrapping RFC3118 Delayed Authentication Using PANA. draft-tschofenig-pana-bootstrap-rfc3118-00 (work in progress).

UDDI n.d. UDDI version 2 specifications. http://www.oasis-open.org/committees/uddi-spec/doc/tcspecs.htm uddiv2.

Valente M, Bigonha R, Bigonha M and Loureiro A 2001 Disconnected operation in a mobile computation system. *Workshop on Software Engineering and Mobility*.

Viterbi A 1995 *CDMA – Principles of Spread Spectrum Communication*. Addison-Wesley, Reading, Massachusetts.

W3C 2003a SOAP Version 1.2. http://www.w3.org . W3C Recommendation.

W3C 2003b The W3C Workshop on Binary Interchange of XML Information Item Sets, September 2003, http://www.w3.org/2003/07/binary-xml-cfp.html.

W3C 2004 Composite Capability/Preference Profiles (CC/PP): Structure and Vocabularies 1.0. W3C Recommendation.

W3C n.d.a File transfer protocol. [RFC 959, http://www.w3c.org/Protocols/rfc959/Overview.html].

W3C n.d.b Hypertext transfer protocol – HTTP/1.1. RFC 2616, http://www.w3c.org/Protocols/rfc2616/rfc2616.html.

W3C n.d.c Web services description language (WSDL) 1.1. http://www.w3.org/TR/wsdl/.

Wahbe R, Lucco S, Anderson TE and Graham SL 1993 Efficient software-based fault isolation. *Proceedings of the Fourteenth ACM Symposium on Operating Systems Principles*, 203–216.

Wahl M, Howes T and Kille S 2000 Lightweight Directory Access Protocol (v3). IETF RFC 2251.

Walker J 2000 Unsafe at any key size: an analysis of the WEP encapsulation. Technical Report, IEEE 802.11 document 2000/362, IEEE.

Walker J and Housley R 2003 The EAP Archie Protocol. draft-jwalker-eap-archie-01.txt (work in progress).

Wallach D, Appel A and Felten E 2000 SAFKASI: a security mechanism for language-based systems. *Trans. Software Eng.* **9**(4), 341–378.

Wallach D, Balfanz D, Dean D and Felten E 1997 Extensible security architectures for Java. *Symposium on Operating Systems Principles*, Saint-Malo, France.

Watanabe F et al 2002 Physical access network topology map in hyper operator overlay architecture. Technical Report, IEEE VTC 2002 Fall, IEEE.

Watanabe F, Hagen A and Wu G 2003 *Connectivity Problem*. IEEE 802.11-03-097r1.

Wee SJ and Apostolopoulos JG 2001 Secure scalable streaming enabling transcoding without decryption. *Proceedings of IEEE International Conference on Image Processing*.

Wessles D and Claffy K n.d. Internet Cache Protocol (icp), version 2. IETF RFC 2186. http://www.ietf.org/rfc/rfc2186.txt?number=2186.

Wi-Fi Protected Access

Wi-Fi Protected Access n.d. http://www.weca.net/OpenSection/protected_access.asp?

Williams S 2004 Efficient structured XML. http://www.esxml.org/.

Wilson, M. 2002 MWIF Network Reference Architecture. MWIF 2001.053.3, Mobile Wireless Internet Forum (MWIF).

Wisely D, Aghvami H, Gwyn S, Zahariadis T, Manner J, Gazis V, Houssos N and Alonistioti N 2003 Transparent ip radio access for next-generation mobile networks. *IEEE Wireless Commun.* **10**(4), 26–35.

WSFL n.d. Web services flow language (WSFL 1.0). http://www-3.ibm.com/software/solutions/Webservices/pdf/WSFL.pdf.

Xi H and Pfenning F 1998 Eliminating array bound checking through dependent types. *Programming Language Design and Implementation*, Montreal, Canada.

Xia S and Hook J 2003a Abstraction-carrying code: a new method to certify temporal properties. *Formal Techniques for Java-like Programs*, Darmstadt, Germany.

Xia S and Hook J 2003b Experience with abstraction-carrying code. *Electronic Notes in Theoretical Computer Science.* Software Model Checking Workshop.

Xia S and Hook J 2003c Implementation of abstraction-carrying code. *Foundations of Computer Security*, Ottawa, Canada.

Xia Y and Rosdahl J 2002 *Throughput Limit for IEEE 802.11.* IEEE 802.11-02/291r0.

Xiao Y and Rosdahl J 2002 Throughput and delay limits of IEEE 802.11. *IEEE Commun. Lett.* **6**(8), 355–357.

Yang L and Hanzo L 2003 Multicarrier DS-CDMA: a multiple access scheme for ubiquitous broadband wireless communications. *IEEE Commun. Mag.* 116–124.

Yasuda Y, Nishio N and Tokuda H 2001 End-to-edge QoS system integration: Integrated resource reservation framework for mobile Internet. *Proceedings of International Workshop on Quality of Service (IWQoS).*

Yokote A, Yegin E, Tariq M, Fu G, Williams C and Takeshita A 2002 Mobile IP API draft-yokote-mobileip-api-01.txt.

Yoshimura T, Yonemoto Y, Ohya T, Etoh M and Wee S 2002 Mobile Streaming Media. CDN Enabled by Dynamic SMIL.

Yu H and Vahdat A 2000 Design and evaluation of a continuous consistency model for replicated services. *Proceedings of 4th USENIX Symposium on Operating Systems Design and Implementation*, 305–318.

Yumiba H, Imai K and Yabusaki M 2001 IP-based IMT network platform. *IEEE Pers. Commun.* **8**(6), 18–23.

Zerfos P, Zhong G, Cheng J, Luo H, Lu S and Li JJR 2003 DIRAC: A software-based wireless router system. *Mobicom 2003.*

Zhang J, Li J, Winstein S and Tu N 2002 Virtual operator based AAA in wireless LAN hot spots with ad hoc networking support. *ACM SIGMOBILE Mobile Comput. Commun. Rev.* **6**(13).

Zhou D, Pande S and Schwan K 2003 Method partitioning – runtime customization of pervasive programs without design-time application knowledge *Proceedings of ICDCS 2003.*

Zuidweg J 2002 *Next Generation Intelligent Networks.* Artech House Publisher.

Zwicker E and Fastl H 1999 *Psycho-acoustics.* Springer.

Index